Electro-mechanical Prime Movers

Edited by Peter C Bell BSc

Macmillan Engineering Evaluations

Published by
The Macmillan Press Limited
Technical and Industrial Publishing Unit

Managing Editor William F Waller
AMITPP AssIRefEng

Advertisement Representative Alan Kay
General Manager Barry Gibbs

The Macmillan Press Limited
Brunel Road Basingstoke Hampshire UK

Price £3.50

© The Macmillan Press Limited 1971
SBN 333 125 436

Printed Photolitho by Page Bros (Norwich) Ltd.

Foreword

The use by industry of the electric motor as a source of mechanical power has grown extensively since the beginning of the 20th century, when practical electrical machines were developed and electrical power supplies were made readily available. The machines discussed in this book are termed electromechanical prime movers, in keeping with the concept that electricity is available as a fuel to the engineer in industry.

The types of motor are divided broadly between d.c. and a.c. machines, of which the general design and development are discussed respectively in chapters 1 and 2. Advances in design are often due to improved materials, and those used in motor construction are covered in chapter 3.

The use of d.c. or a.c. motors varies with the changing pattern of technology; the d.c. motor field is dealt with according to size of equipment, from miniature (chapter 5) through small/medium (chapter 10) to large units (chapter 11). Alternating current equipment for 'miniature' applications is described in chapter 4, and other types included are single-phase motors (chapter 6), 3-phase induction motors (chapter 7), synchronous motors (chapter 8), a.c. commutator motors (chapter 9).

Specialised applications involve the use of stepper motors (chapter 12) and servomotors (chapter 13), whilst a type that is of current interest and of fairly recent development is the linear induction motor (chapter 14).

For the efficient use of electric motors in industry one must consider many ancillary factors in addition to the first choice of equipment. Some of these factors are presented in practical terms as a guide to the equipment user: starting and speed control of a.c. motors (chapter 16) and of d.c. motors (chapter 15); motor protection devices (chapter 17); installation and maintenance (chapter 18); and advice on the international relatiionship of various enclosures and motor sizes (chapter 19).

A glossary of terms associated with the field of electromechanical prime movers and a guide to some of the major manufacturers and suppliers of the equipment complete the practical coverage of the subject.

Contents

Chapter		Page
1	**Development of d.c. Motors**	7
	R A Bruce *BSc (Eng)*	
	Principal Development Engineer	
	CAV Ltd	
2	**Development of a.c. Motors**	15
	P B Greenwood *BSc CEng FIEE*	
	Chief Designer	
	Brook Motors Ltd	
3	**Materials of Construction**	21
	G B Sugden *MSc CEng MIEE*	
	Design Engineer Electrical Group	
	Lucas Aerospace Ltd	
4	**Miniature a.c. Motors**	29
	E H Werninck *AIEE MIMC*	
	Consultant	
5	**Miniature d.c. Motors**	37
	D N Stanley *CEng MIEE*	
	Development Design Engineer	
	Vactric Control Equipment Ltd	
6	**Single-phase Motors**	45
	J T Appleby *CEng MIEE*	
	Manager Research and Development	
	Newman Industries Ltd.	
7	**Three-phase Induction Motors**	55
	C R M Heath *MA*	
	Machine Design Engineer	
	Electrical Power Engineering (Birmingham) Ltd	
8	**Synchronous Motors**	65
	H T Price *DLC CEng MIEE*	
	Engineering Manager Rotating Machines Division	
	Brush Electrical Engineering Co. Ltd	
9	**A.C. Commutator Motors**	73
	R Mederer *CEng BSc MIEE*	
	Head of A.C. Variable Speed Dept	
	Laurence, Scott & Electromotors Ltd	
10	**Small/medium d.c. Motors**	81
	D Ramsden *CEng MIEE*	
	Technical Director	
	Bull Motors Ltd	
11	**Large d.c. Motors**	93
	B Skenfield *BSc CEng MIEE*	
	Chief Designer Medium D.C. Machines	
	English Electric—AEI Machines Ltd	
12	**Stepper Motors**	105
	E H Werninck *AIEE MIMC*	
	Consultant	
13	**Servomotors**	111
	P Vernon *CEng MIEE*	
	Senior Development Engineer	
	Muirhead Ltd	
14	**Linear Induction Motors**	117
	P J Markey *BSc CEng MIEE*	
	Asst Chief Engineer Medium Industrial Machines Div	
	English Electric—AEI Machines Ltd	
15	**Starting and Speed Control of d.c. Motors**	125
	R M Carter *BSc*	
	AEI Semiconductors Ltd	
16	**Starting and Speed Control of a.c. Motors**	133
	J H C Bone *BSc MIMechE FIEE*	
	Chief Engineer Rotating Machines	
	Laurence, Scott & Electromotors Ltd	
17	**Motor Protection Devices**	143
	D Ramsden *CEng MIEE*	
	Technical Director	
	Bull Motors Ltd	
18	**Installation and Maintenance**	155
	T J Williamson	
	Design Department	
	Mather & Platt Ltd	
19	**Enclosures and Motor Sizes**	163
	T J Williamson	
	Design Department	
	Mather & Platt Ltd.	

Guide to Equipment Makers — 170

Glossary of Terms — 188

Chapter 1

Development of d.c. Motors

R A Bruce *BSc(Eng)*
CAV Limited

If an armature, free to rotate and having conductors in which a current flows, is placed in a magnetic field, forces are produced which cause the armature to rotate. Two basic equations can be derived which define the nature of this phenomenon: they are called the induced voltage equation and the torque equation.

Torque production

Principles of operation and performance discussed in this chapter are concerned only with the generally familiar, heteropolar, direct-current motor in its widely accepted, contemporary modes of construction. Assume the use of copper for conductors, and low-carbon steels for the magnetic circuit. The need to keep air gaps in the magnetic circuit reasonably small, and the desirability of protecting the insulated copper conductors from magnetic and mechanical forces, decrees that these conductors are housed in slots on the armature periphery. The mechanism of torque production, and deriving the torque equation, are explained in terms that avoid making use of the classic law that the force on a conductor is given by the product of the conductor length, the current it carries, and the flux density of the field surrounding it. In a machine with armature conductors in slots the torque is due not to mechanical forces on the conductors themselves but to tangential magnetic pulls on the armature teeth. The current in the conductor gives rise to an armature flux which flows up one tooth, across the open top of the slot and down the adjacent tooth so that the density of the bunches of flux crossing the airgap varies across each tooth and, in addition, the flux lines are inclined to the radial direction. The effect is shown in Fig. 1, which indicates that tangential forces are produced which cause the armature to rotate in the direction indicated.

Induced voltage equation

The stationary field system sets up a stationary magnetic field around the armature, adjacent poles being of opposite polarity. The commutator, rotating under stationary brushes of fixed polarity, ensures that the current flowing in each

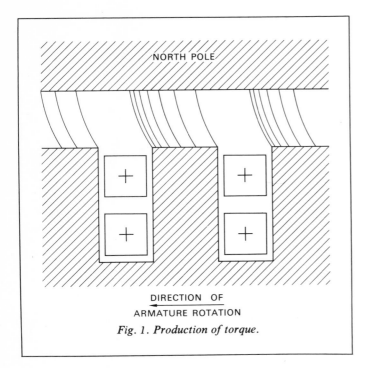

Fig. 1. Production of torque.

armature conductor changes direction as the conductor moves from one pole to the next. In this way the torque produced always causes motion in the same direction.

The current-carrying conductors moving in a magnetic field have voltages induced in them which, according to Lenz's law, are in such a direction as to oppose the current flow. Because of this the induced voltage in a motor is often called the 'back emf'. Induced voltage is defined as being equal to the rate of change of flux, so that: the average voltage induced in a conductor = the total flux it cuts in unit time.

Let e be average induced voltage per conductor; Φ total effective flux per pole in webers; P number of poles; N rotational speed in revolution per second; then $e = \Phi PN$.

For Z_s armature conductors connected in series the total average induced voltage is $E = eZ_s$; if Z is the total number of armature conductors, and A the number of parallel paths in the armature circuit, then $Z_s = Z/A$; hence $E = eZ/A$, which gives us the induced voltage equation:

$$E = Z\Phi NP/A \qquad \text{(i)}$$

If V is armature applied voltage; ΔB is total effective brush drop in volts; I_a is armature current in amp; R_a is armature resistance in ohms; then for equilibrium in the armature winding

$$V = \Delta B + I_a R_a + E \qquad \text{(ii)}$$

By taking equation (i), above, and multiplying throughout by the armature current, I_a, we arrive at

$$VI_a = \Delta B I_a + I_a^2 R_a + EI_a \qquad \text{(iii)}$$

This is the power equation for a motor, the components having the following significance: VI_a is the total power supplied to the armature; $\Delta B I_a$ is the commutation electrical loss; $I_a^2 R_a$ is the power lost as heat in the armature conductors; EI_a is the power converted into mechanical power.

From the converted power EI_a we can derive the torque equation: let T be the torque produced in Newton-metres, then the work done per second is $2\pi NT$ watts. Since this must be the converted power

$$2\pi NT = EI_a$$

Substituting for E and re-arranging gives the torque equation

$$T = (1/2\pi) I_a Z\Phi P/A \qquad \text{(iv)}$$

CONSTRUCTION

Figures 2 and 3 show some details of construction of d.c. motors.

Field System

The field system can take many forms, from a simple two pole horseshoe magnet to an elaborate compound-wound multipole-pair field with interpoles and pole face compensating windings. Most types of field have one thing in common, the yoke, which has the prime, magnetic, purpose of linking all the field poles and interpoles and is usually a hollow cylinder of soft iron or mild steel around the inside circumference of which are ranged the other elements of the field system. The yoke usually has the secondary, constructional, purpose of forming the basic structural member. Sometimes a four pole motor has a square yoke but normally the cross-section is circular and is made from tube stock or wrapped and welded plate, or from cast iron if the weight penalty of its inferior permeability can be accepted.

Permanent magnet motors dispense with the need for field windings and have a simple field construction. In all except the very smallest two pole motors the magnets are attached to the inside of the cylindrical yoke. Ceramic permanent magnet materials such as the barium or strontium ferrites are usually cast to the required shape and fixed to the yoke with an epoxy adhesive.

Metallic permanent magnets can be brazed or silver soldered to the yoke. Other fixing methods for all types of magnet

Fig. 2. Component parts of a 1.5-kW d.c. motor.

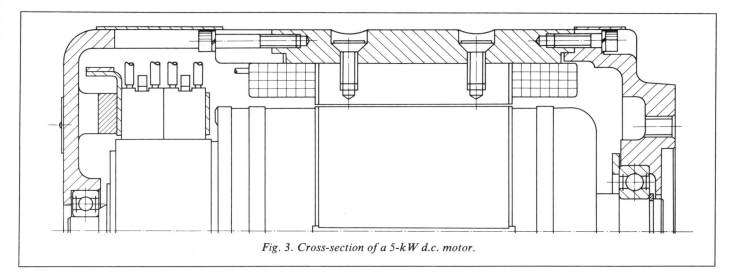

Fig. 3. Cross-section of a 5-kW d.c. motor.

include the use of non-magnetic clamps, and bolting magnets to the yoke with non-magnetic bolts which pass through cast holes in the magnets. In wound field motors the main poles are either solid and made of mild steel, or built up from stamped-out laminations which are riveted or welded together, or bonded into a stack by means of an epoxy adhesive. Interpoles are almost always solid mild steel. If the field supply is controlled by means of a solid state device operating at high frequency, it may be advisable to improve the overall motor efficiency by a special design of the yoke and poles to reduce eddy current losses. Lamination of the entire field iron circuit is one way of achieving this, but much ingenuity is needed to avoid excessive constructional costs.

The construction of field and interpole coils varies widely with the type of size of motor, conductor size being determined by the current and available cooling. Round wires with diameters up to 2 mm are commonly used, sometimes two or three strands in parallel, the smaller sizes being often random wound, the larger sizes layer wound. Where higher currents call for heavier conductors, rectangular cross-sections are used. Some enamel coverings exhibit poor adhesion and a tendency to peel on bends if strip manufacturers' recommended width-to-thickness ratios are not used.

It is common practice to form the wound coil to the inside profile of the yoke, either before or after taping with a suitable insulation. It is often sufficient to impregnate the taped coil with varnish before assembly into the yoke, but for some applications where thermal or mechanical movements can lead to frettage in service it is necessary to impregnate the complete assembled field system with varnish.

Armature

The armature assembly has the motor shaft as its 'backbone'. The shaft provides the drive for the load, and carries the bearings which support the armature and position it within the field system. Within the span of the bearing locations the shaft carries a stack of armature laminations which contain slots for carrying the armature conductors, and also the commutator to which all the armature conductors are connected. The lamination stack, in small or medium sized motors, is usually force-fitted to the knurled shaft. In larger motors it may be bolted to a suitable spider. The armature slots are lined with insulating material, as are the end faces of the stack, and the insulated conductors lie in the slots.

Only in low power or high voltage motors is the armature wound in round wire, currents usually being high enough to necessitate rectangular strip. Connection of armature conductors to the commutator is usually made by soldering, but for high operating temperatures or high mechanical stresses connections may be brazed or welded. Non-magnetic wedges are used to retain the conductors in their slots, and the finished armature is impregnated with varnish to exclude moisture and brush dust. Except in very small motors it is necessary to balance the armature, which usually involves adding suitable masses to end windings until a dynamic balance is achieved. Frequently the mass of copper in the winding overhangs is too great to be self-supporting, especially if high rotational speeds have to be catered for, and the overhang at each end is encircled with a strengthening band. This band, formerly of soldered piano wire with added solder for balancing, now commonly consists of an epoxy-impregnated glass-fibre bandage applied, in the partly cured state at controlled tension, to each overhang. The bandage is finally oven-cured and balancing is effected by forcing self-curing epoxy putty under the bands in appropriate positions.

Brushgear and endshields

A circular cap or endshield is used at each end of the motor to support the outer races of the armature shaft bearings and to locate on the ends of the yoke. In fan ventilated motors the fan is usually mounted on the shaft either inside or outside the drive end shield and draws cooling air through the machine, the air on entry impinging on the commutator.

The commutator endshield is usually the most convenient mounting place for the brushgear (the assembly of brush holders, brushsprings, brushes and interconnectors) which has as its main function the accurate positioning of the brushes in relation to the field poles. The usual types of winding require the ends of each armature coil to be bent through one half a pole pitch after leaving the slot and before being connected to commutator bars. It follows that to commutate the current in a conductor lying in the neutral interpolar zone the brushes must be positioned in line with the main pole centrelines. In unidirectional motors commutation conditions can be improved if the brush position is displaced a small angle from the pole centreline.

DESIGN

The starting point of any design is a performance requirement, usually calculated in terms of terminal input and shaft output after careful sifting of the application data. For traction applications shaft speeds must be derived from vehicle speeds after due appreciation of wheel sizes and gear ratios; torque at the motor shaft will be related to vehicle acceleration rates, rolling resistances, gradients to be negotiated, and gear efficiencies. If the motor is to power a hydraulic system one must relate flow rates and pressures with pump efficiencies, and consider the effects of duty cycle and environment.

The next step is to decide on the type and basic configuration of the motor. Motor performance is discussed below, and guidance is given which should allow this step to be taken with confidence.

The main criteria of any design are that the performance requirements should be met in the most overall efficient way, and that the resultant motor should represent the most economical approach. The selection of motor type and configuration greatly influences both of these considerations, and it remains to ensure that the optimum magnetic and thermal design is achieved.

Space will not permit a detailed exposition of the methods by which initial machine dimensions are chosen, the reader is referred to one of the specialist texts on machine design recommended in the bibliography. Much profit can be gained by special study of the magnetic circuit and of the electrical loading of field and armature windings.

The magnetic circuit

The power required to energise a magnetic field is directly related to the magneto-motive force (mmf) which is in turn a function of the flux density and the length of the flux path.

More precisely: power $\propto$ (mmf)^2 and mmf $\propto \sum_n$(pathlength $\times$ a function of flux density) where n is the number of portions of the complete magnetic circuit.

The problems of magnetic circuit design lie in the nature of the flux density function, which is linear for the air gap, but for the iron is notoriously non-linear. Figure 4 shows a typical characteristic of lamination material. Wherever the flux path comes into conflict with the field or armature conductors, design problems become more acute, and Fig. 5 shows that the conflicts occur in the armature tooth-slot area and in the pole shank region of the field. It is essential that operation in the saturated region (Fig. 4) is limited to only short stretches of the magnetic path. In arriving at the initial machine dimensions application of the torque and induced voltage equations produce derived values of magnetic flux. It is most useful to work with at least two values of flux, one corresponding to the nominal full load operating point, the other to the operating conditions requiring maximum flux. Using the initially chosen motor dimensions the magnetic conditions in the portions of the complete circuit can be evaluated and tabulated for each flux value. It is important that leakage fluxes associated particularly with the critical parts of the circuit be evaluated and added to the main flux in each element of the circuit. Leakage fluxes depend greatly on geometrical configuration, so that considerable mathematical analysis or flux plotting may be necessary before their accurate estimation is possible. A fresh assessment of the magnetic circuit, with particular regard to the changing pattern of leakage fluxes, is essential for accurate evaluation of each operating point. Of special importance is the field distortion caused by armature currents. The laborious nature of these calculations makes the use of digital computers for the analysis of design and performance particularly attractive, and programs written for this purpose can often form the basis of design optimisation routines.

Reference to Fig. 4 makes clear the inadvisability of operating in the saturated region except for short parts of the magnetic circuit such as the armature teeth and the pole shank. It is wise therefore to operate in the 'knee' part of the curve (Fig. 4) for the full load operating point and to ensure that at the operating point at which the maximum flux is required the field excitation is sufficient, and saturation sufficiently restricted, to provide the required flux.

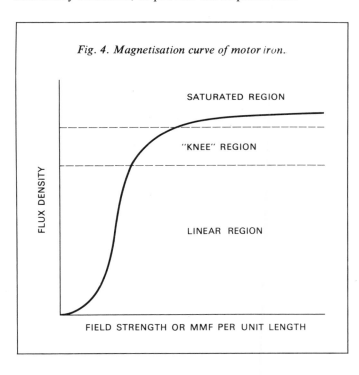

Fig. 4. Magnetisation curve of motor iron.

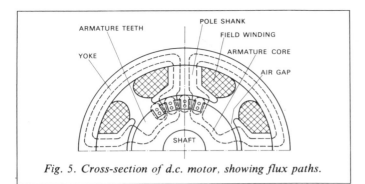

Fig. 5. Cross-section of d.c. motor, showing flux paths.

Permanent magnet fields

Considerations so far have implied the presence of an electromagnetic field system. Permanent magnet motors, however, present their special problems of magnetic circuit design. For stability of operation and ability to withstand demagnetising effects, barium or strontium ferrites are unsurpassed and are particularly suitable for d.c. motors. The high maximum energy levels of *eg* Alnico metal magnets are not attainable in ferrites but the linearity of the demagnetising characteristic and their greatly superior coercive force are factors which result in ferrites attaining working energy levels which match those achievable in Alnico magnets after short-circuit stabilisation. The principal point of design difference between wound field and permanent magnet motors is in the average air gap flux density. It is common for values in permanent magnet machines to be only half of those used in wound field motors. This has two important effects, firstly that the armature slot width to tooth width ratio can be considerably increased, and secondly that air gap lengths can be increased to minimise the effect of variations in magnet performance between motors without paying too great a penalty in excitation mmf. The mmf of a permanent magnet is proportional to its length in the direction of the main flux flow.

Electrical loading

Two principal factors determine the electrical loading of both field and armature: the cooling available and the permissible temperature rise. It is convenient to express the armature electrical loading in ampere conductors per metre of periphery. Its value varies slightly with armature diameter, except in the case of very small motors, but is greatly affected by the type of cooling. If the electrical loading for a fan ventilated motor is taken as one per unit, then removal of the fan and reliance on natural convection and radiation will reduce the loading to about 0·8 per unit. If the air vents in the endshields are now closed there will be a further fall in permissible loading to 0·6 per unit.

The maximum operating temperature of the motor is determined, principally, by the material used to insulate the armature and field conductors, and the commutator segments. A later paragraph discusses economical factors in the choice of materials; it is sufficient here to point out that motor efficiency is likely to be one of the overall design criteria and that this influences strongly the choice of maximum operating temperature.

Subtraction of the operating ambient temperature results in a figure for the allowable temperature rise for the conductors and commutator. It is found in practice that armature conductor temperature rise is roughly proportional to the square of the electrical loading. With an ambient temperature of 35°C and Class A conductor insulation the permitted temperature rise is about 70°C. Using insulation of class F capability, with its permitted maximum of 155°C, increases the allowable temperature rise to 120°C and allows an increase in the electrical loading of about 30%. One of the aims of good design should be to match armature and field winding temperatures and avoid hot spots. This aim is more easily achieved in series motors than in compound or shunt wound machines since shunt winding temperatures are almost independent of motor load, the slight fall off in current density at high motor load being compensated for by increased radiation from the armature. An accurate analysis of thermal conditions within a motor is only practicable if digital or analogue computer techniques are used.

PERFORMANCE

In selecting a motor type to suit an application it is vital that the whole of the operating range be considered. Our main concern is with the full load operating conditions but the behaviour required of the motor at conditions of very light or very heavy loadings must be determined.

Permanent magnet motors

The great advantages of permanent magnet motors are their simplicity and efficiency; the disadvantages are the comparatively poor utilisation of some parts of the magnetic circuit and the lack of design flexibility, that is, the difficulty of varying the performance characteristic.

The flux, Φ, is sensibly constant over the entire operating range, and since Z, P and A are constants, equation (i) reduces to $E = k_1 N$, where k_1 is a constant. Hence from equation (ii) we can arrive at

$$V = \Delta B + I_a R_a + k_1 N$$

When operating at constant voltage, V, and taking into account the fact that both R_a and ΔB are sensibly constant

$$\frac{V}{R_a} = \frac{\Delta B}{R_a} + I_a + \frac{k_1}{R_a} N$$

or
$$I_a + lN + m = 0$$

where l and m are constants which gives a linear relationship of speed with current as shown in Fig. 6. From the torque equation (iv) we see immediately that $T = k_2 I_a$ where k_2 is a constant.

The approximately linear speed-current, torque-current and torque-speed characteristics of the permanent magnet motor make it ideally suited to servomotor and control system applications. Since for this type of use fast response to load changes is desirable, it is common for the armatures to have a high length to diameter ratio. In this case the high torque to armature inertia ratio gives a very high rate of armature acceleration.

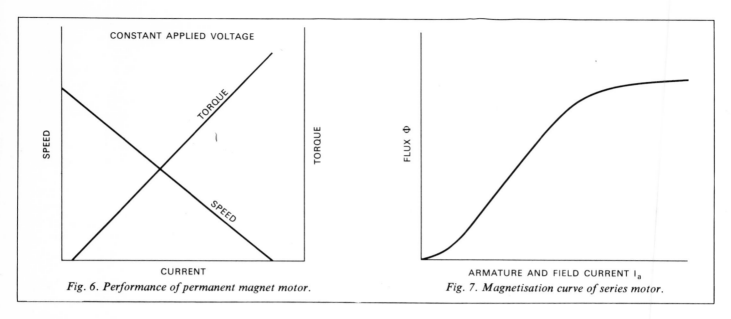

Fig. 6. Performance of permanent magnet motor.

Fig. 7. Magnetisation curve of series motor.

Series field motors

In series motors the flux is very variable since the field mmf is proportional to armature current. Figure 7 shows the flux-current curve for a typical series motor, from which it will be seen that the flux is initially proportional to current but at saturation becomes sensibly constant.

From equation (i) since Z, P and A are constant

$$E = k_3 \Phi N$$

From equation (ii)

$$E = V - \Delta B - I_a R_a$$
$$= m_1 - l_1 I_a$$

since V, ΔB and R_a are sensibly constant; m_1 and l_1 are constants.

Combining we have

$$k_3 \Phi N = m_1 - l_1 I_a$$

or

$$N = \frac{m_2 - l_2 I_a}{\Phi}$$

which gives rise to the speed-current relationship shown in Fig. 8. From the torque equation (iv) we have $T = k_4 I_a \Phi$. Initially $\Phi \propto I_a$, so that $T \propto I_a^2$, but when Φ becomes constant then $T \propto I_a$, which leads to the torque-current characteristic of Fig. 8.

Stable operation under high loadings is the particular feature of the series motor which makes it a favourite for traction applications. The constancy of the ratio of field to armature ampere-turns ensures comparatively good commutation at high loadings, and as with the permanent magnet motor, the starting torque is very high. The main disadvantage is the tendency, especially in the larger machines, for the speed to rise dangerously high on light loads. In traction applications even the minimum load is sufficient to restrain the speed, but during testing precautions are necessary. A method of overcoming the problem of high no-load speed is described under 'compound wound motors', below.

Shunt field motors

A motor in which the field winding is connected in parallel with the armature is called a shunt motor, and its field current is approximately constant and independent of the armature current. The flux is therefore constant so that linear speed-current and torque-current characteristics are achieved. At high loads the low ratio of field to armature mmf results in severe flux distortion, excessive sparking at the commutator and consequent instability of operation. Even with interpoles fitted shunt motors are not satisfactory on very heavy overloads and care must be taken on starting up to ensure that the field winding is connected to the full supply voltage, and time is given for flux build-up before the armature is connected to the supply through a starting resistance.

The performance characteristics are shown, for the stable operating region, in Fig. 9. In practice the fall-off in speed over the operating range up to full load can be made quite small, so that shunt machines are used where a relatively constant speed is required over the whole operating range from zero to full load but where heavy overloads will not be encountered.

Compound wound motors

By employing a combination of shunt and series fields, hybrid characteristics of shunt and series motors can be achieved. If a shunt winding is added to a series motor so as to reinforce the flux, then the sharp rise of speed at very light loads can be curtailed and a safe no-load speed ensured. If a series winding is added to a shunt motor to reinforce the flux, then the gradient of the speed-current curve will be increased, but the ability to cope with heavy overloads will be greatly improved. Such a motor might be used, for example, for driving a heavy machine tool where a steady linear speed-load curve is required but where sudden deep cuts could cause heavy overloads.

Although 'cumulatively' compounded motors, in which the two field windings assist one another, are by far the most common types of compound-wound motor it is sometimes

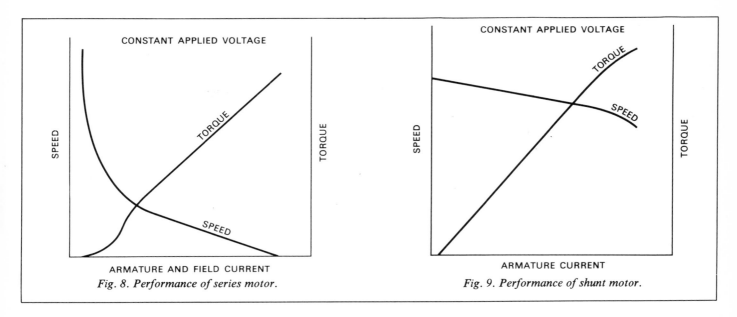

Fig. 8. Performance of series motor.

Fig. 9. Performance of shunt motor.

advantageous to make the two field windings oppose one another; by this 'differential' compounding one can arrange the relative strengths of the fields so that the full load speed is equal to the no load speed, the flux being weakened to offset the fall in induced voltage. The speed characteristic is as in Fig. 10. Such a motor has two main disadvantages: firstly under overload conditions the resultant field weakening leads rapidly to unstable operation and poor commutation; secondly, unless the series field is shorted out during starting, the motor will start up in the wrong direction, because the series field builds up more quickly than the shunt field when the supply is switched on.

DEVELOPMENT

The conventional d.c. motor, except in very small or very large sizes is mechanically simple in its construction, which thus causes few problems in development. Its mechanical simplicity also allows few possibilities of innovation. From the electrical standpoint, however, the d.c. motor is of great complexity, and there is scope for rewarding improvement.

Equipment

For successful development it is vital that tests and test results are precisely repeatable. The primary requirement is for a stable flexible finely-adjustable and accurately metered power supply. The supply is normally a motor-generator set, but for small motors could be a rectified variable-voltage a.c. supply. For consistency and repeatability the only suitable load is a generator. Rope brakes, disc brakes, or even water-wheel dynamometers are subject to variations in performance, mainly due to effects of dissipating the load directly in the form of heat, so that they are unsuitable for development work.

When the loading device is a generator the load can be accurately set and smoothly varied over the desired range, the waste power being dispersed in some remote device such as a water-cooled resistor. Motor load torque is best measured by means of a calibrated strain-gauge coupling between the shafts of motor and dynamometer. Speed should be measured either stroboscopically, using a precision signal generator as comparator, or by means of a pulse-counting device operating from a toothed wheel on the motor shaft and providing digital read-out. Electromagnetic or mechanical tachometers are neither sufficiently accurate nor sufficiently consistent for speed measurement in development work. It is most important that voltmeters, ammeters, and ammeter shunts are of suitable range and regularly calibrated.

Thermal measurements

Except for very large motors in which 'end effects' are of acceptably small significance and build variations negligible, the accuracy and reliability of thermal analysis by calculation is usually unsatisfactory, despite the use of digital computer programs of considerable sophistication. A development exercise to measure and optimise thermal conditions within a motor usually yields beneficial results.

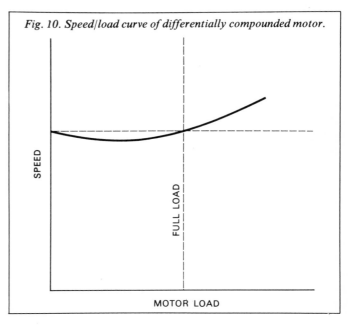

Fig. 10. Speed/load curve of differentially compounded motor.

Measurement of the temperatures and temperature rises of stationary parts is straightforward and best done by the strategic placing of thermocouples. Typically these are fixed to brushes and brush boxes, to endshields, the yoke, and to main poles and interpoles. Thermocouples are taped into field coils and fixed between the turns of interpole windings. The thermocouples are connected through a selector switch mechanism to a direct reading millivoltmeter or to a continuous pen recorder.

Thermal measurements of the armature and commutator present a strong challenge to the ingenuity of the development engineer. Temperature-sensitive paints change colour under the combined influence of time and temperature; they are applied with a small brush to such places as the face of the commutator riser, the overhang of the winding, and the lamination pack, so that after timed tests at controlled loads the colour changes can be recorded. It is usually necessary to carry out calibration checks on a stationary armature with thermocouples attached to observe the precise effect of the nature and colour of the host surface on the colour changes of the paints. When access is unrestricted, reasonably accurate temperature measurements on parts of the armature assembly can be made with a hand-held contact thermocouple device, readings being taken immediately the motor is stopped.

Aims of development

It is most likely that development aims will include the optimum increase of a machine's output, or the reduction of size or weight or cost for a specified output, or the improvement of some aspect of performance, such as efficiency and commutation. Most profit will be derived, in the fulfilment of all these aims, if the tests and investigations are used to provide information for detailed re-appraisals of the design; only in the last resort should 'cut and try' methods be used.

ECONOMICS AND CURRENT TRENDS

Current trends in motor design lead in two distinct directions, both being inevitably governed by economic considerations. The first trend is towards the exploitation of developments in materials and techniques to improve the performance and economics of manufacture of conventional existing designs; the second is towards the establishment of more or less radical changes in the design and construction of motors.

Improved materials

Advances in magnetic and constructional materials for motors are of a minor nature except for permanent magnets. There is a tendency to use sheet steel fabrication for the yokes and housings of large motors, and to use cast aluminium alloys for the endshields of small and medium motors, in both cases replacing cast iron or steel. Continuing development of permanent magnets is along two quite separate lines. In metallic magnets improvements to domain size and orientation give materials with greater energy products, but economic factors usually inhibit their use. For ceramic magnets, the economic climate is brighter, and much benefit is being gained from the use of new barium and strontium ferrites in which higher coercive forces and straight line demagnetisation characteristics are attractive features.

Insulating materials show the most significant improvements. New substances based on modifications or combinations of polyesters, imides, amides, and nylons produce varnishes, laminates (both flexible and rigid), sheets and papers to permit conductors to operate at higher temperatures, and provide better thermal conductivity and mechanical strength to a degree outweighing increased cost. Class A, E, and B machines are rapidly being replaced by new motors operating at Class F, H, or even C temperatures.

New techniques

Two of the most significant new techniques concern the manufacture of commutators and the methods of joining conductors. Commutator manufacture now tends to employ plastic injection moulding techniques to hold rigid (in accurate juxtaposition) the copper bars, mica insulation, and steel supporting members, in preference to the previously used methods of riveting, bolting or screwing. The economic advantages are considerable, and the introduction of plastic injection moulding has liberated designers from previous constraints and led to the adoption of 'face type' commutators, for some small motors, in which the brushes are mounted axially and the working surface of the commutator is normal to the axis.

The acceptance of higher operating temperatures in motors has led to dissatisfaction with the solders used to join conductors to commutator bars, or to make field connections. More satisfactory methods such as brazing or welding are being introduced and are proving not only technically but economically superior to soldering.

Changes in design concept

A steady and continuing improvement in the economics and performance of semiconductors, particularly power transistors and thyristors, has occurred since the 1960s. The maintenance and performance of brushes and commutator in all sizes of d.c. motor has long been the most serious problem area, so that much effort is being expended in discovering suitable methods of replacing mechanical with electronic commutation. Developments are proceeding along two main lines; in one, semiconductors are used to invert the d.c. supply so that a suitable a.c. induction or reluctance motor can complement the inverter and form a system to replace the d.c. motor; in the other, the basic design concept of the d.c. motor is retained, while using semiconductors to commutate electronically. At present (1971) neither line of development offers either performance or economic benefits over conventional designs, but advances in semiconductor design and manufacture bring the realisation of such benefits steadily nearer.

BIBLIOGRAPHY

A E Clayton. The performance and design of direct current machines. *Pitman* 3rd Edition 1959.

J Hindmarsh. Electrical machines and their applications. *Pergamon Press* 2nd Edition 1970.

Chapter 2

Development of a.c. Motors

P B Greenwood BSc CEng FIEE
Brook Motors Ltd

The man generally accepted as the inventor of the induction motor was Nikola Tesla in the year 1887. Equal credit might be accorded to Galileo Ferraris, who about two years earlier had discovered that two coils placed at right angles to each other, and carrying alternating currents displaced in phase by 90° produced a rotating magnetic field. Ferraris in fact showed that a solid cylinder placed in the rotating field was caused to rotate. Tesla however, demonstrated how polyphase voltages could be generated in one machine, and applied to the separate phase windings of a motor. He employed ring type windings, and his rotor had short circuited coils on an iron core.

The next stage was the commercial development of practical motor designs, and this came within three years of Tesla's invention, as a result of the collaboration of, M von Dolivo-Dobrowolsky and C E L Brown. The former invented the squirrel cage rotor, and this was incorporated into the motors designed by Brown.

Wound-rotor induction motors with slip-rings, enabling resistances to be inserted into the rotor circuit during starting were also soon being manufactured. Dobrowolsky even proposed a double cage rotor design, the outer cage to provide an effectively high rotor resistance with corresponding high torque at starting, while the inner cage had low resistance and carried the major part of the rotor current when running. Though not manufactured for many years, it anticipated the double cage rotor designs later attributed to Boucherot.

Undoubtedly the development of the a.c. motor would not have progressed so rapidly in its early years, following its invention in 1887, if the first cut and try methods had not been quickly replaced by a scientific analytical approach to design.

The two first great contributors to the science of a.c. motor theory were André Blondel and Gisbert Kapp. By separate contributions, they mathematically demonstrated the

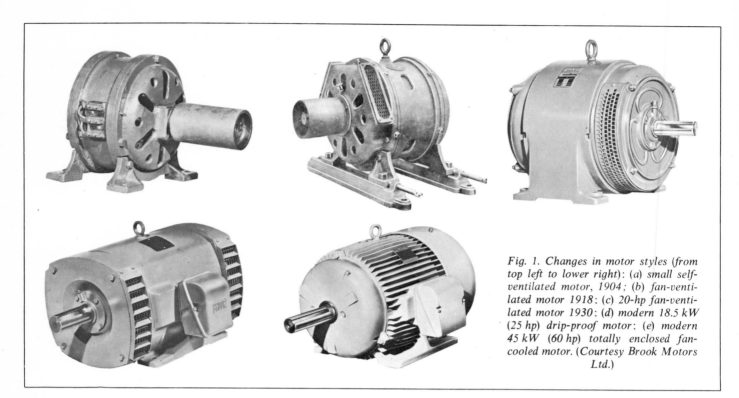

Fig. 1. Changes in motor styles (from top left to lower right): (a) small self-ventilated motor, 1904; (b) fan-ventilated motor 1918; (c) 20-hp fan-ventilated motor 1930; (d) modern 18.5 kW (25 hp) drip-proof motor; (e) modern 45 kW (60 hp) totally enclosed fan-cooled motor. (Courtesy Brook Motors Ltd.)

similarity between the induction motor operating in two different modes, namely running on load, and with rotor locked but with added resistance in the rotor circuit. Hence the possibility of using the same theory for an induction motor on load as for a transformer, the stator winding being considered as the primary winding, and the rotor as the secondary. The concept of leakage fluxes and their associated leakage reactances then led to the use of the equivalent circuit, which enabled the performance of a motor to be simply and accurately predicted, once the parameters of the circuit (namely resistances and reactances) were known.

Heyland first realised that the locus of the primary current vector of the polyphase induction motor is a circle, but the complete construction of the circle diagram, and its interpretation to present in a simple manner a visual aid to understand motor performance was mainly due to B A Behrend in 1896.

Contemporary with the above discoveries in polyphase induction motors, mathematical theories were also being developed to explain the action of the single phase induction motor. Thus, in only ten years from its invention the basic theory of the induction motor as we know it today was fully understood. That is not to say that no further discoveries were to be made. Indeed our knowledge is still increasing, but by the year 1900 the essential product had left the mould, and needed only to be smoothed and polished.

Undoubtedly this early rapid progress, the commercial exploitation of the 3-phase induction motor, and its acceptance in preference to the d.c. motor as the prime mover in practically all industrial drives were all inter-related, as was the decision to use 3-phase a.c. for power transmission influenced by its easier use in providing rotary motion.

TRENDS IN MOTOR DESIGN

Because the basic induction motor had already reached such a technically advanced stage over seventy years ago, it is natural that growth in the industry tends to be steady but undramatic, and changes in design evolutionary rather than revolutionary. Nevertheless, when viewed over a long period of time, the changes have been great.

Figure 1 shows the change in appearance of small to medium sized motors (since the open construction of 1890 on a bedplate) to the modern drip-proof and totally-enclosed fan-cooled forms.

The styling of large motors of 400 mm centre height and above has tended in recent years towards rectangular welded steel unit construction; Fig. 2 shows a typical design in which the frame is split horizontally. The base unit is common, irrespective of the type of cooling, and supports the stator and the rotor bearings. Various alternative types of top cover can be fitted, giving any of the following types of enclosures: drip-proof, pipe ventilated, splash-proof, hose-proof, closed air circuit, air or water cooled *etc*.

The improvements in utilisation of materials, resulting in steady increases in power/weight ratios have been particularly impressive; Fig. 3 shows examples of the relative sizes of motors of equal output rating manufactured by one company at various periods of its history.

Insofar as external dimensions are concerned, the shaft dimensions, centre heights, and fixing holes of the popular ranges of induction motors are now standardised by international agreement. However, it is envisaged that a new agreement involving further reductions in the sizes of most

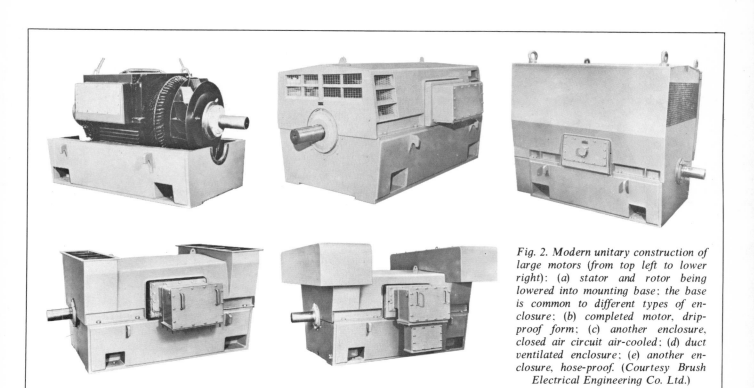

Fig. 2. Modern unitary construction of large motors (from top left to lower right): (a) stator and rotor being lowered into mounting base; the base is common to different types of enclosure; (b) completed motor, drip-proof form; (c) another enclosure, closed air circuit air-cooled; (d) duct ventilated enclosure; (e) another enclosure, hose-proof. (Courtesy Brush Electrical Engineering Co. Ltd.)

motors will take effect in 1980. In the meantime, reductions in the active material contained inside standardised frame sizes are almost continuous. This progressive reduction in size and weight of motors has been due to three principal factors: better cooling, materials, and design techniques.

Improved cooling systems

The earliest motors were open machines without cooling fans. It was soon recognised however that the limitation of a motor's output rating was the temperature attained by the insulation, and that increased output was possible by forced ventilation. Since 1954, the market has tended to expect the standard range of ventilated motors to have drip-proof enclosure. This at first imposed restriction on the efficiency of the ventilation system, but continuous development has resulted in the extremely efficient ventilation systems of modern drip-proof motors.

In the period between the two world wars, totally enclosed motors gained steadily in popularity, but until 1954 were still in the minority in the UK. The advent of the first range of British Standard dimension motors in totally-enclosed fan-cooled frames in that year gave a fresh impetus to the popularity of this type of enclosure, a trend which has continued with the adoption of metric frame sizes in 1966, and the inter-related influence of Continental Europe, where a.c. motors had for many years been mainly of totally-enclosed fan-cooled construction in outputs up to about 50 kW.

In this type of cooling also, development has led to progressive reductions in size. In the most modern designs in sizes up to about 7.5 kW output multi-ribbed yokes of aluminium alloy have largely replaced cast iron, giving improved thermal dissipation. So effective is this form of cooling in small motors that it is now considered by many manufacturers uneconomical to manufacture drip-proof ventilated motors below about 10 kW output.

Advances in materials

Improvements in the active materials have made but a minor contribution to reducing sizes of motors. The substitution of aluminium for copper as the conductor material for squirrel cage rotors has not helped to reduce motor sizes, although it has considerably assisted in keeping down costs.

Improvements in magnetic steels and in techniques for final heat treatment and decarburising have made small but useful contributions towards reducing motor sizes.

The main advances in materials however have been in insulations. These are described in another chapter, but it is sufficient to say here that the modern motor has a temperature rise limited generally to 75°C or 80°C (measured by rise of resistance) compared with 40°C (measured by thermometer) in motors manufactured prior to 1955. This increase in permissible temperature rise of modern insulation has played a considerable part in enabling motor sizes to be reduced, and it is likely that the trend will continue with the probable adoption of insulation systems in future designs of general purpose motors which will operate within a permissible temperature rise of 100°C.

Improved design techniques

The digital computer has had a profound impact on design methods for rotating machinery. First used in the late 1950s,

Fig. 3. Size reductions of 7.5 kW (10 hp) 1500 rev/min totally-enclosed fan-cooled motors (from left): 1948 non-standard; 1954 BS2083; 1960 BS2960; 1966 BS3979. Comparable reductions in a range of 18.5 kW (25 hp) 1500 rev/min ventilated motors are indicated by the weight/date of units: 418-kg 1920; 204-kg 1930; 154-kg 1958; 122-kg 1970. (Courtesy Brook Motors Ltd.)

it is now employed as a design tool by practically every company engaged in electric motor manufacturing. Undoubtedly, by enabling far more accurate design calculations to be made, and because of the great speed and facility of computation, every new design can be analysed far more rigorously than hitherto, and optimised to meet the required specification with the minimum cost of active materials.

There are many ways of using a computer in the design process, and one problem is whether to use an analysis or a synthesis approach.

In an analysis procedure, the motor is designed first. The geometry of the motor, *ie* its entire dimensions right down to minute details of slot size, coil details, *etc*, is then fed into the computer, which is programmed to calculate the expected performance, weights or costs of materials and any specific loadings or other information which the designer may require, or which the expertise at his disposal enables him to include in the program.

A synthesis approach is one in which the desired performance is fed into the computer, which is programmed to determine the best design geometry. Any synthesis technique is bound to utilise more store and time in the computer and is more difficult to program. Furthermore, there are so many options possible that artificial restraints have to be built into the program at various stages. Some design departments are using partial synthesis techniques, but a pure synthesis design program is probably academic.

Most prefer a straight analysis process, in which the designer checks the calculated performance and other output data against the specified requirements, makes whatever changes he deems necessary to the motor geometry, then repeats the analysis. This was indeed the traditional approach using hand calculation in pre-computer days, but the tedious hours of slide-rule calculations are now replaced by a minute or so of computer time. Whereas formerly one or two variations of the motor geometry could limit the efforts of the designer he can now go on analysing repeatedly until he has the best or most economical design.

Furthermore, because he has such a powerful calculating tool at his disposal, he can now select the most accurate calculating procedures available from the entire literature on the subject, whereas with hand calculations, less rigorous methods had to be used. This new ability to make practical use of the most accurate formulae in performance calculations, no matter how complicated the mathematics might be, has in turn stimulated considerable research in recent years into mathematical analysis in areas of motor performance previously little known. For example, stray losses are now more clearly understood, and are quantitatively predictable from the motor geometry.

MOTOR CONSTRUCTION

Manufacturing techniques have improved rapidly in recent years. The practice of winding motor coils by hand directly into closed slots in the stator had largely ceased by about 1930, in a.c. motors below 150 kW in output. It was replaced by various methods of pre-forming the coils prior to insertion into open or semi-closed slots.

High voltage motors, and medium voltage motors larger than 500 kW, have conductors of rectangular cross section thus obtaining a high slot factor, *ie* a high utilisation of the stator slot area. Coils of rectangular conductors are also more rigid than round wires, an important consideration in large motors, as the coils are subject to higher electromagnetic forces than in small motors, and generally require to be mechanically braced. In smaller motors, the coils may be random-wound on formers from round section wires, complete phase groups or even complete phases being gang wound to minimise the number of joints.

Machine-winding of stators developed during the 1950s starting with small power single-phase motors, but now being extended to 3-phase motors up to about 40 kW. Two basic types of machines are used, namely needle-winders in which the wire is wound turn by turn into the slots, and the more recently developed coil-inserting or 'shooting' machines. The latter in effect perform automatically the processes of preforming the coils and then inserting them into the slots, as previously done by hand.

A typical modern winding machine for winding motors of 3.0–7.5 kW rating costs about £25000 but winds, in a matter of minutes, a stator which would take more than an hour to wind by hand. These high-speed winding machines however can only wind coils in concentric groups, which results in a slightly inferior performance compared with double layer lap windings, unless compensation is made in the design.

The tendency for copper to fluctuate widely in price in recent years has made motor manufacturers look closely at aluminium wire as a possible alternative for the stator windings of a.c. motors. Aluminium wires need 64% greater sectional area than copper wires of equal conductivity. Attempting to obtain standardised output ratings from specific frame sizes when using aluminium wire, therefore, imposes considerable design problems. Moreover, the resultant designs are not necessarily lower in cost than corresponding designs employing copper, as, in order to accommodate coils of greater section, slot sizes have to be increased. This results in a reduction in the iron in each lamination, which must be compensated by increased core length.

Although aluminium wire is being used for domestic appliance motors, the tendency may be to retain use of copper as the principal winding material in most a.c. motors, but with a reduction in the ratio of copper to iron. However, much depends on the relative prices of copper and aluminium.

In cage rotor design, however, the extra space required by aluminium compared with copper does not present such a serious problem. Squirrel cage rotors up to 1940 were mainly manufactured with conductors of copper bar brazed to copper end rings. Small-power motors were the first to have an entire aluminium squirrel cage diecast into the slots. It is now common practice to use this technique in sizes up to 100 kW.

In larger sizes, the high cost of dies relative to the smaller quantities produced, the limitations to size of diecasting machines, and the added complication of ventilating ducts in the rotor cores of ventilated motors are all prohibiting factors. Nevertheless the low cost of aluminium relative to copper remains attractive, and there is a tendency to exploit this in large squirrel cage rotors by the use of extruded aluminium conductors, welded to cast aluminium end rings.

Variable speed a.c. motors
The fact that the a.c. motor on a fixed frequency supply is by nature a constant speed machine has presented a constant challenge to engineers to persuade its speed to vary. A slip ring motor operates at reduced speed when external resistances are connected across its slip rings, but the slip energy is wasted in the form of heat. Schemes to recover the slip energy and at the same time permit the speed to be varied are numerous and generally require the use of commutators, either on an auxiliary machine or as part of the motor. A.C. commutator motors have been manufactured since about 1910 and are treated in detail in a later chapter.

In recent years, induction motors have been used in conjunction with thyristors to obtain variable speed. Three different types of drive briefly mentioned here are:
(*a*) The use of a variable frequency thyristor inverter to supply power to the stator of a squirrel cage motor.
(*b*) The use of a variable frequency thyristor inverter in the rotor circuit of a slip ring induction motor, either recovering the slip energy at sub-synchronous speeds or supplying power at super-synchronous speeds.
(*c*) The use of a variable voltage thyristor drive to the stator of a squirrel cage motor, a method which is particularly suited to applications in which the torque/speed relationship of the load follows a quadratic law, *eg* fan and centrifugal pump drives.

Although these are essentially control system developments, they do require the development of motors with special characteristics, especially in the case of type (*c*).

Change-speed a.c. motors
A squirrel cage motor can be provided with separate windings (dual-wound) to give two speeds in inverse ratio to the number of poles in the two windings. On the other hand, the Dahlander method of obtaining two speeds in the ratio 2:1 from one winding has been in common use for more than 70 years.

A significant advance was made in 1957 by the discovery of the principle of Pole-Amplitude-Modulation (PAM) by Professor Rawcliffe and his colleagues at Bristol University. The phrase was coined because of the similarity of the principle to the theory of frequency modulation in the field of telecommunications. In essence, a rotating field having 'p' pole pairs can be modulated by means of a field having 'k' pole pairs, resulting in a field having $p \pm k$ pole pairs. (The sequence of the modulating field determines whether the sign is + or −.)

In practice, the modulation is applied by reversal of half of the windings in each phase. Figure 4 shows in a simple way how this can be done for one phase. The coils of the complete 3-phase winding are, however, irregularly grouped in an apparently chaotic manner, but resulting always in electrical balance comparable with that of orthodox single speed windings. The external connection changes which are necessary to change the speed are in fact identical to the connection diagrams of the 2:1 Dahlander motor.

The first PAM motors were close-ratio two-speed motors, *ie* 4/6, 6/8, 8/10 pole motors *etc*, and these combinations are still the most popular, and compare most favourably with the alternative dual-wound motors. However, wide-ratio pole combinations are also possible, in all ratios except 3:1. PAM motors giving three speeds from one winding are also possible.

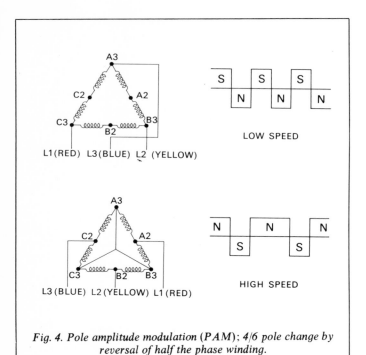

Fig. 4. Pole amplitude modulation (PAM); 4/6 pole change by reversal of half the phase winding.

The advantage of PAM motors is that the whole of the copper is active on each speed, whereas in the alternative dual-wound 2-speed motors the slots have to accommodate two windings, only one of which is active at one time. Hence PAM motors are physically smaller and less costly than dual-wound motors of similar outputs.

Synchronous motors

Considerable design and development has taken place in the fifteen years to 1971 in small synchronous motors of the reluctance type. Developed originally from the squirrel cage motor, the earliest motors were made by simply milling channels in the periphery of squirrel cage motors, so that the magnetic reluctance along one axis (the direct axis) was kept low, whilst the reluctance along an axis at 90 electrical degrees (the quadrature axis) was increased. This resulted in a rotor which would align itself to the rotating magnetic field and run in exact synchronism, but with very modest performance.

The performance of synchronous motors can be improved by increasing the ratio of the reactances X_d and X_q along the direct and quadrature axes. This was achieved by introducing flux barriers or flux guides into the rotor core laminations, or by complete segmentation of the rotor. Much research and development has been conducted in America, and in Britain (notably at Leeds University) on variations of rotor geometry to optimise performance.

Small synchronous motors having permanent magnets embedded in the rotor are also a recent development. Although having superior performance to reluctance motors, they are far more costly to manufacture.

There is a small but growing market for small synchronous motors, wherever exact synchronism is required between two or more drives. Examples are to be found in the manufacture of synthetic fibres, in plastics, the glass industry, and in paper manufacture.

Chapter 3

Materials of Construction

G B Sugden *MSc CEng MIEE*
Lucas Aerospace Ltd

Materials used in the construction of electric motors fall broadly into the following categories: (1) Materials of the magnetic circuit. (2) Materials of the electric circuit. (3) Insulation materials. (4) Materials used for mechanical construction.

All motors comprise an electric circuit and a magnetic circuit electromagnetically linked together. The purpose of the insulation material is to confine the flow of electric current within the electric circuit only.

Numerous factors govern the choice of materials used in electrical machines, and the decision to use a particular material is usually based on a compromise of all such factors. This chapter describes the various materials which are employed in the construction of electric motors, and the important factors governing the choice of these materials.

MAGNETIC CIRCUIT MATERIALS

The magnetic circuit of almost all electrical machines consists of iron in some form. In machines where the field system is stationary, the frame constitutes the yoke which magnetically connects the poles. Forged steel, cast steel, and cast iron yokes are frequently used, but special exceptions include aircraft machines and very small motors, in which entirely laminated field circuits are used. Field systems which rotate are frequently made from a solid steel forging except where sheer size precludes this method and other means must be used; for example, poles assembled separately onto the yoke. In special cases a laminated construction is used, or the field system is machined from solid material.

Armatures, *ie* the part of the magnetic circuit which is associated with pulsations of the main body of flux, are almost always of laminated construction. If iron or steel carries a flux which is varying, hysteresis and eddy current losses are produced. The former are due to molecular action, while the latter are the losses due to currents flowing

Nomenclature
- f = frequency, Hz
- B = flux density, Wb/m^2
- H = magnetising force, At/m
- t = thickness dimension, m
- B_m = maximum flux density, Wb/m^2
- B_{rem} = remanent flux density, Wb/m^2
- H_c = coercive magnetising force, At/m

in the metal due to induced emfs. Investigations carried out on laminated material with sinusoidally-alternating excitation have shown that the losses are given in terms of the maximum flux density B_m by

Hysteresis loss $= k_h f B_m^x$ watts per unit weight or volume

Eddy current loss $= k_e t^2 f^2 B_m^2$ watts per unit weight or volume

The factors k_h and k_e depend on the magnetic characteristics and electrical resistivity of the metal. The exponent x of the maximum flux density B_m in the hysteresis loss usually lies between 1.5 and 2.3.

The thickness of the plate is denoted by t. Since the eddy current loss varies directly with the square of the plate thickness, obviously this loss may be reduced considerably by using thinner plates. Typical values of plate thickness lie between 0.35 mm and 0.5 mm although in some special applications 0.2 mm material has been used. To simplify calculations, the hysteresis and eddy current losses are usually combined together and presented as a family of curves showing losses against flux density for given material of thickness t at frequency f.

In electrical machines a sinusoidally varying flux is rarely achieved, the change is more complex and the iron loss is roughly proportional to B_m^2 but even for the same B_m, the specific loss is greater in the teeth than in the core. In fact, losses occurring in a machine may be more than twice as much as the values obtained from laboratory tests on carefully prepared specimens. Fig. 1 shows typical results obtained on various materials when tested in actual machines.

The permeability of magnetic materials is not a constant value but varies with flux density. This variation makes it necessary in practice to use magnetisation curves relating flux density to the magnetising force, for the rapid determination of the necessary excitation. Fig. 2 gives typical curves for some commonly used materials in electrical machines, although strictly speaking each specimen has its own magnetisation curve, but in practice it is usual to employ a single average curve for each class of material.

Lamination materials include sheet steel of several qualities, transformer silicon steel, special nickel-steel alloys such as 'Permalloy', 'Mu-metal', and cobalt steel alloys. The latter are being increasingly used in aircraft applications because of their high permeability. Various grades of silicon steels are widely used; they possess quite a wide range of different properties with regard to iron losses, saturation flux density, and mechanical qualities. Fig. 1 and 2 depict characteristics of one such silicon steel. The various grades are categorised in British Standard 601 parts 1 and 2.

Typical values of working flux densities for the materials shown in Fig. 1 and 2 are: (1: for industrial applications) cast iron, yokes and poles etc., 0.7 Wb/m^2; cast steel, poles 1.9–1.5 Wb/m^2; alloy steel plates, teeth 1.8–2.1 Wb/m^2 at 50 Hz, cores 0.7–1.0 Wb/m^2 at 50 Hz. (2: aircraft applications) 50% cobalt steel, teeth 1.8–2.2 Wb/m^2 at 400 Hz, cores 1.7–2.1 Wb/m^2 at 400 Hz.

Insulation

Insulating material must of course separate the laminations. One form consists of thin sheets of paper between adjacent sheets of metal. More commonly an insulating treatment or high resistance coating is applied to the individual laminations. The actual voltages encountered will vary, from less than 0.04 volt per centimetre of core length on small machines to more than 2 V/cm of core length on large high speed motors or generators.

The most widely used form of magnetic lamination insulation is core plate enamel, an organic varnish baked onto the laminations. This is relatively thick, and adds about 0.005 mm to the thickness of a lamination per side. Other types of lamination insulation include waterglass coating, and phenolic thermosetting resin filled with an organic silicate. In the latter type, if the resin is destroyed by overheating, the organic silicate decomposes to mineral silica which continues to provide separation between laminations.

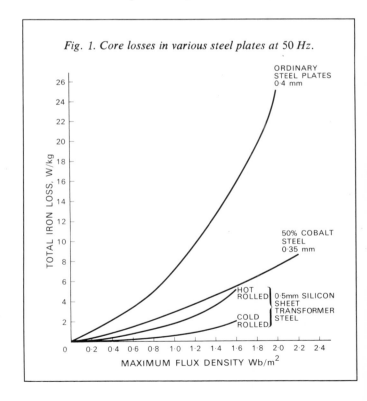

Fig. 1. Core losses in various steel plates at 50 Hz.

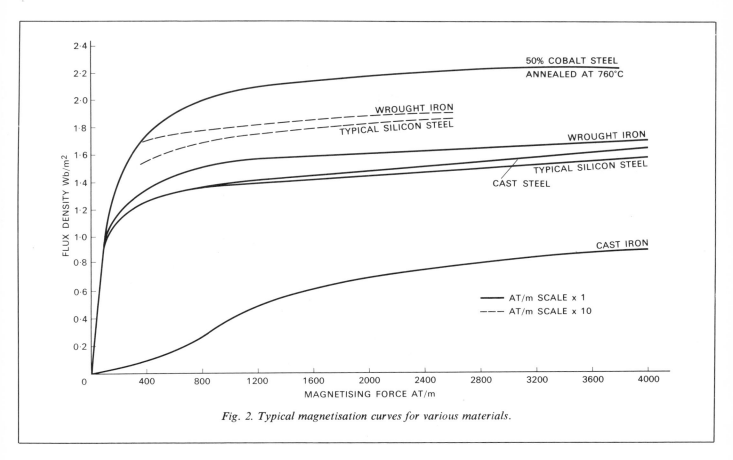

Fig. 2. Typical magnetisation curves for various materials.

Sometimes insulation is provided by oxidation of the lamination surface by heating in a controlled atmosphere; this gives quite a thin coating, adding about 0.5% to the thickness of a 0.5 mm lamination. On some low speed machines no intentional coating is applied, the natural film of oxide present on the material providing sufficient barrier. Space factors or 'stacking factors' for laminated material vary from 0.9 to as high as 0.98 for special purpose and aircraft electrical machines.

Permanent magnets
Permanent magnet materials are used extensively in a great variety of motors. Freedom from separately-fed excitation windings offers advantages. Permanent magnet fields are often used in small d.c. motors and fractional-power synchronous motors, for automotive, aircraft, toy and industrial applications. However, compared with similarly sized machines with wound fields, they have inferior power factor, efficiency and power output. These shortcomings may be offset by the economic advantages of employing permanent magnet materials.

There are numerous methods of arrangement of magnets to produce desired field arrangements; for example, a single annular two-pole magnet, axially magnetised and contained within an interdigitated assembly, is often used to produce a multipole field. Such an assembly is illustrated in Fig. 3. Particular arrangements are usually selected to give the cheapest fabrication costs. Another commonly-used process incorporates the magnets in an aluminium die casting; this forms a convenient method of assembly and also provides mechanical strengthening for comparatively brittle magnets which are to be run at high speeds. Various magnet field assemblies are illustrated in Fig. 4.

There are a wide variety of permanent magnet materials used in electric motors. Table 1 gives the properties of some commercially available materials. The properties listed are average values determined by statistical means. In practice, variations in performance of at least 10% between machines of identical design using permanent magnet materials is quite common.

A wide range of ceramic bonded barium ferrite magnet materials have recently become available. This material possesses unusual magnetic characteristics having low remanent flux density but high coercive force. Bonded ceramic magnets are becoming quite widely accepted for use in permanent magnet machines such as small d.c. motors.

ELECTRIC CIRCUIT MATERIALS

The conducting material most frequently used for armature windings is copper, because of all commercial metals it occupies the least space for a given current carrying capacity. In addition, it has many excellent mechanical qualities. It is easily drawn to wire and its ductility enables it to be bent without breaking; furthermore, it can be readily welded and soldered, and the action of air does not cause deleterious oxidation. The price of copper is relatively high, and aluminium wire is recognised as an alternative.

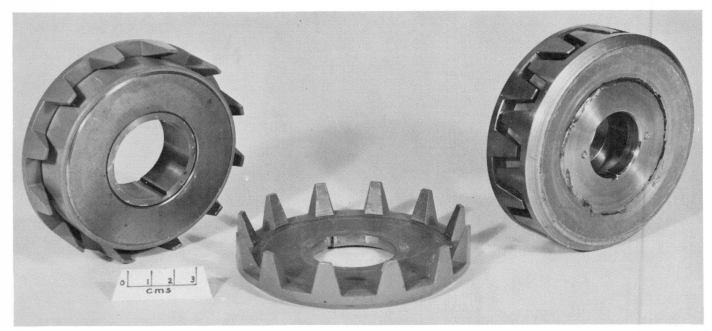

Fig. 3. Multipole interdigitated field assembly showing annular two-pole magnet.

The electrical conductivities of aluminium and copper are in the ratio 1 to 1.66 at 15°C so that if aluminium wire of the same resistance is used it should have a cross section of 1.66 times that of the copper wire which it replaces. It is very difficult to give reliable figures on the comparative cost of copper and aluminium windings, because the labour cost is often a large proportion of the total and this differs widely in different manufacturing organisations. The larger perimeter of conductor required in using aluminium results in improved heat conductivity and a lower mean temperature. The resistance is therefore correspondingly lower, so that it is often possible to use a cross section of less than 1.66 times that of copper wire without exceeding the specified temperature rise.

Table 1. Properties of permanent magnet materials.

Material	Remanence Wb/m²	(BH)max J/m³	Coercivity AT/m × 10⁴	B at (BH)max Wb/m²	H at (BH)max AT/m × 10⁴	How produced	Machinability
Columax	1.35	59 700	5.89	1.17	5.09		
Alcomax IV SC	1.22	41 400	6.21	0.90	4.62		
Alcomax III SC	1.32	48 500	5.57	1.07	4.54		
Alcomax II SC	1.37	47 000	4.78	1.16	4.06		
Alcomax IV	1.15	35 800	5.97	0.835	4.30		
Alcomax III	1.26	43 000	5.17	1.02	4.22	Castings (hard and brittle)	Grinding only
Alcomax II	1.30	43 000	4.62	1.10	3.90		
Alnico (high remanence)	0.800	13 500	3.98	0.52	2.60		
Alnico (normal)	0.725	13 500	4.46	0.47	2.88		
Alnico (high coercivity)	0.650	13 500	4.93	0.425	3.18		
Alni (high remanence)	0.620	9950	3.82	0.403	2.47		
Alni (normal)	0.560	9950	4.61	0.348	2.86		
Alni (high coercivity)	0.500	9950	5.41	0.304	3.26		
35% Cobalt	0.900	7560	1.99	0.593	1.27		Difficult to tap
15% Cobalt	0.820	4930	1.43	0.525	0.938		
9% Cobalt	0.780	3980	1.27	0.500	0.796		
6% Cobalt	0.750	3500	1.15	0.468	0.748	Rolled or cast	Suitable for all operations when annealed
3% Cobalt	0.720	2780	1.03	0.422	0.662		
Chromium steel	0.980	2270	0.557	0.620	0.366		
1.25% Tungsten	0.98	1980	0.477	0.640	0.310	Rolled	
Carbon steel	0.90	1580	0.398	0.618	0.256		
Bonded barium ferrite	0.2	8000	14.0	0.10	8.00	Ceramic bonding process	Grinding

Fig. 4. Various permanent magnet field assemblies.

Handling of aluminium coils during manufacture is by comparison easier because the density of aluminium is approximately half that of copper, but it should be remembered that the mechanical qualities of aluminium are not so good as those of copper and there is not quite the same ease in making satisfactory electrical joints.

On relatively small machines armature coils are wound from round wire; however, as the size of the conductor increases the poor space factor inherent in the use of round wire leads to the use of square and rectangular bars. The corners of the bars are usually rounded off during the process of rolling so as not to cut the insulation, and to permit uniform covering on the corners where enamel is used. Large rectangular sections have conductor widths several times larger than the thickness and the term 'strap' is used to describe them.

Rectangular section conductors have the added advantage of better heat dissipation and it is often possible to control the arrangement of the armature conductors so as to keep down the voltage between adjacent turns. Field windings frequently employ rectangular straps where the current is above, say, 100 A. For smaller currents rectangular wire is used down to about 5.5 mm^2 cross-sectional area; below this size it is more convenient to use round wire. It should be emphasized that these are generalisations, and there are likely to be many exceptions found in practice.

For the electrical circuit of squirrel-cage induction motor rotors, various materials are used. It is often an advantage to increase the resistance of these conductors without reducing their size, where resistance is necessary in order to give the motor the right characteristics, and where considerable mass is needed to limit the rate of heating. In such cases brass conductors have been employed, or copper conductors used in conjunction with endrings of high resistance.

Other materials used for squirrel cage rotor bars include phosphor bronze, tin copper alloy, and zinc alloys. For endrings, manganese bronze, cast brass and cast gun metal. Aluminium is quite extensively used; and with this material it is possible to cast the bars and rings onto the rotor as a single unit, often shaped with fan blades for cooling purposes.

D.C. motors using superconducting field windings have recently been constructed and tested, and these have certain advantages over machines of conventional construction, particularly for marine applications. Superconducting materials are materials which possess zero resistance at temperatures close to absolute zero, niobium-tin/copper composites have been successfully used. The application of superconductors to electrical machines is still in its infancy, and many problems, such as the size and reliability of associated refrigeration plants, must be resolved before the full advantages of superconductors can be reaped.

INSULATION MATERIALS

Insulation is a vital link in the application of electricity. The design of insulating parts and the character of insulation is an important factor in electric equipment design. Furthermore, insulation is a key factor in relation to the reliability of electrical equipment. All solid materials are conductors of electricity to some degree; insulators are conductors having a high resistivity such that the flow of current through them may be neglected for power equipment.

The dividing line between conductors and insulators is difficult to define, and they are separated by the class of semiconductors. Insulating materials are very variable in many significant properties and it is necessary to use statistical methods and a large number of observations to determine a usable value of the property being studied. Insulation resistance tests (often at 500 V d.c.) are commonly made to check insulation condition. The property of a particular insulation varies quite widely with temperature, humidity, cleanliness, as well as the magnitude of the applied voltage and the length of time that the voltage is applied. When these variables are properly understood many apparent inconsistencies of insulation resistance measurements may be explained.

Choice of insulation materials in electrical machines has a significant effect on size, weight, cost and reliability of a particular machine. It is not necessarily the case that the cheapest insulation produces the cheapest machine for the same machine performance. A more costly insulation which permits a thinner insulation wall having better thermal conductivity, or higher temperature operation, may result in a lower overall cost, or a machine with more competitive performance.

Selection of the insulation material for a particular application is not an exact science, it requires judgment and compromise based on at least the following considerations.
(1) Dielectric strength requirements.
(2) Mechanical characteristics required.
(3) Temperature classification.
(4) Special service conditions, *eg* chemical contamination, moisture.
(5) Compatibility with other materials used in the system.
(6) Manufacturing expertise available.

A summary follows of typical materials which are frequently used in electrical machines: (1) *Conductor insulation*: Paper, cotton, silk, asbestos, PVA, enamel, polyester enamel, glass fibre tape, mica tape, polyimide tape, polyimide enamel, polyester imide enamel. Conductor insulation is usually provided by some standard form of covering such as enamel, but with larger and special conductor sizes, mica tape may be applied directly to the strap before forming. If forming in such a way would lead to damage of the insulation, the bare copper is formed to shape, and taped afterwards by hand or hand-guided taping machine. Taping may be applied in several ways: (a) Butted, in which the adjacent edges of consecutive turns touch to give a smooth finish; (b) spaced, in which the adjacent edges of consecutive turns do not quite touch; (c) overlapped, in which there is always at least one thickness of tape along the conductor and a minimum of about one-third of the tape width overlapped. Tape is almost always overlapped and this is often specified as 'half lapped'.

(2) *Ground insulation* (*between winding and core*): Resin films, varnish treated cloth, paper backed mica, glass backed mica, continuous mica tape, fabricated cells using laminated and moulded materials, polyimide film. The simplest form of ground insulation is the slot cell, most commonly used on small random wound motors. Two conflicting characteristics are needed: high dielectric strength and maintenance of shape after forming. To overcome this, many slot cells are made of two materials, an outer component such as paper which retains preforming, and an inner flexible material, usually film, of high dielectric strength which is little affected by bending.

(3) *Structural binding*: Cotton tapes, glass fibre tapes, asbestos tapes, synthetic resin fabric tapes. Usually, large a.c. motors have to be designed to withstand sudden and appreciable changes in load during normal operation, and during starting the armature current may be five or more times full-load current. These currents in the armature coils produce electromagnetic forces that tend to distort the coil ends. The overhangs must therefore be supported so that the insulation is not damaged by these forces. Typical methods of bracing overhangs include: (a) multiple rings and spacing blocks between coils, the blocks being tied from coil to coil to give a rigid assembly; (b) one support ring to which every coil is lashed; these may be floating or attached to the frame. Rings farthest from the core may be made of steel, but near the core nonmagnetic material is used to reduce the effects of leakage flux and minimise eddy currents.

Impregnants
Another vital part of an electrical insulation system is the impregnation. The function of impregnants is to coat the surfaces and fill the insulation voids of electrical windings. For this purpose special varnishes and resins are used. Impregnation increases the resistance of the insulation to penetration of moisture or other contaminants and also minimises the effect of conducting compounds on vulnerable creepage surfaces. Varnishes also provide adhesion of windings in the slots and produce rigidity in the insulation system to withstand mechanical shocks and vibration. Filling of internal voids improves the thermal conductivity of the insulation and minimises the effects of ionisation. Proper sealing of the insulation surfaces also promotes increased insulation life.

Various compounds are used for impregnation, usually carried in a solvent which holds the solid resins and gums in fluid form so that they may be applied to required positions before the solvent is evaporated to form uniform thin films. Thinners, which may or may not be true solvents of the solids, are often lower-cost liquids used to dilute varnishes to the desired viscosity for a particular process.

Typical resins used in impregnating varnishes are: epoxides, silicones, phenolics, polyesters, and polyimides; other resins include polyethylene and fluorocarbons.

Insulation material is classified according to operating temperature, in British Standard 2757:1956 'Classification of insulating materials for electrical machinery and apparatus'. There are seven classes of insulating materials, for which the relevant operating temperatures (°C) are shown here in parentheses: Y (90°); A (105); E (120); B (130); F (155); H (180); C (above 180).

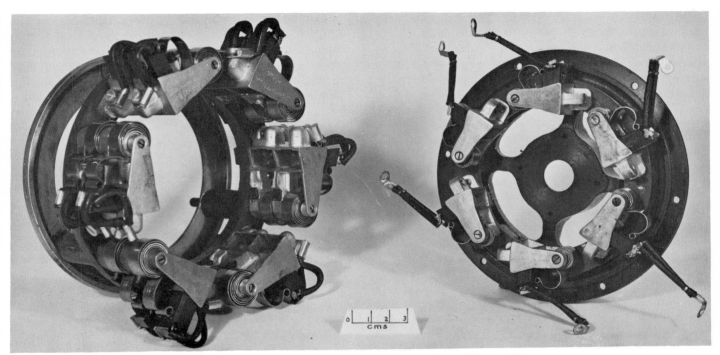

Fig. 5. Commutator brush gear assemblies for high speed d.c. machines.

A summary of materials which belong to these classes is now given; (a comprehensive classification of most insulating materials is given as a guide in the Appendix of BS2757).
Class Y. Materials or combinations of materials such as cotton, silk and paper without impregnation.
Class A. Materials such as cotton, silk and paper when suitably impregnated or coated with dielectric liquid.
Class E. Wire enamels based on polyvinylformal. Paper laminates, cotton fabric laminates.
Class B. Glass fibre, asbestos, built-up mica, glass fibre, and asbestos laminates.
Class F. Materials or combinations of materials such as mica, glass fibre, asbestos, with suitable bonding such as alkyd, epoxy, and polyurethane resins.
Class H. Silicone elastomer and combinations of materials such as mica, glass fibre, asbestos bonded or impregnated with materials such as silicone resins.
Class C. Mica, glass, quartz, and asbestos. Polyimide films and varnishes.

COMMUTATORS

Commutators are assembled by alternately stacking bars of hard-drawn copper and sheets of insulation such as mica or micanite, in a circular form within a metal ring whose inside diameter is somewhat larger than that of the finished commutator. During assembly the thickness of each sheet of insulation is measured and if necessary adjusted to within specified limits. Use of mica is convenient because it is easily split to size, typical separation thickness is 0.8 mm. After assembly the segments are clamped tightly together, and are usually held in place by vee-shaped clamping rings, one at each end. The copper segments are insulated from the vee-rings and supporting hub by material such as micanite, which has been carefully moulded to the correct shape. In some cases sufficient clearance can be left to avoid the need to insulate the hub. Vee-rings are made of cast iron for low speed machines, but steel is necessary for high speed machines.

In large machines the 'risers' connecting the commutator segments to the armature coils are made of copper strips, these being riveted and soldered into slots in the commutator bars before the commutator is assembled. For smaller machines the risers are formed by simply having on each bar a lug, which is slotted to take connections from the armature coils.

Small commutators sometimes use a soft grade of mica between bar segments which wear down at roughly the same rate as the copper segments. More often hard mica is used and the top of each mica segment is removed, a process known as 'undercutting'. For smaller sizes, commutators are produced by a moulding process using materials such as phenolic, epoxide or alkyd resin, the segments being assembled as described above but the vee-rings are replaced by the moulding material. Even smaller sizes use moulded face commutators.

BEARINGS AND LUBRICANTS

Bearings of electric motors are usually of the ball, roller, sleeve or needle types. Normally an armature is supported by bearings at both ends of the shaft; information about bearings for particular applications is readily available from bearing manufacturers. Motors which are mounted in other than horizontal positions require thrust bearings, which take up the thrust caused by the added weight of the armature.

Considerations in choice of bearings include: (a) Type of lubrication, *eg* grease or oil feed. (b) Speed and load on bearing. (c) Required life and type of duty *eg* intermittent with long periods of standing. (d) Temperature range of operation. (e) Nature of environment, *eg* presence of contaminants or lubricants from other areas, possibility of shock or vibration. (f) Size of shaft and other dimensions limiting space. (g) Degree of alignment required in a particular application and operating position; design limitations with regard to accessibility of bearings.

Where there is danger of penetration into the bearing or into the motor by outside contaminants a sealing arrangement may be required. Such conditions influence bearing selection and the mechanical configuration.

Oil is the generally preferred lubricant for most high speed and high temperature applications. Mineral oils are frequently used, but for wide temperature ranges, synthetic oils are employed. Good general purpose greases are commercially available and are widely used. A grease should be clean, containing no abrasive impurities, and should not oxidise during long periods of storage and operation. It should possess low resistance to shear and low fluid friction and be serviceable over the required temperature range. Prelubricated shielded and sealed bearings are extensively used where space limitations rule out the use of a grease-filled housing, where housings cannot be kept free of grit or other contaminants, or where relubrication is not practicable.

BRUSHES AND HOLDERS

Natural graphite was originally chosen for sliding contacts because of its properties as a solid lubricant. Its continued use has shown that graphite, in common with other forms of carbon, exhibits other valuable properties. Prominent among these properties are:
(1) Resistance to the effects of high temperature. Instantaneous very high local temperatures occur under all sliding contacts; carbon retains its properties because it remains solid at these temperatures.
(2) Carbon has a low density and therefore has low inertia so that the carbon brush follows the irregularities of the moving surface.
(3) Carbon does not weld to metals under conditions where metals would weld together, such as the heat of an electric arc.
(4) Carbon has a low coefficient of friction against metals.

There are various forms of natural carbon produced by the decomposition of carbon compounds, and each has physical characteristics which differ considerably. They retain these characteristics up to about 2500°C at which they turn to graphite which is the stable crystalline form of carbon.

The whole range of brushes can be broadly classified into six groups: (1) Natural graphite. (2) Hard carbon. (3) Electro-graphite. (4) Metal graphite. (5) Metal carbons. (6) 'Treated' carbons.

Brushes have been treated with the following materials and combinations of materials; barium fluoride, molybdenum disulphide, PTFE, and zinc oxide. The brushes are either impregnated with these materials or the material is embedded in cores. In aircraft electrical machines, treated brushes are frequently used to improve brush wear rates at altitude. Brush wear rate is greatly impaired at the low air densities and extreme dryness that occur above about 6000 metres. Using these special materials, brush wear rates at altitude may be as good as obtained at ground level with a natural graphite brush, provided that a 'skin' is properly formed on the commutator. To produce this skin using treated brushes necessitates running at ground level for 12 to 15 hours with the machine operating at approximately half full-load. Recently, machines using brushes made of carbon fibre have been tried, and these appear to be quite promising, particularly in the field of superconducting motors.

The most common type of brush holder is the slide type in which the brush is fitted in a box-shaped guide. Pressure is exerted on the brush by a spring acting on the top of the brush. The construction of a brush box must be as rigid and solid as possible. The most satisfactory materials used in brush box construction are brass, pressed steel, bronze or some other copper-based alloy.

Helical coil springs in compression, and clock type spiral springs are used to apply pressures on the brushes; they are generally made of phosphor bronze or spring steel. To form a good connection between a brush and its holder a plaited or twisted copper flexible connector is often used, this being securely fixed into the brush during manufacture. The other end of the flexible connector is secured to the bush holder using a tag. Typical commutator brush gear assemblies used in high speed d.c. machines are illustrated in Fig. 5.

BIBLIOGRAPHY

1. A E Clayton and N W Hancock. The performance and design of direct current machines. *Pitman* Engineering Degree Series.
2. G L Moses. Insulation engineering fundamentals. *Lake Publishing Company* (reprints of articles in *Insulation* magazines).
3. A J Dekker. Electrical engineering materials. *Prentice-Hall* Electrical Engineering Series.
4. Classification of insulating materials for electrical machinery and apparatus. British Standard 2757:1956, *British Standards Institution*.
5. F G Spreadbury. Fractional h.p. electric motors. *Sir Isaac Pitman & Sons Ltd.*
6. C G Veinott. Fractional and sub-fractional horse-power electric motors. *McGraw-Hill Book Company.*
7. Steel sheets for magnetic circuits of power electrical apparatus. British Standard 601: pt 1 and 2, *British Standards Institution.*
8. A D Appleton. Superconducting machines. Reprint from *Science Journal* April 1969, v. 5, no.4. International Research and Development Co Ltd.
9. R J Parker and R J Studders. Permanent magnets and their application. *John Wiley and Sons, Inc.*
10. Permanent magnets. *Permanent Magnet Association publication.*

Chapter 4

Miniature a.c. Motors

E H Werninck *AIEE MIMC*
Consultant

Small electric prime movers find so many applications in nearly every field of human activity that they are manufactured in a great variety of designs, and in very large numbers. To achieve economic production, interchangeability, and basic performance standards, a range of electric motors generally known as fractional horsepower motors has been defined. This group, which comprises motors with outputs up to 746 watts (1 hp) at 1000 rev/min, as well as miniature motors which have outputs expressed in milliwatts are dealt with here. The types range from three-phase induction motors (which are scaled-down versions of their larger counterparts) and special versions suitable for single-phase running, to series wound types and synchronous motors which, because they are small, permit the use of other electromotive principles.

The single-phase induction motor is particularly important since the most universally and readily available supply is single-phase a.c. generally with a frequency of 50 Hz. A notable exception is in the United States and Canada, where the frequency is 60 Hz. To start induction motors satisfactorily when supplied with single-phase requires some means of phase splitting, which in the shaded-pole motor is achieved by placing a short-circuited conductor around part of the stator poles.

The maximum speed of induction motors is limited by the frequency of the supply, and a 2-pole 50-Hz induction type motor cannot be made to run at more than 3000 rev/min. In some applications, such as portable tools and domestic appliances, where weight has to be kept to a minimum and starting torques several times normal running torque are required, an a.c. series motor is therefore used. This type of motor was developed from the d.c. series motor and extensively used in vacuum cleaners and became known as a 'vacuum cleaner motor'. The current, and more useful description is 'universal motor', because such motors also operate satisfactorily from d.c. supplies.

It is a characteristic of small motors that many are defined and classified by their principal application, *eg*: blower,

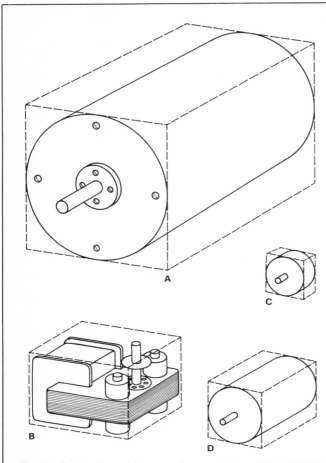

Fig. 1. Comparison of motor characteristics. (A) BS2048 B56 frame; 180–250 W ($\frac{1}{4}$–$\frac{1}{3}$ hp), fan-cooled, 1400 rev/min; single-phase, split-phase or capacitor start; dimensions overall 6.5 × 6.5 in × 8 in long. (B) Shaded pole motor; 1.8–2.5 W, 2200 rev/min, 50 Hz; 2.25 in high, 2.5 in deep, 3 in wide. (C) Clock motor; 100 mW, 200 rev/min, 50 Hz; 1.5 × 1.5 in, 1 in deep. (D) 180 W ($\frac{1}{4}$ hp), 22000 rev/min, 400 Hz (intermittent rating); 2.25 × 2.25 in, 4.5 in deep.

oil-burner, refrigerator, compressor, record-player, and clock-motors. Some manufacturers also use these applications for defining the rating under certain specified conditions.

A type of single-phase unit known as the repulsion–induction motor combines the high starting torque of the series motor with the constant speed running characteristics of an induction motor. Since this motor requires a special arrangement which short-circuits the commutator when it has reached a certain speed, it is the most costly of single-phase motors and its choice will be influenced by other possibilities of starting a load requiring high initial torques.

Fig. 1 shows graphically various types of motors, typical ratings, and the effect of speed on size for a given output. Thus Fig. 1A is the outline of a standard 186 watt (0.25 hp) 4-pole induction motor (typical load speed 1430 rev/min); 1B represents a shaded pole motor, with an output of about 3 watts at about 1400 rev/min. A motor with an output around 0.1 W can, when geared down to say 1 rev/h, drive a chart-drive mechanism or a timer at synchronous speed;

it originates from the electric clock industry and is shown in 1C. To show the benefits obtained by using a high frequency supply the outline of a 186 watt (0.25 hp) aircraft 400-cycle motor is shown in 1D. A universal motor would have about the same diameter but, to accommodate the commutator requires to be about 20 mm (0.75 in) longer.

The aircraft motor shown (Fig. 1D) is built in one of the standard frame sizes which originated in the military equipment field around 1950, and which are known as MIL or NATO sizes. As the same fixing arrangements are also used for synchros, resolvers, servomotors, and potentiometers it makes a contribution to standardisation in special equipments. Fig. 2 and Table 1 show brief details of these sizes.

Definite purpose motors such as record-player, vacuum-cleaner, and clock motors soon found so many other applications that specialist fractional horsepower motor manufacturers included them in their range and can now often offer a wide variety of special purpose designs from which a preliminary selection can be made. The manufacturer will primarily classify the motors according to their main constructional features, as shown below. In some cases, however, to give guidance on the operating conditions for which the motor was originally designed, typical applications may be quoted. These indicate starting characteristics, cooling air requirements, and noise level. Most motors supplied to equipment manufacturers are either of the housed or skeleton type. Where stator and

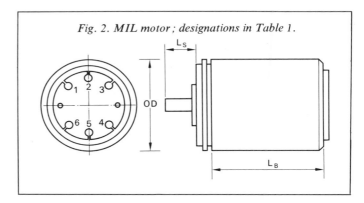

Fig. 2. MIL motor; designations in Table 1.

Table 1. Typical dimensions of MIL motors.

Designation	OD nominal in	L_B in	L_S in
08	0.75	0.8	0.365
11	1.06	1.6	0.50
15	1.45	1.6	0.50
18	1.75	2.0	0.60
23	2.25	3.1	1.00

rotor units are supplied there must be close liaison to ensure that performance of the final assembly is not affected by mechanical deviations in alignment and airgap.

Classification

The following type classification is used here to survey the more common types:

(1) *Universal or a.c. series motors*: fractional uncompensated windings in sizes up to 300 W. These are usually very high speed units with load speeds of 5000 to about 15 000 rev/min and are an obvious choice where a high power to weight ratio is essential, and where it is uneconomical to provide a high-frequency supply for induction motors.

(2) *Squirrel cage induction motors*: Apart from the type of enclosure (see construction) this group can also be sub-divided according to supply and method of starting. (a) *Three-phase motors*—these can usually only be used for industrial applications where they are built into a machine or remain fixed. (b) *Single-phase induction motors*—these are subdivided according to starting method and usually appear as: (i) *split-phase motors*; (ii) *capacitor start*; (iii) *capacitor start and run*; (iv) *shaded pole*; (v) *repulsion induction*; (vi) *synchronous*, a feature that generally overrides the above classification as nearly all such motors start as induction motors, and could thus also come under any of the above headings (i) to (iv).

(3) *Special rotor induction motors*: Apart from the normal clock motor, special synchronous motors such as the hysteresis motor use solid iron rotors. Other applications are high-speed variable-frequency motors such as used in dental drills. One should also mention that eddy currents induced in copper discs or cylinders also provide 'fleapower' motors as evidenced in the electricity meter.

PRINCIPLES OF OPERATION

Universal motors

Because of the few d.c. supplies still in use, it is now usually of little importance to design universal motors so that their performance is, as nearly as possible, equal on a.c. and d.c., but their operation is best understood if they are considered as a d.c. series motor. The design aim is to keep armature current and field flux in phase so that at any instant the unit behaves exactly like a d.c. series motor. Inductance in the field and a.c. losses are kept to a minimum by using laminations and relatively few turns. The deviation of d.c. performance is also due to mutual inductance and setting up of transformer currents, causing a.c. speeds and starting torque to be lower than that obtained on d.c. For motors above 300 W output it becomes necessary to use compensating windings or a distributed field, to obtain acceptable commutation. Improvements in materials and techniques of manufacture have made it possible to market portable tools and appliances which have motor load speeds of some 14 000 to 20 000 rev/min.

Squirrel cage induction motors

The rotating field created by a three-phase supply gives a very useful constant-speed motor characteristic. Further-

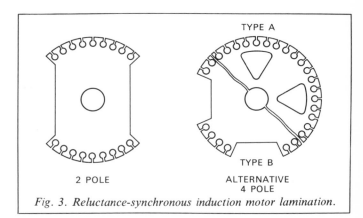

Fig. 3. Reluctance-synchronous induction motor lamination.

more, by careful rotor design such induction motors can be made to give very satisfactory starting torques with reasonable currents when started direct on-line.

When the supply is single-phase (more usual for small motors) and this is applied to a stator the resulting pulsating field produces no torque. If, however, the motor is made to approach its synchronous speed, the reaction of the rotor current helps to set up a rotating field, and as long as the single-phase supply is maintained, the motor will not only continue to run, but actually deliver power. In practice, single-phase motors are provided with a second winding in which the current is caused to be out of phase with that in the main winding, either by using many more turns or by putting a capacitor in series. This produces a flux which in the ideal case would be in quadrature with the main winding flux, thus turning the motor into a two-phase machine which would behave rather like the three-phase motor described. The various types of single-phase motors are discussed further when considering their starting and running characteristics.

The normal squirrel cage induction motor has a finite number of rotor conductors, and so-called slot effects cause harmonic torques and noise. To overcome this the bars are usually skewed, which results in further loss of torque and efficiency. If part of the iron circuit is removed, as shown in Fig. 3, the rotor polarises when near synchronism and actually pulls in if the load torque is not too high. A two-pole rotor has two possible synchronising positions, whereas a four-pole rotor has four.

Hysteresis motors have rotors made of hard magnetic material, which becomes polarised in any rotor position so that the rotor can pull in any position around its periphery. Single-phase stators must be very carefully designed to give a good rotating field for starting.

The clock motor, which usually has a magnetised multipolar rotor, is so designed that when the stator is energised, the rotor tries to move towards the nearest pole. If the direction is incorrect, it is prevented from building up speed by a mechanical stop which causes the rotor to recoil and start to rotate in the correct direction. For this reason clock motors and most shaded pole motors with short-circuited copper loops are not reversible.

CONSTRUCTION

Constructional features have a considerable influence on the size, performance, and cost of small electric motors. Extensive research and development has done much to make motors more efficient, quieter, more reliable, and longer lived, on the one hand, and also more economical to produce. Standardisation necessary for long production runs has had to be balanced by introducing alternative features at various stages in manufacture. Thus the same stator assembly can be wound for different voltages and ratings and brought together at the assembly stages with end frames containing either plain or ball bearings and ventilating openings, or totally enclosing the motor. In Britain BS2048 is used for the fixing dimensions, and enables equipment manufacturers considerably to reduce the purchasing specification for their motors and to use alternate suppliers.

Taking one of the frame sizes (B56) of BS2048 and the catalogue of one of the leading manufacturers as an example, one finds that, in this frame size, outputs range from 93 W (0.125 hp) at 960 rev/min to 560 W (0.75 hp) at 2850 rev/min for drip-proof ventilated split-phase and capacitor start single-phase induction motors. Typically these motors are suitable for up to 60 starts per hour, and can be provided with a special resilient mounting which greatly reduces the direct and indirect noise during starting and running; the mounting bracket can also incorporate automatic belt tensioning. The stator is enclosed in a steel tube, and the endframes which are die-cast in aluminium alloy can provide a variety of features. In addition to alternative bearing and ventilating arrangements they sometimes incorporate a mounting flange. It should also be noted that the BS2048 Pt.1 B56 base is interchangeable with the American NEMA56 dimensions. Split-phase and capacitor-start motors may also contain a centrifugal switch, which under average operating conditions should not limit the life of the motor.

Sleeve bearings are provided with felt-packed reservoirs which hold sufficient oil to last the life-time of the motor under intermittent operating conditions, and require the addition of only a few drops of oil once a year if on continuous duty. End-thrust and axial oscillations that may be set up by the drive are absorbed by thrust washers at either end. The method of fitting of ball bearings when assembling the motor has considerable influence on their life and quiet running. Ideally one bearing should be fixed axially whilst the second one floats in the housing to allow for thermal expansion and optimum alignment of inner and outer tracks. One of the difficulties in this is that, due to rapid acceleration on starting, and vibration during running, the outer race begins to creep, eventually rotates, and causes the bearing to overheat and fail prematurely. A special patented liner which, it is claimed, eliminates this problem, is incorporated in some designs (eg of English Electric—AEI). Diecast aluminium rotors usually have fan fins cast at each end for ventilation and are virtually indestructible in service.

Smaller induction motors, including the shaded pole family and universal (series-) motors, can also be obtained in fairly conventional housings, though some mounting arrangements are specially adapted to fit particular requirements. The absence of space limitations and particularly favourable ambient conditions have led to the use of open or skeleton constructions. In such motors the stator stack itself forms the body and the endframes are diecast or metal-pressings. A typical example is the shaded pole motor specially developed for low-cost record-players. (Fig. 1B). In contrast, a special type of external rotor synchronous motor has been produced for high quality turntables to ensure smooth running (absence of 'wow') and minimum stray magnetic flux (detected by sensitive pick-ups and reproduced as hum).

For many applications a short axial length, can not only have a very considerable influence on the design of the driven device, but bring with it other benefits such as quieter running and better cooling. In axial airgap motors stator and rotor resemble two discs in which one side forms the 'active' or airgap part, whilst the other provides the flux return path (Fig. 4). This type of motor is primarily suited to flange mounting and the fixing flange usually contains the stator stack and bearing housings as shown in Fig. 5. This illustration also shows that the large exposed rotor surface makes it easier to dissipate rotor losses, and thus prevent overheating when the motor is started and stopped frequently or speed-controlled. Another compact drive arrangement with automatic belt tensioning is shown in Fig. 6. In many applications in industry as well as in the domestic field (eg washing machines and spin dryers) it is desirable to bring the motor to rest as quickly as possible, and several special constructions of conventional motors (such as conical rotors) have been devised to achieve this. In the axial airgap motor no such modifications are

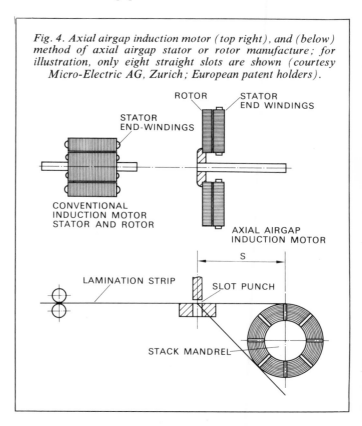

Fig. 4. Axial airgap induction motor (top right), and (below) method of axial airgap stator or rotor manufacture; for illustration, only eight straight slots are shown (courtesy Micro-Electric AG, Zurich; European patent holders).

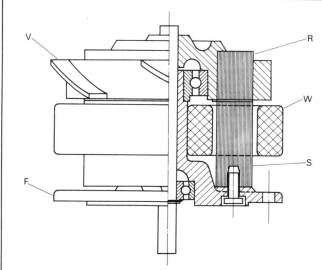
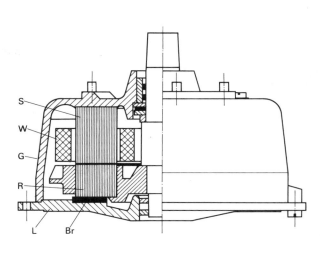

Fig. 5. Flange mounted axial airgap motor; F fixing flange; S stator stack; V integral fan blades on rotor; W stator winding.

Fig. 6. Axial airgap brake motor: L bearing endframe; Br brake lining; G housing; R rotor iron; S stator iron.

necessary, and it suffices to replace the normal deep grooved ball bearing by needle or roller types and to attach a braking lining to the face of the rotor.

Though machines for winding small stators are now on the market, they only become economical when very large throughputs of the same type are possible. The more difficult and costly manufacture of the axial airgap motor stack (shown diagrammatically in Fig. 4) is balanced to some extent by the greater ease of winding.

Shaded pole motors with outputs up to about 15 W often have special forms of stators (sometimes known as horse-shoe types) with the winding reduced to a single coil. Though this reduces manufacturing costs to a minimum, it results in lower efficiency and makes it more difficult to avoid noise-producing stray fields and torque harmonics.

CHARACTERISTICS AND PERFORMANCE

Universal or series motors

Universal (series) motors have maximum torque at starting and a speed which is very load-dependent, thus requiring careful estimate and control of the load if some consistency in speed is required. For such applications as portable tools, vacuum cleaners, mixers and so on, speed is not critical and the normal series characteristic motor is most suitable. The high speeds make it essential to balance armatures and carefully turn commutators concentric with the bearings. Such motors normally operate satisfactorily from supplies which may vary by plus or minus 10% from the nominal value. If units are to be exported, the regulations obtaining in the destination country for maximum heat rise under overvoltage conditions, clearance of live parts, and radio-interference suppression must be studied and carefully observed. Most universal motors are rated with forced air cooling by the fan fitted in the motor, and care must be taken that, under all operating conditions, air can enter and exhaust freely. Because on no-load such motors can reach extremely high speeds (up to 30000 rev/min) and also operate at speeds where air and some mechanical noise is inevitable, they are unsuitable for locations where quiet running is of paramount importance.

The limiting factor for running periods before services is almost always brush-wear; 1000 hours is a typical value for normal operation. Frequent starting and intermittent heavy duty can shorten this to a few hundred hours, which for some domestic appliances can still represent several years use (daily use for half an hour is 182.5 hours a year).

As with any engineered product, the maximum output from a given frame size may for reasons of economic manufacture be considerably reduced and the graphs shown here are typical of motors of fairly high quality. Fig. 7 shows the performance of a typical universal motor.

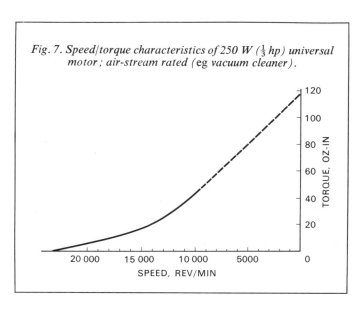

Fig. 7. Speed/torque characteristics of 250 W ($\frac{1}{3}$ hp) universal motor; air-stream rated (eg vacuum cleaner).

Induction motors

Three-phase versions: Most requirements in this category are met by the standard versions of mains frequency motors offered by manufacturers. These are usually wound for: 290/320–500/550 V; 220/250–380/440 V; 200/220–346/340 V 50 Hz.

Preferred outputs are $\frac{1}{8}, \frac{1}{4}, \frac{1}{3}, \frac{1}{2}, \frac{3}{4}$ hp (approximately 93–560 W) 4-pole having a rated load speed of about 1425 rev/min. These motors may be started direct on line, at a starting torque of the order of three times full load torque and starting current six times rated load current.

For factory installations, and special machines such as are used in the woodworking industries, where it is economical to install high-frequency generators, motors for 200 cycles may be required. Aircraft and military polyphase motors are, however, wound for 400 Hz, which is preferable as the most useful speed of around 11 000 rev/min can be obtained from 4-pole machines with more economical and therefore efficient windings.

In addition to the absence of starting problems, the three-phase motors, because they work with a true rotating field, are not only inherently quieter, but far less likely to set up resonances due to torque pulsations.

Single-phase versions: It is practical to operate small three-phase units from a single phase supply by connecting a capacitor between the third phase terminal and, according to direction of rotation required, one or the other of the phases to which the supply is connected. The capacitor required is usually larger than that for a two-phase wound motor, but has the advantage of giving the motor a slightly higher output, particularly if an additional tuning capacitor is put across the leading phase. A typical instrument motor gives an output of 25 W three-phase and 19 W single-phase with 2.5 microfarad capacitor. On three-phase, the motor has a high resistance rotor characteristic with a maximum torque of 0.3 Newton-metres (43 oz in) which occurs at starting. On single-phase the maximum torque of 0.18 Nm (25 oz in) occurs at 1000 rev/min, whilst the starting torque of 0.29 Nm is slightly less than a third of the three-phase value. In both cases the rating on full load is continuous as determined by the total losses of 28 W, which in this frame-size was found to be the maximum which could be allowed for the usual 55°C temperature rise.

Split-phase motors

The auxiliary phase required for starting a single-phase induction motor can only produce the necessary phase displaced flux if the current flowing in it is also phase displaced from the main winding current. If a true two-phase winding is used this could, in theory be achieved by putting a resistance in series with one winding and placing them both in parallel across the supply. The starting torques obtainable by such an arrangement would be far too small to be of practical value. In the split-phase motors which are widely used in such applications as washing machines and oil-burners, the auxiliary phase is used for starting only and is disconnected when the motor has reached about 70% of the theoretical synchronous speed. This allows most of the winding space to be taken up by the main winding and the two windings to be wound with substantially different resistances. Typical starting torques for this type lie around 150% rated full-load torque.

For very low powers and starting torques more of the order of 50% permanent split-phase motors were at one time used, particularly for small fans. Nowadays the shaded pole motor is almost always found to be cheaper and more reliable.

Capacitor start motors

By varying the value of the capacitor and rating of the starting winding, a wide range of starting torques can be designed into capacitor start motors. The most usual value offered commercially is around 300% full load torque, which is of the same order as that for three-phase versions. The values of starting capacitances used can be of the order of 700 μF for 110 V, and 130 μF for 220/240 V, and many years of development have resulted in special fairly cheap electrolytics which can be mounted on the motor.

Capacitor start and run

A distinction must be made between single and two value capacitor motors. In small instrument sizes and larger low starting torque units it is possible to leave the capacitor in

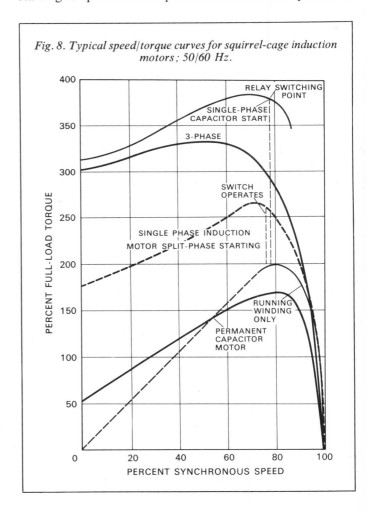

Fig. 8. Typical speed/torque curves for squirrel-cage induction motors; 50/60 Hz.

circuit with the suitably rated auxiliary winding. These motors are also referred to as permanent split capacitor motors and use paper-type capacitors, usually separately mounted.

Capacitor start/capacitor run motors in one commercial range have the running capacitor, typically 12 to 15 μF, connected permanently in series with the main winding. The action of the centrifugal switch is thus to take the starting capacitor out of circuit. Fig. 8 summarises and contrasts typical induction motor torque speed curves.

Shaded pole motors

There is usually only one main winding in shaded pole motors, and this is connected to the supply. The auxiliary winding in most cases consists of one or more short-circuited copper loops embracing up to one-third of the pole arc. The current induced in the shading loops produces a flux which lags the main flux and thus provides the conditions necessary for setting up a rotating field. Elaborate stator winding machines capable of winding multislot small stators are a recent introduction, so that formerly much effort was expended to develop single coil small motors with acceptable performance. The design of the special horseshoe lamination shown in Fig. 1B went the full circuit when one well known American manufacturer replaced the shading coils by small former wound coils, which could be used for starting the motor with a capacitor, or controlling it from a valve amplifier, thus making it into a high starting torque or servo-type unit.

The fact that the flux in the shaded pole always lags behind the main flux fixes the direction of rotation, which in some applications is a disadvantage. Some shaded pole motors with two or more main coils have therefore been designed and produced. Another version of the shaded pole motor, referred to as the shaded loop motor, uses a distributed winding as the main winding and a shaded winding consisting of one or at the most a few turns of thick copper wire, shorted either by soldering or by a shorting bar externally.

Fans, record-players, displays and many other devices do not require high starting torques or very powerful motors and so still provide a very large market for the shaded pole motor. For fan applications, in particular, the shaded pole motor has the further advantage that it can be speed-controlled over a fairly wide range by a simple series resistance.

Repulsion induction motors

Repulsion induction motors have a wound rotor, a form of commutator, brushes, and a short-circuiting device. On starting, current is induced in the rotor (which is completely isolated from the mains) and this flows through part of the winding joined by the brushes. The brushes are short-circuited, and their position is critical for obtaining maximum torque, which of the order of 300% rated torque. When the motor reaches about 75% of its synchronous speed centrifugal force actuates the shorting device and the rotor is turned into a squirrel-cage motor. The brushes, which at this stage have no useful function to perform, may also be lifted off to prevent excessive wear. It need hardly be stressed that this motor is not only costly but very susceptible to adverse operating conditions, so that its selection is likely to be restricted to very special applications.

Synchronous motors

The special construction of the 'clock' type synchronous motor described earlier gives some clue to the very poor efficiency, which may be less than 1%. While this may seem incongruous in a world striving for greater efficiency in everything that consumes power, one must consider that the power involved is only 2 W to 3 W and the cost of manufacture a fraction of that of a more efficient motor. The subsynchronous motor has the advantage that for clocks it requires less gearing and still can deliver adequate power. For larger industrial clocks and instrument applications various forms gradually approaching conventional induction motors are manufactured and usually offered with a variety of integral reduction gears.

Considering the larger motors (up to 750 W at 1000 rev/min) one can distinguish between salient pole, induction, and hysteresis types. The first, which again are scaled-down versions of widely used large synchronous motors, have no inherent starting torque and must be provided with special windings or external starters. In very small sizes, various special designs involving permanent magnets have been developed but by far the most economical and useful version is the specially adapted induction motor. This is variously described as variable reluctance, salient pole, or synchronous reaction motor. Up to about 95% of synchronous speed the motor behaves exactly like an induction motor and can therefore be any of the types described. Provided the load is relatively small and does not have excessive inertia it will then accelerate further up to synchronous speed. The rotor polarisation is derived from the rotating field and there are as many definite rotor positions as there are poles. Thus the two-pole motor has two distinct positions 180 degrees apart. This and the fact that load pulsations can cause the rotor to oscillate by a few degrees can be a distinct disadvantage in certain scientific and recording instruments.

One company, which markets a wide range of synchronous induction motors has also developed a range of hysteresis synchronous motors for applications where more than one speed or the synchronisation of large inertia loads are required. On starting the hysteresis rotor behaves like a high resistance induction motor with an infinite number of rotor bars. The run-up under the action of eddy currents and hysteresis effect gives almost constant torque up to synchronous speed. Once synchronised, the hysteresis motor provides a particularly smooth drive over its rated torque range. One company which for many years has specialised in external rotor motors for, among other applications, high quality sound recording and reproducing equipment list a variety of hysteresis motors. Some of these are provided with cages for additional damping whereas others have composite rotors, part hysteresis part reluctance. The latter was specially developed for capstan motors in studio equipment and each motor is tested for wow and noise before despatch.

APPLICATIONS

There are many special small motors manufactured for specific purposes which, as has been indicated, find application in other spheres. The equipment designer should therefore at the earliest possible stage of a new project endeavour to define his requirements so that he can be advised on feasibility and on existing designs. Considering that some equipments used for measurement, analysis, data processing, and the like may have as many as five small motors, it is clear that the care with which these are selected has a marked influence on the technical and commercial quality of the final product.

ACKNOWLEDGMENTS

The author is indebted to many more people than can be mentioned here for providing information. Thanks are due in particular to English Electric—AEI, Brook Motors Ltd, Croydon Engineering Co Ltd, Micro-Electric Ltd, Switzerland, and Papst-Motoren AG, Germany, who gave permission to use their latest technical information.

BIBLIOGRAPHY

C G Veinott. Fractional Horsepower Electric Motors. *McGraw-Hill.*

F G Spreadbury. Fractional Horsepower Electric Motors. *Pitman.*

E K Bottle. Fractional Horsepower Motors. *Griffin* 1948.

R Beyart. Les Petits Moteurs Electriques. *Dunod,* Paris.

S F Philpott. Fractional Horsepower Motors. *Chapman & Hall* 1951.

A Skrobisch. How to apply sub-fractional horsepower motors. *Product Engineering* March 1946.

H M Fulmer. Applying special small motors for powering engineered products. *Product Engineering* Jan 1948.

Ing. H Jaun (Zürich). Elektrische Maschinen in axialer bauart. *Technica* Nr. 16 1964.

Dr Ing P Vaske. Einphasenmotoren für Kleingeräte. *Technica* Nr. 5 1964.

H J Cadoux. Selection of fractional horsepower motors. *Engineering Materials & Design* Nov 1958.

E H Werninck. Miniature motors. *Electrical Review* 25 12 1959.

British Standard Specifications: 170, 2048, 3861, 3456.

Chapter 5

Miniature d. c. Motors

D N Stanley CEng MIEE
Vactric Control Equipment Ltd

In general, the term 'miniature d.c. motor' refers to machines having an output in the order of 375 W (0.05 hp) and less, and in particular to those of 10–20 W. However, the scope of this chapter has been extended to embrace motors having outputs of up to 1.0 hp (745 W). Motors within the range 0–1.0 hp vary in construction and exhibit vastly differing characteristics depending upon the application. This chapter gives a brief insight into the characteristics and construction of most of the types in use.

WOUND FIELD MOTORS

Wound field motors are produced in varying sizes but mainly from 0.1 to 1.0 hp (75-745 W) and are, in general, for commercial industrial markets. The characteristics depend upon the method of connecting the field winding with respect to the armature and supply. Fig. 1(a), (b), (c) show the connections for shunt, series and compound motors respectively. The basic equation used in d.c. motors is the back emf equation:

$$E = (2Z/a)(N\Phi/60)p$$

where E = back emf; Z = total armature conductors; a = no. parallel paths in armature winding; N = speed in rev/min; p = no. pole pairs; Φ = air gap flux (Webers).
For a given motor

$$E = K_1 N\Phi$$

where $K_1 = \dfrac{2Z}{a} \cdot \dfrac{p}{60}$ = constant

The supply voltage V applied to the motor has to overcome: (a) the back emf; (b) the resistive drop in the armature circuit. Hence:

$$V = E + Ia \cdot Ra$$

where Ia = armature current.
Thus

$$V = K_1 N\phi + Ia \cdot Ra$$

and

$$N = \frac{V - Ia \cdot Ra}{\phi K_1} \tag{1}$$

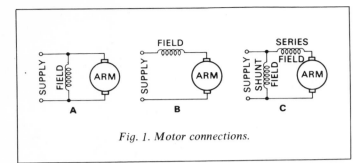

Fig. 1. Motor connections.

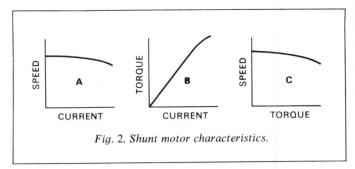

Fig. 2. Shunt motor characteristics.

Also torque

$$T(\text{Nm}) = \frac{\text{armature watts} \times 60}{2\pi N} \text{ where } N = \text{rev/min}$$

ie

$$T = \frac{EIa}{N} \cdot \frac{60}{2\pi} \quad \text{but } E = K_1 N\phi$$

therefore

$$T = K_1 \phi \, Ia \cdot (60/2\pi)$$
$$T = K_2 \phi \, Ia \qquad (2)$$

It is now possible to predict variations of speed and torque with load current for the three types mentioned.

Shunt motor
Since the supply voltage V is constant, the current in the field remains constant (ignoring changes in resistance due to temperature changes) and therefore flux ϕ remains constant. Normally $Ia \cdot Ra$ is fairly small compared to V and so the speed falls slightly for increases in current. In fact the smaller the motor, the nearer will $Ia \cdot Ra$ approach V, producing a greater speed variation with current, *ie* larger speed regulation.

Also since ϕ is constant, $T = KIa$ and the resulting torque/current curve is linear. However, at higher loads the effect of armature reaction (opposing flux set up by current in armature winding) will reduce the total flux and cause the curve to bend over slightly. (Fig. 2 A, B, C). The speed/torque curve shows that the speed falls slightly as the load torque is increased from zero to full load (typically 10–20%), but this motor is regarded as a constant speed machine, and is therefore used for such drives as line shaft drives, lathes, drilling and grinding machines, etc.

Series motor
Since $Ia \cdot Ra$ is normally small compared to V then $N = K/\phi$ and since the current Ia is flowing in the field, the flux produced will follow the magnetisation curve (Fig. 3A). The speed will therefore approximately follow the inverse of the magnetisation curve. In practice, it will deviate due to the effects of armature reaction and the fact that $Ia \cdot Ra$ has a finite value.

From eqn (2) $T = K\phi Ia$, therefore up to saturation $T = KIa^2$; after saturation $T = KIa$ and the curve takes the shape shown.

The speed/torque curve shows that the speed falls rapidly with load torque increases. At low torques the speed tends to race and care should be taken not to remove the entire load from a series motor. Such motors are used: (a) when high starting torques are required; (b) when the load is subject to heavy fluctuations and a reduced speed is desired to compensate for the high torques provided. Examples of applications include fans, vacuum cleaners, air compressors.

Compound motors
Compound motors combine the advantages and disadvantages of both shunt and series motors, giving characteristics between the two, in varying degrees. Such motors are used when a good starting torque is required, but the load may fall so low that a series motor would race. Applications such as punching machines, pumps, fans, small rolling and stamping machines are typical.

Most applications in this category of motors are those for various sections of industry in which climatic conditions are not unduly adverse. Quite low classes of insulation are therefore acceptable and, in the main, class (E) insulating systems are used enabling a maximum hot spot temperature of 120°C to be reached.

Since the early 1960s, the use of polyester film has increased and is now used extensively for slot liners and field coil insulation. When combined with polyester wire covering and impregnated (either vacuum or trickle feed) with polyester varnishes, this may be used quite safely up to 150°C maximum hot spots. This has almost entirely replaced the older class A classification which allowed a maximum temperature of 105°C, using cellulosic materials such as presspaper and leatheroid.

In some cases epoxy resins are used in place of slot liners. The material is fixed to the inner surfaces of the slot either by employing an adhesive and cold spraying, or by immersing a heated stack of laminations in a bath of resin, having previously screened off the parts not to be covered. This type of insulation is referred to as fluidised bedding, and can be adapted to insulate poles from field windings. Certain resins used are capable of withstanding temperatures of up to approximately 160°C when used in this application.

Two basic forms of armature winding are used (lap and wave wound), and the choice of winding determines the number of slots in the armature and the number of commutator segments. Obviously these factors also determine the number and placing of the brushes, which is a consideration of the original design. The characteristics of winding selection and brush design result in the following summarised factors; for lap windings the number of brushes is equal to the number of poles. For wave windings the number of brushes required is 2, displaced by 1-pole pitch, although more may be used, up the the number of poles, each displaced by 1-pole pitch, without affecting the number of parallel paths in the winding.

From the foregoing, it will be appreciated that the choice of number of commutator segments is large provided the above rules are observed. There remains another factor, however, which restricts certain numbers, namely, the commutation ability of the machine (*ie* sparking at the brushes) which is determined by what is known as the reactance voltage E_R. There is no clear-cut formula for calculating this since factors such as brush life required, duty cycle and rating, and indeed application, have an influence upon just what values are permissible and it is up to the designer to make his own judgment. However, it will suffice to state that:

$$E_R = K_1 Ia\, N\, S\, (TPC)^2$$

where Z = total no. of armature conductors; Ia = armature current; N = rev/min; S = no. commutator segments; TPC = turns per coil.

but $\qquad TPC = Z/S$

hence $\qquad E_R = K_1 Ia N Z^2/S$

Given that Ia, N, and Z are constant, the lower the number of segments, the higher will be the value of reactance voltage and the poorer the commutation ability of the motor, resulting in a shorter brush life. The brush life on some common high-speed appliance motors is usually only 100–200 hours, but the better engineered motors for industrial purposes in which the speed is much lower, have a brush life in excess of 3000 hours.

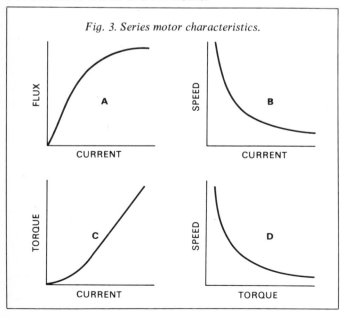

Fig. 3. Series motor characteristics.

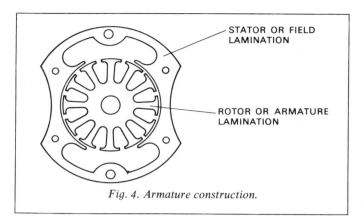

Fig. 4. Armature construction.

Another factor closely associated with commutation, which can sometimes limit the current rating of a d.c. motor, is the armature ampere-turns which distorts and weakens the main flux and causes the true magnetic neutral to swing away from the geometric neutral which effectively means that the brushes are now incorrectly placed. In the case of a motor, the brushes must be moved against the direction of rotation of the armature to correct this (not possible for reversible motors). The distorting effect of armature reaction is cancelled out by the correct design of interpoles placed between the main poles and in series with the armature. However, it will be appreciated that the cost of the motor is greatly increased if these are used.

Looking briefly at the effect of increasing the number of poles in a d.c. machine, it may be generally stated that the higher the pole number the less the volume of active iron necessary in the yoke and armature core, and a decrease in the copper overhang. Hence, a saving in materials, although the frequency of magnetisation increases and can cause the tooth loss to become excessive. However, except in larger machines, the saving in materials is exceeded by the increased labour costs, and for this reason small motors are usually manufactured as 2-pole non-interpoled machines. To assist commutation which is, in many cases, the limiting factor, the brush grade is carefully selected and is usually of the electrographitic class, having a relatively high specific resistance.

The method of construction may take the form of: (a) a steel body having the laminated poles screwed on; (b) a completely laminated 'stator' stack serving as the motor body as shown in Fig. 4. The former method, normally adopted for the higher output range, uses bobbin wound field coils placed around the poles, whereas the latter construction makes it more readily adaptable for machine winding and has the added advantage of employing a fully laminated magnetic circuit.

A copper alloy (85–90% copper, 5–10% tin) moulded commutator of 'barrel' construction is almost always used, and the brushes are held in position by rocker or cartridge type brush holders. Uni-directional motors are usually provided with moveable brushes in order to obtain the correct brush position when installed. The brush pressure varies for different applications and is normally 14–21 kgf/mm² (2–3 lbf/in²). The laminated armature stack is

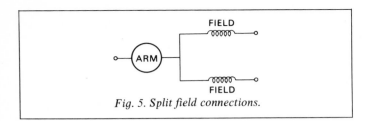

Fig. 5. Split field connections.

press fitted or stuck to the shaft and the winding normally machine wound; the recent trend has been to 'hot-stake' the connections to the commutator, a form of resistance welding which gives a more reliable joint than soldering and can be carried out with a degree of automation.

For certain applications the slots of the armature stack are skewed by up to 1 slot pitch in order to reduce the effect of cogging (*ie* fluctuations in the torque produced by the varying reluctance paths of the slots and teeth). In the main, the armature is ground to obtain the desired air gap length, typically 0.38–0.64 mm (0.015–0.025 in) before winding. For high speed applications a degree of balancing is necessary, usually done by placing lead weights in the slots after winding and impregnating.

Various grades of silicon steels are used for the lamination materials (up to 3% silicon) enabling the magnetic circuits to be worked at densities of 1.6–2.0 Wb/m². In this style of motor the thickness normally used is 0.5–0.64 mm (0.020–0.025 in) and each lamination has its mating surfaces insulated by various types of oxidation, thereby increasing the interlaminar resistance to eddy currents. A typical iron loss for this type of material and thickness is 5.5–7.7 W/kg (2.5–3.5 W/lb) at a density of 1.5 Wb/m² and a frequency of magnetisation of 50 Hz.

SPLIT FIELD MOTORS

A split field motor is a wound field motor in which the field is split into two sections so that reverse speeds may be obtained by exciting one half or the other. Sometimes there are only two field coils, one on each pole, and half the winding is excited for each direction of rotation. The latter arrangement produces somewhat better balanced flux conditions, but involves extra complications in the construction. They are normally series wound and exhibit typical series motor characteristics. Fig. 5 shows the connections. Normally these are used in position control systems and are powered by push-pull amplifiers.

PERMANENT MAGNET MOTORS

As the name implies, these are motors in which the field excitation is supplied by a permanent magnet or magnets in the magnetic circuit. They are generally manufactured up to 225 W ($\frac{1}{3}$ hp), but mainly up to 10 W and run at speeds from 2 000–20 000 rev/min: they are usually two pole machines. Their construction and materials differ depending upon the market to which they are aimed: (a) for aircraft and military applications where they are used in all kinds of control devices and servo mechanisms generally up to 5 W, 4 000–8 000 rev/min: (b) the consumer market where use is found in a variety of applications from toy cars to battery powered razors and tape recorders where outputs may be 1–15 W at speeds of 2 000–20 000 rev/min.

Before discussing the various constructions in detail, a brief look at the theory of magnets is included. A complete hysteresis loop is shown in Fig. 6. Quadrant II is the area of interest when dealing with permanent magnets; this is shown in Fig. 7.

The value of B_r is the residual induction retained by the magnet after being magnetised to saturation and then has the external magnetising force reduced to zero. The total induction B_n is equal to the sum of two magnetic inductions; μH generated by the external magnetising field and B_i, intrinsic induction, the contribution of the magnetic material placed in the field. Therefore $B_n = B_i + \mu H$ where B is in lines/cm², $\mu = 1$ and H is in oersteds. Reversing the external field causes B_n to decrease along its demagnetising path, B_i lags and eventually returns to zero completely demagnetising the magnet. The reverse field to achieve this is the intrinsic coercive force H_{ci}.

In a permanent magnet motor the external force is supplied by the ampere-turns of the armature and so when a motor draws current, the intrinsic induction B_i available to provide the air gap flux settles at a point on the intrinsic magnetisation curve, say point P of Fig. 8. Just where on the curve this point falls depends upon the armature ampere-turns and the geometry of the motor, *ie* air gap and magnet lengths, and respective areas. As the current is reduced to zero, the working point on the curve moves along a recoil loop until $B = B_d$. The slope of the recoil loop depends upon the recoil permeability of the material. Should current be further increased, the working point falls to point Q and then follows a recoil loop from that point.

A measure of the figure of merit of a material is its $(B_i . H_i)$max. product measured in mega-gauss-oersteds (10^{-4} gauss = 1 Tesla; $10^3/4\pi$ oersted = 1 A/m). This is required to be high for best advantage, since in a motor, magnet pole area depends upon the air gap flux density, whereas the magnet radial thickness depends upon the peak armature external field; thus for minimum area and minimum radial thickness, B_i and H_i should be the maximum possible.

Magnets are produced in several ways, out of various materials, and are normally one of the following. Cast out of magnetic material such as Alnico which has a typical $(B_i . H_i)$max. of 1–2 and is isotropic (able to be magnetised in any direction), and Alcomax with a typical $(B_i . H_i)$max. of 4.5–6.0 and which is anisotropic (*ie* its properties differ in different directions; value given for preferred direction). Instead of casting, these materials may be sintered (*ie* particles of material pressed into shape and bonded with an adhesive) in which case their respective $(B_i . H_i)$max. values fall to approximately 60% of the cast value. The

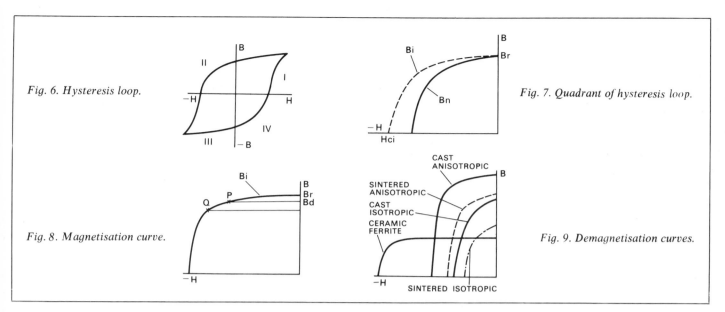

Fig. 6. Hysteresis loop.

Fig. 7. Quadrant of hysteresis loop.

Fig. 8. Magnetisation curve.

Fig. 9. Demagnetisation curves.

third group are ceramic ferrites and are magnetic iron-oxide compounds with a typical $(B_i . H_i)$max. value of 4-10. The respective intrinsic demagnetisation curves are shown in Fig. 9.

From Fig. 9 the following observations are apparent. Ceramic ferrites would produce a lower flux level than cast anisotropic materials but could hold this level whilst being subjected to much stronger demagnetising fields. Since most permanent magnet motors are assembled before the magnet is magnetised, should the armature be removed during its working life, the resultant increase in circuit reluctance causes the magnet to demagnetise, and when re-assembled, will work on a new lower recoil loop. Thus ceramic ferrite magnets are best used for applications where removal of the armature is an advantage or when the magnet is required to be magnetised before assembly.

Units for the aircraft market normally employ anisotropic material cast in the form of a continuous ring usually bonded into a stainless steel or anodised aluminium case dimensioned to international standards. In other types ceramic ferrite magnets are used in the form of segments (*ie* two pieces) bonded to a steel frame which also serves as the return magnetic path.

In nearly all permanent magnet motors the magnet axial length is usually about 10% longer than the armature core to reduce the leakage flux and increase the useful flux crossing the air gap.

Because of the environment, the armature stack is made from rustproof nickel iron such as Radiometal or Permalloy insulated laminations in thicknesses of 0.35 mm (0.014 in) capable of densities of up to 1.4–1.6 Wb/m² and having a very low loss — typically 1.1 W/kg (0.5 W/lb) at a density of 1.5 Wb/m² and frequency 50 Hz. In most cases the armature slots, usually an odd number ranging from 5 to 11, are skewed to reduce cogging and to destroy any preferred axis within the stack, the laminations are staggered during build.

The armature winding is a simple 2-pole lap type machine wound, having the connections to the commutator hot staked. In this application these motors are required to operate in higher temperatures, namely, maximum hot spots of 155°C (Class F), 180°C (Class H) and even to 200–220°C. In such cases polyimide materials are employed, both as wire covering and impregnating varnishes, and in sheet form for slot liners. It should be noted that some forms of polyimide materials are used successfully in temperatures as high as 400°C, however these applications never involve d.c. motors. For those insulated to F, fluidised bedding is usually used and in some cases, since the life required of these units is normally less than 1000 hours, polyester insulations may be used at temperatures outside its normal range but holding good for the life of the motor.

Normally a copper alloy face type commutator is used with the brushes pressing axially against it, the brushes being mounted through the plastic end frame. Because of the high altitude conditions met in these applications, special grade brushes are used. Much work has been carried out by brush manufacturers on the problem of brush performance at high altitude, and it has been determined that the main cause of rapid brush wear is the low absolute humidity that can exist at altitudes above 1500 m (5000 ft). It has also been determined that the resistance to wear of brushes, which are basically established industrial grades, can be greatly improved by adding such materials as PTFE and molybdenum disulphide into the brush mix in quantities of about 10% by weight or placing cores of the material in the brush. These materials provide lubrication and greatly reduce the brush wear. Brush life is usually guaranteed for about 1000 hours, but in some cases can be extended to 2000–5000 hours in good condition.

The same basic equations for wound field motors hold good for permanent magnet motors:

$$N = \frac{V - Ia\,Ra}{\phi K_1} \quad \text{and} \quad T = K_2 \phi\, Ia$$

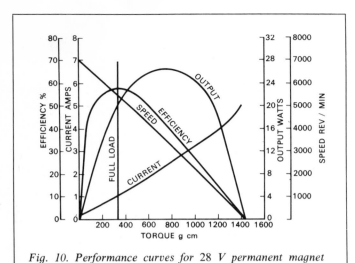

Fig. 10. Performance curves for 28 V permanent magnet motor.

However, since on small motors the resistance is quite high, the effect is to cause the speed to drop rapidly as load is increased. As the motor is loaded, since $T = K_2 \phi Ia$ and ϕ decreases slightly due to armature reaction effects, the current increases and speed falls to a value such that its back emf balances the difference between the supply voltage and $Ia.Ra$ drop and results in a very linear speed/torque characteristic. These motors are normally designed for standard voltages (6–28 V d.c.); typical performance curves are shown in Fig. 10. Generally speaking, the shape of the curves in Fig. 10 is true for all types of permanent magnet motors.

On the motors manufactured for the consumer market, where the very arduous conditions of vibration and shock, high altitudes and large changes in ambient temperatures, as found in aircraft applications, do not prevail, manufacturing tolerances are relaxed and cheaper materials used.

In most cases, since a high proportion of the loss in the motor is brush friction, where currents are small enough the carbon brush is replaced by small wire wiper arms made from precious metals, trailing on a small rhodium plated commutator. The commutators are normally only three, or perhaps five, segments and may be of the face or barrel construction. The armature stack contains the same number of slots, normally unskewed, and is made from 0.35–0.65 mm (0.014–0.025 in) thick silicon steel laminations. The winding is always machine wound, being generally insulated from the teeth by fluidised bedding. Where currents are too high for wiper arms to handle, very small carbon brushes are used either fixed to the end of flat beryllium copper arms or in the form used in the aircraft type motors. The brush grades usually contain copper or silver to keep the contact drop low.

The magnet construction varies according to the application, but is usually one of the following: ceramic ferrite in a segmented form; cast isotropic materials in block form; or sintered isotropic materials in the form of a continuous ring. The first two methods necessitate the use of a steel case to provide a low reluctance return path for the flux, either in the form of a fabricated steel tube or a simple pressed steel case giving an open type construction; however, some manufacturers mould the ceramic segments and a low reluctance circuit into a plastic body. The ring construction is pressed or moulded into a plastic body. These moulded cases can be made to contain apertures into which plain sintered plastic bearings can be press fitted, and a plastic end cap can be ultrasonically welded to the body to form the complete motor.

TORQUE MOTORS

In general, torque motors are designed essentially for high torque standstill operation in positioning systems, and for high torque at low speeds in speed control systems. They are manufactured in sizes ranging from torques of several ozf-in to several thousand lbf-ft.

Most torque motors are frameless and require very little space, offering great flexibility and adaptability in application. They are thin compared to diameter and have a relatively large axial hole through armature to enable the motor to be directly attached to the load; therefore no gearbox, with the result that backlash is eliminated and a high torque to inertia ratio exists at the load shaft.

Special design features such as high level magnetic saturation, together with the use of a large number of poles, reduce the armature self inductance to low values, thereby producing a very fast time response to changes in voltage. The torque increases directly with input current, independent of speed or angular position and is linear through zero excitation, assuring no dead band due to torque non-linearities.

In construction the torque motor consists of an anisotropic magnet, either cast in ring form and annularly grain oriented, or in block form mounted in a low reluctance steel ring to enable a large number of poles to be formed. The armature laminations are made from high permeability silicon steels containing a large number of slots into which the winding is machine wound and insulated from the iron with fluidised bedded resins: polyimide coverings are normally used, allowing high temperature rises to be attained.

The windings, in general, are wave wound having the connections made at the non-brushgear end of a printed circuit copper commutator which is laid in the slots on top of the winding. The whole armature assembly is then potted with an epoxy-resin to produce a very robust construction with good heat dissipation. The brushgear essentially consists of a moulded glass nylon or plastic ring containing small beryllium copper brush arms onto which the carbon brushes are secured. The complete assembly is finally screwed to the magnet ring.

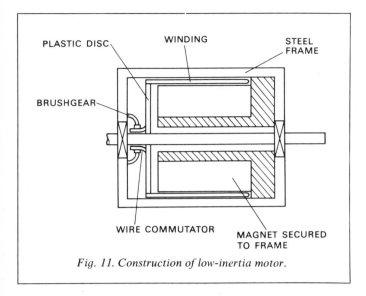
Fig. 11. Construction of low-inertia motor.

LOW INERTIA MOTORS

In many instrument type applications, small d.c. drive motors are required to have very quick response, necessitating a very low rotor inertia. Such a motor incorporates an iron-free rotor consisting of a hollow winding rotating in a small air gap. Fig. 11 shows a typical construction. Excitation is supplied by a high flux permanent magnet and the steel frame acts as the return path.

The winding is basically one of two types: (a) wound in a similar manner to an iron cored armature having straight very shallow slots, with the connections made to a small plated barrel style commutator. This type, however, has a bunched-up end winding creating a higher inertia and requires a long air gap to enable the coil to rotate, thereby reducing the torque produced. To get over this inherent disadvantage, winding (b) is used, this being wound on the same principle as rf coils used in radio work where the winding is continuous and only two wires deep. The coils are skewed in winding to keep the ends the same thickness.

The wire normally has a self-bonding covering or is finally lacquered so that it is rigid when removed from the former and has equi-spaced loops brought out which are connected to a small diameter commutator. This is normally a moulded plastic disc, which fits into the end of the coil, containing a number of precious metal wires, bent to shape and having small precious metal wire brushes.

These types of motor produce very small torques, but by virtue of their very low rotor inertia, accelerate to speed extremely quickly.

BRUSHLESS MOTORS

In practice, the brush friction loss in miniature d.c. motors is normally quite a high proportion of the total losses, and efficiencies seldom exceeding 40-50% are attained. Very high efficiencies would result if this loss were removed, quite apart from the obvious improvements in life, reliability and noise.

In order to duplicate the characteristics of a conventional permanent magnet d.c. motor with a brushless d.c. motor, it is necessary to sense rotor position and switch stationary armature circuits in proper relation to a moving permanent magnet field. The function of the commutator and brushes can be duplicated in a motor by a solid state electronic switching system. The result is a brushless d.c. motor without the deficiencies of a mechanical commutator.

The brushless motor can use a permanent magnet rotor for its field excitation and a slotted stator with a conventional three phase induction motor winding for the armature field. The current in the armature is switched by a solid state three phase bridge switching network. Switching logic and rotor position sensing for the commutation process is accomplished by using a reluctance switch — which merely provides a voltage proportional to angular position — and dictates the switching speeds and sequences required.

Another form of winding possible is an inverted conventional d.c. motor having the normal armature winding placed in slots around the stator. The end connections normally taken to a commutator are fed by a controlled thyristor and diode connected in parallel opposition. This involves quite a large number of devices (depending upon the number of poles, slots, etc) which are fired by the methods previously described. The speed of the motor is controlled by the voltage applied to the armature winding, similarly to the conventional permanent magnet motor in which reducing the voltage causes the speed to fall in order to reduce the back emf and maintain the equation $E = V - Ia.Ra$ where Ia is constant for constant torque.

Since the windings are distributed and the slots may be skewed (as a normal induction motor), fairly low ripple torque is obtained. However, because of their complicated construction and cost, they are mainly restricted to the fractional-power range, and a very narrow specialised field where the disadvantages of the conventional motor cannot be tolerated and a smooth low ripple torque must be obtained.

STEPPER MOTORS

Although not a d.c. motor in the strict sense, steppers warrant mentioning in this chapter if only to serve as an indication of the vastly differing types of motor available. A stepper motor is a device whose output shaft moves in discrete steps when excited from a switched d.c. supply. They are to be found in both military and industrial fields on such applications as digital actuators, incremental drives, numerical control devices and the like. They are produced in two basic forms: (a) variable reluctance, and (b) permanent magnet. These motors are discussed in more detail in another chapter.

SPEED CONTROL OF MOTORS

Basically there are two methods open by which to vary the speed of a d.c. motor: (i) by varying the armature voltage; (ii) by varying the field current.

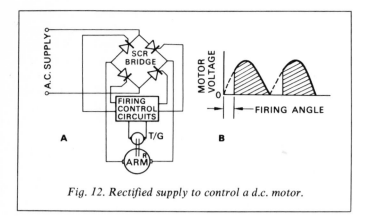

Fig. 12. Rectified supply to control a d.c. motor.

Small permanent magnet motors obviously cannot utilise method (ii) and therefore methods are employed to adjust the armature voltage in order to compensate for speed changes due to load or input voltage changes. A purely mechanical arrangement is to adopt a centrifugal switch connected in the input line to the machine so that, if the speed increases, bob weights react on the contacts and disconnect the motor from the supply. A more accurate method is to adopt a tachogenerator, coupled to or built into the motor, whose output voltage signal is proportional to the motor speed; in its simplest form this output is compared with a reference voltage and the error between the two used to control the base of a transistor in series with the motor, thereby adjusting the voltage at the armature.

The same basic methods can be used to control the speed of wound field motors; however, these can also utilise the fact that reducing the field current will increase the speed. Mechanical governors are available that operate when the speed increases and in operating short circuit a preset resistance in the field circuit.

Electronic controllers employ various methods of feedback and operate more advanced circuitry, but each basically compares a voltage proportional to speed with a constant reference voltage and uses the difference between these to reduce the difference to as low a value as possible. In many instances d.c. motors are fed from rectified supplies and, in particular, use thyristors to control speed. Fig. 12 shows the schematic of such a system employing a single-phase full-wave controlled rectifier system. By varying the firing angle, the mean d.c. voltage to the armature is changed. Again tachogenerator feedback may be employed to automatically adjust the firing angle. When used on rectified supplies the performance of a d.c. motor is greatly affected and unless designed specifically for use with this waveform will result in reduced brush life and, in some cases, it will not produce the required torque for the load.

The torque of a d.c. motor is proportional to the mean d.c. current, the commutation limit dependent upon the peak d.c. current and the I^2R or heating loss proportional to the square of the rms current, thus as the form factor (ratio rms to mean) increases, for the same torque, the commutation worsens and the I^2R loss increases by the square of the form factor. Table 1 indicates the effect on copper loss for various waveforms when operating at constant torque.

In addition to the above effects, the high ripple produces high frequency pulsating fluxes in the poles, resulting in additional iron losses and, in the case of permanent magnet motors, tends to demagnetise the magnet by varying degrees.

BEARINGS

Almost all the types of motors described employ single row radial bearings; most having one bearing locked to shaft and end shield and the other free to take up any end float against some form of spring washer.

The larger slower speed motors usually use grease-lubricated sealed-for-life bearings, the amount of grease usually filling about one third of the free space. The smaller higher speed motors generally employ shielded bearings containing only a few drops of oil. Where very low friction is required, bearings with self-lubricating PTFE cages are employed.

For consumer type motors, bearings having fairly low ABEC numbers are used, whereas for servo type motors where a higher degree of accuracy is required, ABEC 5 and 7 are commonly employed. For shafts up to 10 mm diameter and housings up to 30 mm diameter, the normal diametral tolerances for ABEC 5 bearing fits are 0 to −0.005 mm on the shaft and 0 to +0.005 mm on the housing. On certain small machines, plastic (mainly Nylatron and sometimes PTFE) plain bearings are used; these are machined to produce a diametral interference fit of about 0.05 mm in the housing. When fitted, the actual interference fit on o.d. is reflected 100% onto the i.d., so that allowances must be made to obtain a diametral clearance between shaft and bearing of about 0.05–0.03 mm.

The life of bearings depends upon the speed and loadings encountered; normally ball bearings of this size have a life in excess of 5 000 hours, and more normally 10 000–20 000 h, depending on the type of machine and application. Plain plastic bearings have a very much reduced life and, in general, never exceed 1000 hours.

Table 1. Effect on copper loss for various waveforms in motors operating at constant torque.

Type of supply	Firing angle	Form factor	I^2R loss %
Pure d.c.	—	1·0	100
Full wave single-phase	0	1·11	123
Full wave single-phase	90°	1·57	246
Half wave single-phase	0	1·57	246
Half-wave single-phase	90°	2·22	494

Chapter 6

Single-phase Motors

J T Appleby *CEng MIEE*
Newman Industries Ltd

The performance of single-phase machines is usually inferior to that of three-phase machines, and their cost, for a given output and speed, is higher. For this reason single-phase motors are normally only used where a three-phase supply is not available and their application is mainly restricted to domestic, office and agricultural purposes.

However, a very large number of single-phase fractional-power (less than 1.0 hp, 0.75 kW) machines are used in the home for driving washing machines, vacuum cleaners, refrigerators, fans, central heating circulating pumps, and many other domestic appliances. In offices, single-phase motors of up to about 0.75 kW (1 hp) are used in computers, tabulators, print machines, and other office equipment. For agricultural purposes, single-phase motors of up to about 15 kW (20 hp) are used for grain elevators, grinders, mills, driers, milking machines, and similar applications. Single-phase motors of varying outputs and speeds are used for ventilation and air conditioning purposes in domestic, office and agricultural installations.

In the three-phase motor, a rotating field is set up which produces a rotating torque when the rotor is stationary, thus producing a starting torque. However, in the single-phase motor there is only one stator winding, and this produces a pulsating torque with a net zero torque at zero speed and it is, therefore, not self-starting. For a single-phase machine it is thus necessary to produce a rotating torque at standstill, and the method used to obtain this starting torque is largely responsible for the several forms of single-phase motors available. Different applications require different starting characteristics and these are also provided for by the different methods of connections.

The main types of single-phase motors are listed below, the definitions being taken from the International Electrotechnical Commission publication on Terminology of Rotating Machines.

Fractional horsepower motor: A motor having a continuous rating not exceeding 1 hp (0.75 kW) per 1 000 rev/min.

Split-phase motor: A single-phase induction motor having an auxiliary primary winding, displaced in magnetic position from, and connected in parallel with the main primary winding. There is a phase displacement between the currents in these two windings. (Note: Unless otherwise specified, the auxiliary circuit is assumed to be opened when the motor has attained an appropriate speed.)

Resistance start split-phase motor: A split-phase motor having a resistance connected in series with the auxiliary primary winding. (Note: The auxiliary circuit is opened when the motor has attained an appropriate speed.)

Reactor start split-phase motor: A split-phase motor designed for starting with a reactor normally in series with the main primary winding. (Note: The auxiliary primary circuit is opened and the reactor is short-circuited or otherwise made ineffective when the motor has attained an appropriate speed.)

Capacitor motor: A split-phase motor with a capacitor normally in series with the auxiliary primary winding.

Capacitor start motor: A capacitor motor in which the auxiliary primary winding connected in series with a capacitor is in circuit only during the starting period.

Capacitor start and run motor (sometimes known as a *permanent split capacitor motor*): A capacitor motor in which the auxiliary primary winding and series connected capacitor remain in circuit for both starting and running.

Two-value capacitor motor: A capacitor motor using different values of capacitance for starting and running.

Universal motor: A motor which can be operated by either direct current or single phase alternating current of normal supply frequencies.

Repulsion motor: A single-phase induction motor with a primary winding, generally on the stator, connected to the power source, and a secondary winding, generally on the rotor, connected to a commutator, the brushes of which are short circuited and can occupy different positions.

Repulsion induction motor: A repulsion motor with an additional rotor cage winding.

Repulsion start induction motor: A repulsion motor in which the commutator bars are short circuited or otherwise connected at an appropriate speed to give the equivalent of a cage winding.

Synchronous motor: An alternating current motor in which the speed on load and the frequency of the system to which it is connected are in a constant ratio.

Shaded pole motor: A single-phase induction motor having one or more auxiliary short-circuited windings displaced in magnetic position from the main winding, all these windings being on the primary core, usually the stator.

PRINCIPLES OF OPERATION

Split-phase motor

As given under the definitions, there are two separate windings on the stator of a split-phase motor, the main winding, sometimes referred to as the 'running' winding, and the auxiliary winding, sometimes referred to as the 'start' winding, each being a complete winding in itself. These windings are usually, but not necessarily, displaced 90° (electrical) from each other and carry currents which are out of phase, hence the term 'split-phase'. The rotor is usually of squirrel cage construction; the schematic representation of a split-phase motor is given in Fig. 1.

For starting, the windings are connected in parallel to the supply voltage, and the phase difference between the two windings (which is necessary to produce a rotating torque at standstill) is obtained by making the ratio of resistance to reactance of one winding different from the other. Various methods are used to obtain this difference entailing the use of resistance, inductive reactance, a combination of these and also capacitive reactance (which reduces the inductive reactance of one winding).

As the speed of the motor approaches about 80% of the synchronous speed, the motor can develop about the same torque on the main winding alone as it can with both windings, and above about 85% of the synchronous speed the torque developed with both windings in circuit is less than that developed by the main winding alone. Obviously, it is advantageous to cut out the auxiliary winding when the motor is up to about 75% synchronous speed, and this is carried out by a starting switch. Starting switches are usually of the centrifugal type although voltage, current and thermal operated switches are available and may be preferred for some applications; the four types are described separately.

Another important reason for disconnecting the auxiliary winding is that less current and power is drawn from the supply, thus preventing overheating of the auxiliary winding.

Resistance start split-phase motor

This is the simplest, cheapest, and most common type of split-phase motor. The auxiliary winding is wound with a wire that has a much larger resistance than the main winding, the additional resistance being normally attained by using a finer wire and winding less turns. This saves copper weight and space, and allows the main winding to occupy more than half the total slot area. The additional main winding copper section consequently reduces running losses and heating, thus increasing the efficiency. Improvements in both locked rotor and breakdown torques are also obtained. Since the auxiliary winding is cut out by the starting switch when the motor reaches about 75% synchronous speed, the current density in this winding can be very high, but no severe overheating occurs since the winding is in circuit for only a very short time.

Reactor start split-phase motor

This type of motor is uncommon and not generally available. An external inductive reactor is connected in series with the *main* winding during the starting period. This reduces the starting current and increases the phase angle between the two windings thus producing more torque per ampere of line current than the conventional split-phase motor.

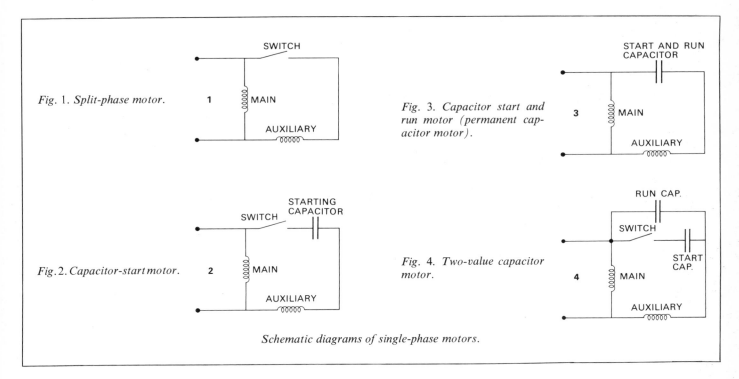

Schematic diagrams of single-phase motors.

Capacitor motor

Although the definition of a capacitor motor is given, the term 'capacitor motor' should never be used by itself in specifying a motor. There are three definite types of capacitor motor, each with very different characteristics, and the term should always be qualified by the type.

Capacitor start motor

The capacitor start motor has two windings similar to the split-phase motor generally spaced 90° (electrical) from each other. A capacitance is inserted in series with the auxiliary winding, and the start switch disconnects both the auxiliary winding and capacitor when the motor reaches about 75% synchronous speed. The windings are specially designed for this motor (the auxiliary winding generally contains more copper than the auxiliary winding of a comparable split-phase motor) and it is not possible merely to fit a capacitor in series with the auxiliary winding of a standard split-phase motor. (Fig. 2.)

Capacitor start and run motor

With each of the four types of motor described above, the auxiliary winding is disconnected from the supply when the motor approaches full load speed and the main winding only is in circuit when the motor is running. However, if the auxiliary winding is retained in circuit in series with a capacitor then the rotating field produced by the two windings will be continuously effective. The performance of the motor will thus be improved and compare more favourably with a three-phase motor. The disadvantage of this type of motor is that the capacitor (which must be continuously rated) is the same value for both starting and running conditions, since no start switch is fitted and the capacitor is in the auxiliary winding circuit for all conditions.

To obtain a satisfactory locked rotor torque it is necessary to have a large value of capacitance, but for running conditions a much lower capacitance is required, otherwise an excessive current is drawn from the supply. The capacitor value best for running conditions is thus generally used, which results in a low locked rotor torque. This motor has characteristics best suited to centrifugal fan and similar applications where the required torque at slow speeds is low.

A virtue of the capacitor start/run motor is that it is speed-controllable by voltage control, and a very large number of these motors are used for ventilation equipment, their speed being controlled automatically by auto-transformer or phase-controlled thyristors. (Fig. 3.)

Two-value capacitor motor

Two windings are provided, as in the split-phase motor, and to overcome the inherent disadvantage of the capacitor start and run motor, two capacitors are provided. A high value capacitor is inserted in series with the auxiliary winding to give a starting performance similar to the capacitor start motor. A starting switch is fitted which cuts out the large value capacitor (which need only be short time rated) leaving in circuit a lower value continuously rated capacitor for running.

A disadvantage of the two-value capacitor motor is the initial cost of the permanent capacitor, which can be expensive for a large motor. However, this initial cost can be easily outweighed by the lower running costs, due to increased power-factor and efficiency, and since both windings are used a given power can be obtained from a smaller frame-size than a normal split-phase motor (where the auxiliary winding is not used in the running condition). (Fig. 4.)

Universal motor

The universal motor is similar in construction to a direct current series motor; it has a stationary field winding, usually salient pole, and a wound armature with commutator and brushes. If an ordinary d.c. series motor is supplied with an a.c. voltage the field winding flux and the current in the armature will reverse almost at the same time producing a unidirectional but pulsating torque. To obtain satisfactory performance with a.c. the whole of the field and armature iron is formed with laminations which are thinner than normal and have a better magnetisation characteristic. In addition, the airgap is smaller and the commutator larger than the corresponding d.c. motor. A series compensating winding may be added to the stator (displaced 90° electrical from the main field winding), to reduce the effect of armature reaction mmf.

The universal motor is designed to work on a supply of any frequency from a specified maximum down to d.c. The motors usually operate at speeds up to 10 000 rev/min and the change of speed with load may be considerable (ie, a series characteristic). For a given load torque the speed is usually somewhat higher on d.c. than on a.c., this difference being kept to a minimum by designing the motors for high speed operation. The speed of the motor may be controlled by any method of varying the terminal voltage, eg auto-transformer, series resistance, phase-controlled thyristors. Constant speed operation within certain limits can be obtained on small universal motors by providing them with a centrifugal switch which operates at a given speed and inserts a resistance in series with the supply, the switch being capable of opening and closing many times per second.

Universal motors are widely used in domestic appliances such as vacuum cleaners, food mixers, hair driers, and in electric drills and other power tools. (Fig. 5.)

Repulsion motor

The repulsion motor consists of a normal single-phase stator winding connected to the supply, and a wound rotor with commutator and brushes which are short-circuited. If the brushes are in line with or at an angle of 90° electrical to the main flux no torque is produced. At an angle of approximately 45° to the main flux, maximum torque is obtained and, by altering the position of the brush axis, the speed of the motor may be varied or reversed. Reversal may also be obtained by using two separate stator windings displaced 90° electrical with a fixed brush position on the axis of one winding. Rotation is reversed by reversing one winding.

The motor has a series type of speed torque characteristic, the no-load speed being usually 1.5 to 2.5 times the synchronous speed. The main advantage of this motor is its high starting torque per unit current. Starting torques of the order of three to four times full load torque with currents of about three times full load current may be obtained. The speed may be controlled by any method of varying the terminal voltage, or by brush shifting. The main disadvantage of the motor is its high initial cost and the increased maintenance associated with brush gear and commutators. (Fig. 6.)

Repulsion induction motor

The repulsion induction motor is of similar construction to the repulsion motor with the addition of a second rotor winding of the squirrel cage type embedded beneath the original winding. On starting, the squirrel cage has little effect due to its high leakage reactance, most of the flux linking only the outer commutator winding. As the motor accelerates the reactance of the squirrel cage winding decreases and produces torque. The result is a torque-speed curve with the good starting characteristics of the repulsion motor and the near constant speed running characteristic of the induction motor. On no-load the motor runs at just above synchronous speed. Reversal may be achieved by an additional stator winding or by brush shifting. (Fig. 7.)

Repulsion-start induction motor

The repulsion-start induction motor is of similar construction to the repulsion motor except that a centrifugal device is fitted which operates at about 70% of full load speed, short-circuiting all the commutator segments and also lifting the brushes clear of the commutator to reduce wear. At starting, the motor has the advantage of the repulsion motor torque characteristic, whereas when the commutator segments are short-circuited the motor functions as a normal induction motor with a near constant speed running characteristic and a no-load speed tending to synchronous speed. The torque-speed curve is not so smooth as that of the repulsion induction motor, there

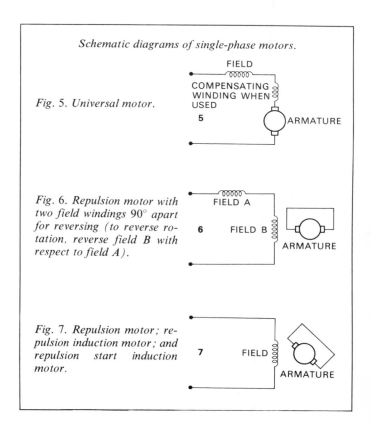

Schematic diagrams of single-phase motors.

Fig. 5. Universal motor.

Fig. 6. Repulsion motor with two field windings 90° apart for reversing (to reverse rotation, reverse field B with respect to field A).

Fig. 7. Repulsion motor; repulsion induction motor; and repulsion start induction motor.

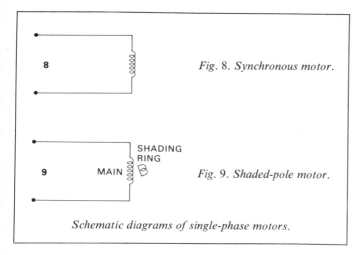

Fig. 8. Synchronous motor.

Fig. 9. Shaded-pole motor.

Schematic diagrams of single-phase motors.

being a dip in the middle of the accelerating range. The motor is not reversible when running.

Both the repulsion induction and the repulsion start induction motors have a high cost compared to the squirrel cage motor and so are normally used only when the high starting torque per unit current of the former is essential.

Synchronous motor

This type of motor operates at a constant speed at loads up to 'pull-out' when supplied with a constant frequency, and is therefore used in small sizes for electric clocks, timers and other applications where a constant speed is essential. The synchronous speed in rev/min is given by: 120 × supply frequency (Hz) divided by the number of poles.

There are several novel constructions for this type of motor, but basically the principle of operation is that the stator provides a single-phase pulsating flux similar to the induction motor. The rotor is not cylindrical, but is shaped to form a number of salient poles so that when running normally the rotor is held in step with the pulsating flux by the concentration of the flux in these rotor poles. If the motor is to be self-starting a rotating field must be provided; this may be done by shading rings as in the shaded-pole induction motor. Other types of synchronous motor incorporate permanent magnetic material in the rotor. (Fig. 8.)

Shaded-pole motor

The shaded-pole motor uses the normal squirrel cage rotor, but the stator has salient poles. Each pole has a slot running axially along it, dividing it into two unequal parts; a band of copper (called a shading ring) surrounds the smaller of the two parts. The main alternating flux induces a current in this shading ring which produces a second alternating flux out of phase in time and position with the main flux. The two fluxes form a weak rotating field sufficient to start the motor.

This type of motor is simple and cheap to manufacture but is used only for the smallest sizes due to its very low efficiency, power factor and starting torque. It is not reversible. (Fig. 9.)

CONSTRUCTION

An exploded view of a typical single-phase fractional-power motor is shown in Fig. 10 and its construction is described here in detail. (Numbers in parentheses refer to key numbers on Fig. 10.) The stator core is pre-built from low loss electrical steel sheet laminations and the stator frame is pressure die-cast around it in aluminium alloy (12). The stator winding (13) is wound with synthetic enamelled covered wire and inserted into the stator slots, the whole of this operation often being carried out automatically.

Before inserting the coils, the stator slots are insulated with a material which has a high mechanical and electrical strength, the type of insulation depending upon the temperature classification of the motor. A wedge of insulation is formed and inserted above the coil in each slot to prevent the wires coming out of the mouth of the slot and protruding into the motor bore. Insulation is also inserted in the end windings between the main and auxiliary windings. The complete stator is immersed in synthetic insulating varnish, after which the impregnation is baked, to render the stator resistant to moisture and to prevent movement of the coils due to vibration, etc. The rotor core is assembled from laminations of similar material to that used for the stator, and the squirrel cage winding is die-cast aluminium, making the rotor very resistant to damage. Fan blades are cast integral with the end-ring. The rotor core (14) is shrunk onto a high grade steel shaft (15) machined all over to very close limits. The complete rotor and shaft is then dynamically balanced.

The end brackets (4 and 20) are pressure die-cast from a similar material to the frame and accurately machined to fit on the spigot of the motor frame. Both frame and end bracket spigots are machined to fine limits to ensure concentricity of rotor and stator to give a uniform air gap.

The ball bearings (18 and 23) are grease lubricated, the non-drive-end (NDE) bearing being retained by an internal bearing cap (24) and retaining tabs (25) which are fitted by screws to the NDE end bracket. The drive-end (DE) bearing (18) is pre-loaded by a spring thrust washer (19) which is inserted into the DE bearing housing before the bearing is fitted. The grease is retained in the bearing by an internal bearing cap (17) and retaining circlip (16) which is fitted external to the housing. After assembly the bearings are protected by the fitting of the NDE (21) and DE (31) external bearing caps. Bearings may be 'greased for life' or facilities may be provided for lubricating the bearings by a nipple or grease screw (3).

The rotating part of the centrifugal switch (8) shown is of a type often fitted. Actuating weights are riveted to arms retained by springs. Fingers on the arms are fitted to a thruster; during rotation, centrifugal force causes the weights to move outward against the spring pressure. As the arms move outwards the fingers move axially towards the rotor, thus moving the thruster, which, in its stationary position, rests against the arms of the static part of the switch

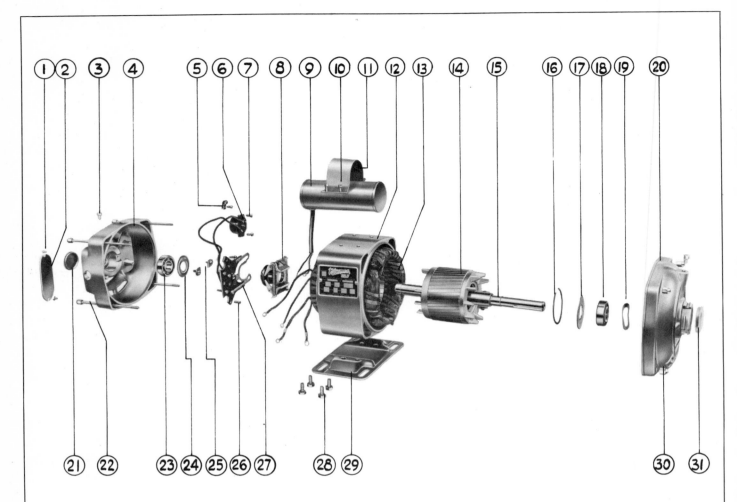

Fig. 10. Exploded view of single-phase fractional-power motor.
(1) Terminal cover fixing screws. (2) Terminal cover. (3) Greaser screw. (4) Non-drive-end (NDE) end bracket.
(5) Cable clip. (6) TOC (thermal operated cutout). (7) TOC fixing screws. (8) Rotating part of centrifugal switch. (9) Capacitor.
(10) Capacitor clip fixing screws. (11) Capacitor clip.
(12) Stator frame. (13) Stator winding. (14) Squirrel cage rotor core. (15) Shaft. (16) Bearing cap retaining circlip. (17) DE bearing integral cap. (18) Drive end (DE) bearing.
(19) Spring thrust washer for pre-loading bearing. (20) DE end bracket.
(21) NDE bearing housing external cap. (22) Tie rods. (23) NDE bearing. (24) NDE bearing internal cap. (25) NDE bearing retaining tabs. (26) Switch fixing screws. (27) Static part of centrifugal switch, incorporating terminals.
(28) Baseplate fixing screws. (29) Baseplate. (30) Tie rod dome nut. (31) DE bearing housing external cap.

(27) whose contacts are closed. When the thruster moves away the contacts open and the thruster, being well clear of the static portion, does not rub on the switch arms (which would cause wear) at normal running speeds. The static part of the centrifugal switch incorporates the terminals for the machine, and when fixed to the NDE end bracket with the fixing screws (26) the terminals are accessible after removing the terminal box lid (1 and 2).

If the motor is of the capacitor start type, the starting capacitor (9) can be fitted to the motor frame by a clip (11) and fixing screws (10). This capacitor is in series with the auxiliary winding during starting, and is cut out of circuit by the centrifugal switch.

If required the motor can be protected by a thermal operated cutout (TOC) which operates if the motor current exceeds a given figure. The TOC (6) is fitted to the NDE bracket by fixing screws (7) and can be automatic or manual reset. If manual reset, the reset button protrudes through a hole in the end bracket.

To prevent the leads from the capacitor, winding and TOC catching in the fan blades, they are secured to the end bracket with a cable clip (5). The end brackets are fitted to the frame and held together with four tie rods (22) fitted with dome nuts (30). If foot mounting is required, a baseplate (29) is bolted to the frame with four fixing screws (28).

Switching of auxiliary winding

The most common method of switching the auxiliary winding is by a centrifugal switch, of the usual type in which rotating weights operate against a spring. At the required speed the centrifugal force on the weights is sufficient to overcome the tension on the spring, the weights fly out, and in so doing operate the switch. With this type of switch the springs and/or weights must be adjusted for different motor speeds.

Relay switching is also used, where it is impossible or impracticable for a centrifugal type switch to be fitted. A typical example is a refrigerator motor where the motor is an integral part of the cooling system and is inaccessible because it is built into the refrigerating unit. There are three types of relay switching: current operated, potential operated, and thermal operated.

The current operated type relay has its coil connected in series with the main winding. When the supply is switched on the initial heavy inrush current of the main winding flows through the relay coil and the relay operates, closing the contacts and introducing the auxiliary winding into circuit. As the motor speed increases the current in the main winding (and relay coil) decreases and at some predetermined current the relay de-energises, the relay contacts open and the auxiliary winding is disconnected.

The potential type relay is connected in parallel with the auxiliary winding. This relay is operated by the voltage across the auxiliary winding which increases as the motor speed increases. When the voltage reaches a predetermined level the relay operates and opens the normally closed contacts (connected in series with the auxiliary winding) thus disconnecting the auxiliary winding. Applications of potential type relay are limited to capacitor start type motors where the characteristics provide enough voltage change across the auxiliary winding to operate the relay. This type of starting relay is uncommon.

The thermal type relay depends upon the motor current to generate heat in a thermal element which actuates contacts. When the supply is switched on, the main and auxiliary windings are in circuit and the motor accelerates. The motor winding current operates the thermal element, which may be directly or indirectly heated, thus switching a normally closed contact (fitted in series with the auxiliary winding) and disconnecting the winding. Whilst current continues to flow through the main winding (and relay) the auxiliary winding is disconnected. When the motor supply is disconnected the thermal element cools and the relay contact closes, this reset time being in the region of 5–15 seconds. Thermal relays should therefore only be used on applications where reset time is not important (*eg* refrigerator and oil burner motors). The reset time delay can, however, be utilised for general time delay and switching applications and motor reversing (the time delay ensuring that the motor has come to rest before it is reversed).

CHARACTERISTICS AND APPLICATIONS

Resistance start split-phase motor. Shunt torque-speed characteristic (approximately constant running speed), not reversible when running but can be reversed when stationary by reversing the starting winding connections. Suitable as a fixed speed drive for light industrial domestic and office use where the higher starting torque of the capacitor-start motor is not necessary, typical starting torque 150% FLT (full-load torque). In common with other types of induction motor, the running speed is restricted by the number of poles and the supply frequency; *eg* maximum possible no-load speed for a 50 Hz supply is 3000 rev/min (2 poles). (Fig. 11.)

Reactor-start split-phase motor. Application details similar to the resistance-start split-phase motor; however, this type is little used.

Capacitor-start motor. Shunt torque-speed characteristic, reversible when stationary but not reversible when running. Good starting torque, typically 250% FLT. Suitable as a fixed speed drive for light industrial, domestic and office use where a better starting torque than the split-phase motor is required. (Fig. 12.)

Two-value capacitor motor. Shunt torque-speed characteristic, good starting torque, efficiency and power factor, reduced noise and vibration, reversible when stationary but not reversible when running. Applications similar to the capacitor-start motor.

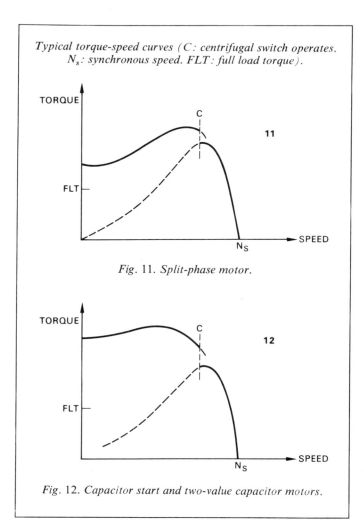

Typical torque-speed curves (C: centrifugal switch operates. N_s: synchronous speed. FLT: full load torque).

Fig. 11. Split-phase motor.

Fig. 12. Capacitor start and two-value capacitor motors.

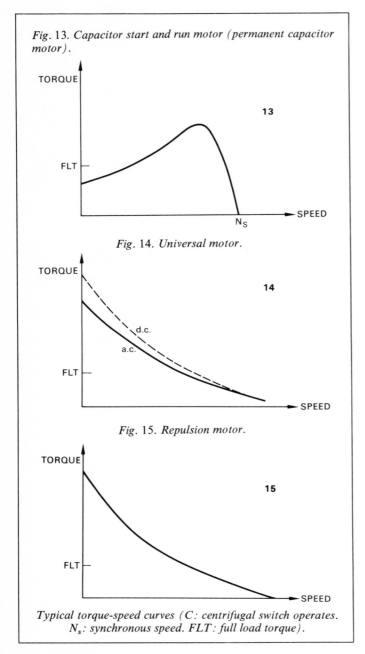

Fig. 13. *Capacitor start and run motor (permanent capacitor motor).*

Fig. 14. *Universal motor.*

Fig. 15. *Repulsion motor.*

Typical torque-speed curves (C: centrifugal switch operates. N_s: synchronous speed. FLT: full load torque).

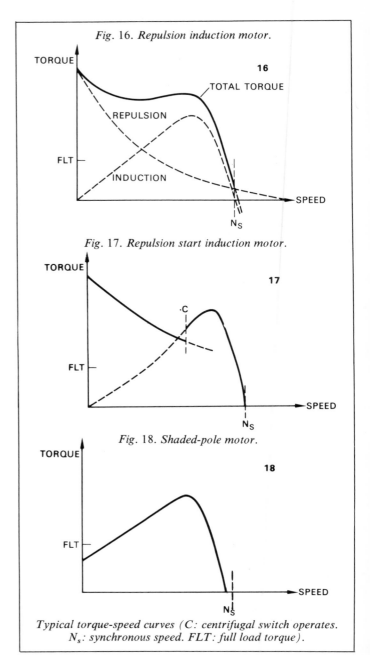

Fig. 16. *Repulsion induction motor.*

Fig. 17. *Repulsion start induction motor.*

Fig. 18. *Shaded-pole motor.*

Typical torque-speed curves (C: centrifugal switch operates. N_s: synchronous speed. FLT: full load torque).

Capacitor start and run motor (*permanent capacitor motor*). Shunt torque-speed characteristic, low starting torque typically 60% FLT; reversible, good power factor. Suitable for small pumps (*eg* central heating circulation), domestic, agricultural, and industrial fans, and other drives where low starting torque is permissible. As the motor runs as a two-phase motor, it is less prone to noise and vibration; reliability is increased by the absence of a centrifugal switch and electrolytic starting capacitor. This type of motor is extensively used for the variable voltage speed control of fans, fan torque varying as the square of the speed. (Fig. 13.)

Universal motor. Series torque-speed characteristic (considerable variation of speed with load torque), good starting torque, operation at speeds up to 10 000 rev/min, speed may be controlled, supply voltage a.c./d.c. not reversible. Used for small domestic equipment such as vacuum cleaners, hair driers, food mixers, also electric drills and portable power tools. (Fig. 14.)

Repulsion motor. Series torque-speed characteristic, no-load speed usually 1.5–2.5 times synchronous speed, may be speed controlled and reversible, good starting torque typically 350% FLT with low starting current; used for lifts, cranes, hoists, spinning, and printing machines. (Fig. 15.)

Repulsion induction motor. Shunt torque-speed characteristic, running speed related to synchronous speed, good power factor and efficiency, reversible by an additional stator winding or brush shifting, very good starting torque per unit current. Suitable for compressors, pumps or any fixed speed drive requiring high starting torque with low starting current. Not greatly used due to its high cost compared to the squirrel cage motor. (Fig. 16.)

Repulsion-start induction motor. Shunt torque-speed characteristic, running speed related to synchronous speed, not reversible when running, very good starting torque per unit current; uses similar to the repulsion induction motor. (Fig. 17.)

Shaded-pole motor. Shunt torque-speed characteristic, low starting torque power factor and efficiency; not reversible. Made only in small sizes; used for desk and ceiling fans, record player and tape recorder drives. (Fig. 18.)

Synchronous motor. For loads up to a 'pull-out' value the speed is absolutely constant when supplied at a constant frequency. In small sizes the motor is used for electric clocks, timers, some record player and tape recorder drives.

Future trends
No great changes in the characteristics of single-phase motors is currently envisaged since all known methods of obtaining maximum performance with minimum cost have already been exploited and incorporated into present-day designs. Improvements will undoubtedly take place in insulation and cooling techniques which will probably increase the output of a given frame thus effectively reducing the cost of power output.

New materials, and in particular, plastics, will also be introduced and these, together with mechanised winding, will be incorporated into production methods and help offset the natural trend of increasing labour and material costs. One increasingly likely change will be the substitution of copper by aluminium for the winding wire.

BIBLIOGRAPHY

Herbert Vickers. The Induction Motor. *Sir Isaac Pitman & Sons Ltd.*
Cyril G Veinott. Theory and Design of Small Induction Motors. *McGraw Hill Book Co Inc.*
G S Brosan and J T Hayden. Advanced Electrical Power and Machines. *Sir Isaac Pitman & Sons Ltd.*

Chapter 7
Three-phase Induction Motors

C R M Heath MA
Electrical Power Engineering (Birmingham) Ltd

The polyphase induction motor invented by Nikola Tesla in 1886 had been developed by 1895 virtually into its present squirrel-cage form. The slip-ring induction motor was developed a short time later. Because of its simplicity and cheapness, the 3-phase squirrel-cage induction motor is now by far the most common type of motor in use for driving industrial plant, particularly of small power ratings (up to 20 kW). The slip-ring induction motor is not as simply constructed as the cage induction motor, but it is possible to control its starting performance and speed by connecting external equipment (commonly resistors) into the rotor circuit via slip-rings. 3-phase induction motors can be made to any desired power ratings, but natural running speeds are related to the power-supply frequency.

CONSTRUCTION

The stator of a 3-phase induction motor is made up of a slotted magnetic core and a 3-phase electrical winding, which is inserted into the slots and must be supplied with 3-phase power. The rotor also has a slotted magnetic core, but the electrical winding can be of two types. The cage-type rotor has bar conductors inserted into the rotor slots, the bars being connected together at each end of the rotor via conducting end rings. The slip-ring rotor has an insulated 3-phase winding which is terminated on slip rings. A typical small cage induction motor is shown in section in Fig. 1.

Magnetic core

The stator and rotor cores are made up of a large number of thin magnetic-material laminations clamped together. Typically the laminations are punched from 0.35–1.0 mm thick insulation-coated silicon steel; this form of construction is necessary to reduce eddy-current iron losses. Typical rotor and stator slot shapes are shown in Fig. 2. Semi-closed slots tend to be used rather than open types, since they give a smaller effective air-gap reluctance and hence a lower magnetising current.

Motors smaller than about 1 m in diameter generally employ one-piece stator-core laminations, the centre part

Fig. 1. Sectioned TEFC cage induction motor: 2 hp (1.5 kW) 4-pole 'Kapak' D90L metric (courtesy English Electric—AEI Machines Ltd). (A) drive end ball bearing; (B) sectioned stator coil; (C) diecast rotor end-ring and cooling fins; (D) non-drive end ball bearing; (E) fan for surface cooling; (F) fan cowl.

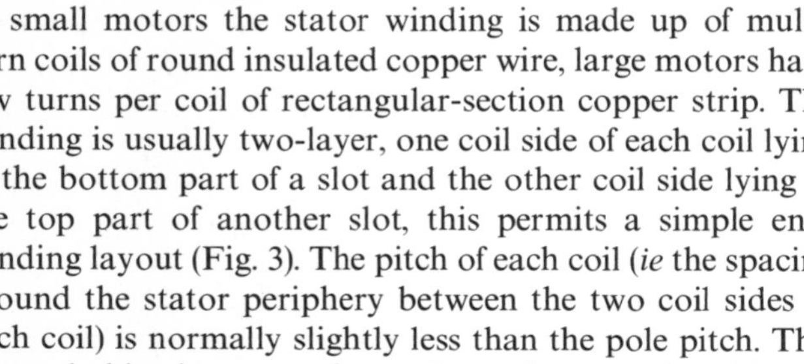

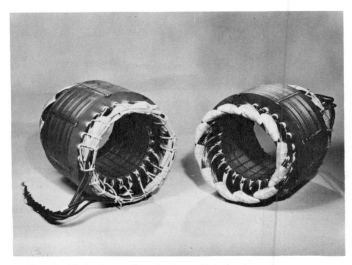

Fig. 3. End views of typical small induction motor stator (courtesy English Electric—AEI Machines Ltd).

being used for the rotor. Larger stators are made up with segments of core periphery, joined by butt and lap joints. The cores and teeth have to be designed to be able to carry the peak operating flux without excessive magnetic saturation effects, *ie* iron losses and magnetising amp-turns.

Electrical windings

In small motors the stator winding is made up of multi-turn coils of round insulated copper wire, large motors have few turns per coil of rectangular-section copper strip. The winding is usually two-layer, one coil side of each coil lying in the bottom part of a slot and the other coil side lying in the top part of another slot, this permits a simple end-winding layout (Fig. 3). The pitch of each coil (*ie* the spacing around the stator periphery between the two coil sides of each coil) is normally slightly less than the pole pitch. This short pitching improves the mmf waveform produced by the winding. Insulation has to be inserted in each slot between the two coils and between each coil and the slot sides (Fig. 2A) and in the end winding between phases. The thickness of the insulation depends on the supply voltage, and the quality on the class of insulation (maximum operating temperature).

Fig. 2. Typical stator and rotor slot shapes. (A) Semi-closed stator slot showing typical layout of 2-layer winding (small motors). (B) Open stator slot (large motors): rectangular strip conductor. (C) Single-cage diecast rotor slot. (D) Double-cage diecast rotor slot. (E) T-bar rotor slot. (F) Double-bar rotor slot.

In large motors mechanical coil support may be necessary for the end windings (Fig. 4) to cater for the forces produced by current surges, especially during starting. The exposed parts of the winding are usually coated with thermosetting varnish or wrapped in bitumen-impregnated tape (large motors) and stoved to harden; this helps to prevent ingress of moisture which could cause insulation breakdown in service. Small anti-condensation heaters can be provided to keep the winding temperature above that of the surroundings when the motor is idle, and hence prevent any moisture problems.

Cage motors up to about 200 kW usually have die-cast aluminium rotor windings which make the rotor extremely robust. The complete assembly, including end rings and internal fans at each end of the rotor, is formed around the core by pressure injection of aluminium into a single mould. Larger cage motors have copper or alloy bars brazed or welded to copper end-rings (Fig. 5). The cage winding usually has no special insulation from the core, and this necessitates careful design in large motors to keep down stray losses. The rotor-slot design depends mainly on starting performance requirements. The rotor slots are often skewed, by one stator-slot pitch over their length, to reduce magnetic noise and crawling tendencies during starting and run up.

The rotor of a slip-ring motor is wound for the same number of poles as the stator winding, but it often has fewer turns/coil; this limits the maximum induced rotor voltage (at standstill) and thus simplifies the rotor insulation (especially important in large high-voltage motors). For normal operation the slip rings must either be short circuited or connected together through an external circuit. If the external equipment is only required for starting, a centrifugal switch can be used to internally short circuit the slip rings at a predetermined speed.

Mechanical construction

Small motors commonly have cast-iron frames, which can be readily made to include special features such as external cooling fins (TEFC types). Larger frames have to be prefabricated by welding. Some parts of small motors are now made of cast-aluminium-alloy which is very light. Cooling fans are often made of die-cast aluminium or more recently of moulded synthetic material such as polypropylene.

The air gap is made as short as possible since this largely determines the motor magnetising (reactive) current and hence the power factor on load. It is limited, however, by manufacturing tolerances and minimum safety clearances necessary to prevent 'pull over' due to gravitational, magnetic (unbalanced magnetic pull due to rotor eccentricity) and rotational forces. The small air gap, typically 0.25–0.75 mm in small motors and up to 10 mm in very large motors, necessitates precise positioning of the shaft within the frame. Ball and roller bearings are used on small motors with grease lubrication, larger motors have to use sleeve bearings with forced oil lubrication. The effect of any transverse loadings on the shaft extension, due to belt pull or overhung weight tends to bend the shaft and reduce the air-gap clearance. Applications involving large loadings of these types may therefore need a bigger than normal diameter shaft to prevent pull-over.

PERFORMANCE

When a 3-phase supply is connected to an induction motor stator winding, a rotating mmf is produced.[1,2,3] The fundamental component of this mmf rotates at a constant

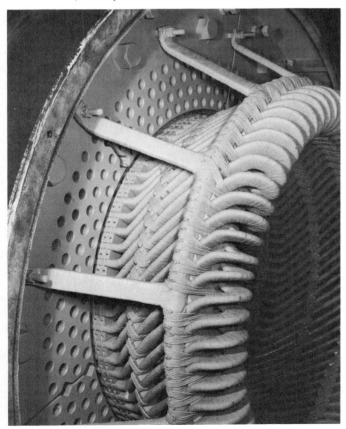

Fig. 4. *Stator end winding of very large motor showing mechanical coil support: 9000 hp (6.7 MW) 6-pole 11 kV (courtesy Parsons–Peebles Ltd, Witton).*

Fig. 5. *Double-cage rotor of large induction motor, before connection of end-ring to outer cage: 650 hp (485 kW) 3.3 kV 6-pole (courtesy Parsons–Peebles Ltd, Witton).*

speed known as the synchronous speed (N_s), given by: N_s(rev/min) = 60 × supply frequency divided by number of pole pairs.

The fundamental component of the resulting magnetic field cuts the conductors of the rotor winding at a speed referred to as the slip speed. This speed is usually expressed in terms of the motor synchronous speed, and is known as the motor slip: s = (synchronous speed minus actual rotor speed) divided by synchronous speed (per unit).

The rotating field induces a voltage in the rotor winding and hence a rotor current, which reacts with the rotating field to produce a torque. The rotor is thus accelerated towards synchronous speed, but can never attain this speed because the rotor voltage and current would then be zero and there would be no sustaining torque. Full-load torque is however produced at comparatively low values of slip: 0.03% to 0.05% for small motors and 0.01% to 0.02% for larger motors.

The frequency of the rotor current is directly related to the motor slip by: rotor frequency = slip (per unit) × supply frequency (Hz). This is a maximum at starting, when it equals the supply frequency.

Although the rotating field has a predominant fundamental component there are many harmonic fields present as well. This is because of the stepping in the net mmf waveform, which results from the winding in separate slots. The relative magnitudes of the different harmonic fields are determined predominantly by the winding details.

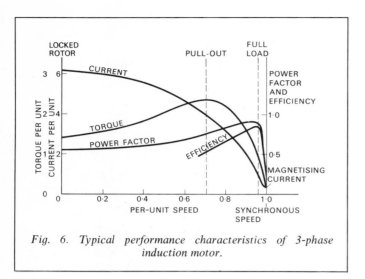

Fig. 6. Typical performance characteristics of 3-phase induction motor.

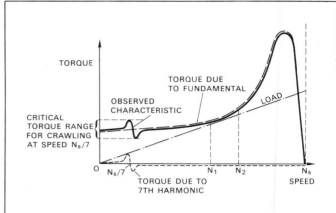

Fig. 8. Effect of seventh harmonic field on motor torque-speed curve.

Characteristics

Typical performance characteristics for a small cage induction motor are shown in Fig. 6. For most performance calculations it is possible to represent an induction motor by the per-phase equivalent circuit of Fig. 7.[1,2] In practice the circuit parameters of Fig. 7 do not remain constant under all conditions. In particular the resistances vary with temperature, and the secondary resistance and inductance change as the slip frequency changes, especially during starting. The latter is due to changing magnetic saturation effects and to 'skin effect' (or 'deep bar effect') which causes the current distribution to be non-uniform at high frequencies (for rotor conductors this effect is usually significant at 50 Hz). By the use of computers the circuit parameters can however be continuously varied to take account of these effects.

In addition motor performance is affected by the presence of mmf harmonics which can cause torque dip and stray load loss. The effect of a seventh harmonic field on a motor torque-speed curve is shown in Fig. 8; this could cause crawling at around one-seventh synchronous speed under certain load conditions. Skewing of the rotor bars largely prevents any crawling, but it introduces additional stray load loss.

Using the equivalent circuit of Fig. 7, and assuming the parameters to be constant (for simple analysis, parameter changes can be allowed for by suitable correction factors), the following expressions can be derived.[3] The effects of the magnetising current branch (X_m) have been neglected: these are significant in small motors particularly those with high pole numbers.[1]

Per-phase locked-rotor current (I_{LO})

$$= \frac{E_1}{\sqrt{[(R_1 + R_2)^2 + (X_1 + X_2)^2]}} \text{ (amps)} \quad (1)$$

Locked-rotor torque (T_{LO})

$$= \frac{28 I_{LO}^2 R_2}{N_s} \text{ (Newton-metres)} \quad (2)$$

Pull-out torque

$$= \frac{14 E_1^2}{\{R_1 + \sqrt{[R_1^2 + (X_1 + X_2)^2]}\} N_s} \text{ (N-m)} \quad (3)$$

Slip at pull-out torque

$$= \frac{R_2}{\sqrt{[R_1^2 + (X_1 + X_2)^2]}} \text{ (per unit).} \quad (4)$$

Slip at around full load

$$= \frac{I_2 R_2}{E_2} \simeq \frac{I_1 R_2}{E_1} \text{ (per unit).} \quad (5)$$

(Note: N_s in rev/min).

The total power input to the secondary is $I_2^2 R_2 / s$. This can be split into heating loss in the rotor, $I_2^2 R_2$ and mechanical output power given by $I_2^2 R_2 (1-s)/s$. Hence, rotor-input-power : rotor-output-power : rotor-copperloss

$$= 1 : (1-s) : s \quad (6)$$

In a slip-ring induction motor the total secondary resistance (R_2) can be varied by the addition of external resistance; this typically gives a range of torque-slip curves as shown in Fig. 9. The effects of increasing R_2 (more correctly R_2/X_2) can be summarised as: (1) reduction in starting current (eqn. 1) and increase in starting torque (eqn. 2), ie improved starting performance; (2) reduction in the speed at which pull-out torque occurs (eqn. 4) although no change to the value of the pull-out (eqn. 3), ie poorer speed regulation above this speed; (3) an increase in the full-load slip (eqn. 5),

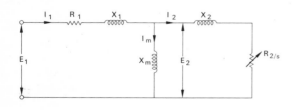

Fig. 7. Per-phase equivalent circuit of induction motor. R_1 primary resistance. R_2 secondary resistance (referred to primary). X_1 primary leakage resistance. X_2 secondary leakage resistance (referred to primary). X_m magnetising reactance. S per-unit slip.

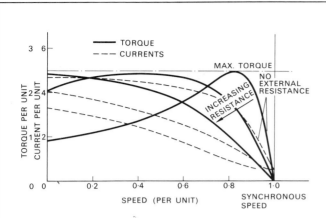

Fig. 9. Typical curves showing effect of adding external rotor resistance to a slip-ring motor.

ie giving poorer efficiency and greater rotor-circuit losses (eqn. 6) (the losses in the rotor itself remain much the same), but also a simple means of load-dependent speed control.

These effects also apply to the design of cage induction motors, but once a cage motor is constructed its performance cannot be varied. The design is therefore a compromise between the ideals of starting performance (item (1)) and running performance, ie full-load slip (item (3)); the latter is important since it affects both efficiency and motor heating (for a cage motor all of the rotor circuit losses are dissipated within the motor).

Two special types of cage rotor designs are commonly used to obtain higher ratios of starting torque to starting current for the same full-load slip. T-bar rotors (Fig. 2E) make use of skin effect. Double-cage rotors (Fig. 2F) are designed so that at high slip the lower conductors have a much higher reactance than the upper conductors. Both of these produce a non-uniform current distribution and a higher effective ratio of R_2/X_2 during starting. For normal running (low slip) the current flows fairly uniformly in all of the conductors.

All induction motors, however, have a starting current on full voltage which is many times the full-load current (usually over five times). Reduced-voltage starting is therefore commonly used for squirrel-cage motors to limit the starting current, but this seriously reduces the starting torque ($I_{LO} \propto E_1$ but $T_{LO} \propto E_1^2$ from equations (1) and (2)). For star-delta starting the locked-rotor torque is reduced to about one-third of its full-voltage value.

In BS2613:1970 code letters define the starting performance of cage induction motors when connected direct on line to a 50 Hz supply. Each letter implies a particular range of values of 'starting (locked-rotor) kVA' and 'locked-rotor torque'. From the former it is possible to calculate the locked-rotor current by:

locked-rotor line current
$$= \frac{\text{kVA/kW} \times \text{motor rating in kW} \times 1000}{\sqrt{3} \times V_{LINE}}$$

Also, since

full-load line current
$$= \frac{\text{motor rating in kW} \times 1000}{\sqrt{3} \times V_{LINE} \times \eta \times \cos \phi},$$

then $\text{kVA/kW} = \dfrac{\text{locked-rotor current}}{\text{full-load current}} \times \dfrac{1}{\eta \times \cos \phi}$,

where V_{LINE} = line to line voltage, η = full-load efficiency, $\cos \phi$ = full-load power factor.

Small/medium sized cage-motors are usually B or C types; C types usually having a special double-cage rotor or a high-resistance rotor in small motors, to give the high starting torque. Larger motors tend to be D or E types, to keep down the power surge on starting.

Magnetising current and power factor

The power factor of a 3-phase load is defined by:

$$\text{power factor} = \frac{\text{total watts taken from supply}}{\sqrt{3} \times V_{LINE} \times I_{LINE}}.$$

The magnitude of total current depends on the reactive current as well as the power component of current as shown in Fig. 10. At normal loads the reactive current component is virtually the same as the magnetising current, which is very nearly equal to the no-load current ($I_{(NL)}$), as the no-load losses and hence $I_{R(NL)}$ are usually small. The frame size of a particular machine is largely determined by the full-load torque requirement and not the speed. For a given frame size, therefore, the continuously-rated output-power capability, and hence the power component of current ($I_{R(FL)}$), is roughly proportional to the speed, ie the inverse of the number of poles. But the magnetising current for the same air-gap flux density increases with increasing pole number; hence full-load power factors worsen with increasing pole number, as shown in Fig. 11.

The low-speed induction motor therefore has the following disadvantages: poor power factor, large motor size and high capital cost per kW; and it is often better to employ a high-speed motor (4-pole is most common) with belts or reduction gears.

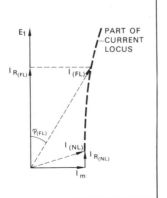

Fig. 10. Vector diagram showing per-phase no-load and full-load current components. I_R power component of current (resistive). I_M magnetising current (reactive). (NL) no-load. (FL) full-load.

Losses and efficiency

It is difficult accurately to measure losses and hence efficiency under full-load conditions (*ie* by input-output measurements), particularly for large induction motors. All standard methods of loss testing and calculation are therefore based on the assumption that losses on load can be sub-divided into the following parts, each of which can be separately determined.
(i) Stator iron loss: eddy-current and hysteresis losses due to flux pulsations in the stator core and teeth.
(ii) Friction and windage losses and rotor iron loss.
(iii) Stator copper loss: I^2R losses in the stator winding.
(iv) Rotor copper loss: I^2R losses in the rotor which are given by slip × total rotor input power (eqn. 6).
(v) Stray load losses: additional losses caused by the load currents, due to changes of flux distribution and eddy-currents.

A method of calculating losses is given in BS269. The motor efficiency is given by: the ratio of the output power to the sum of the output power and the motor losses. Losses (i)–(ii) are virtually independent of loading and are obtained from no-load tests. Except in small motors these generally form a fairly small part of the total losses on full load. Losses (iii)–(v) depend largely on the motor power component of current and hence the loading, and generally make up the major part of the full-load losses. Hence, as the motor output drops, these losses decrease, and the motor efficiency does not therefore change very much from rated full load to half load as shown in Fig. 12. Motors of the same power output but different pole number tend to have similar efficiencies. BS269 allocates a standard value for the stray load loss of 0.5% of the rated output; but the actual stray load loss depends on the individual motor design, and can be several times this value.

Reversing and electrical braking

A 3-phase induction motor can be reversed by interchanging any two of the supply leads; this causes the stator field to rotate in the opposite direction.

Electrical braking may be required either for rapidly stopping a motor, or for retarding a load which is tending to accelerate a motor, such as in cranes and mine hoists. There are three main methods of electrical braking.[5]
(i) *Plugging*. If two of the supply leads are interchanged during normal motor running the motor is rapidly and severely braked. The supply must be disconnected when standstill is reached to prevent the motor accelerating in the opposite direction. The rotor and stator currents during plugging are very high for motors over a few kW, unless limited by external resistors either connected in series with the stator winding or to the rotor sliprings.
(ii) *Regenerative braking*. This only applies when the load is tending to accelerate the motor. With normal mains supply connected, the motor runs at above synchronous speed and feeds power back into the supply (although still taking reactive magnetising current). For a slip-ring motor the external resistance may be used as a braking regulator.

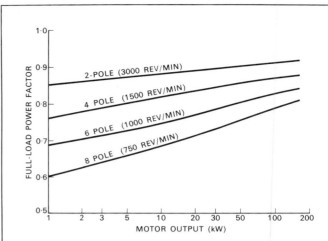

Fig. 11. Typical power factors of 3-phase TEFC cage induction motors, showing effect of pole number (design-B types, BS2613: 1970).

(iii) *Dynamic braking*. This involves disconnecting the 3-phase power supply and supplying part of the stator winding from a d.c. source (low voltage to give tolerable I^2R losses). Energy is dissipated in the rotor circuit as the rotor conductors cut the now stationary air-gap field, and the motor either rapidly but smoothly slows down to standstill, or produces a steady retarding force, depending on the type of loading.

Testing

Manufacturers carry out complete performance tests on prototype motors. For motors of repeat design and construction, however, only tests to ensure that the performance is consistent are necessary. The exact tests depend on manufacturing methods and on the quantity of similar motors being produced. In general, however, testing can be divided into three parts.
(i) Tests to check the overall torque/current-speed characteristics. In large motors complete curves at reduced voltage may be taken. For small cage motors measurements of starting torque and current, pull-up torque and pull-out torque are usually sufficient. For slip-ring motors only the pull-out torque is measured (this equals the maximum starting torque).
(ii) Tests to check the normal running performance. These can include the no-load current in each phase, the no-load watts, full-load current, and speed, and a temperature run on full-load. The temperature rise is now usually calculated from measurements of winding resistance at the start and end of a temperature run. In very large motors embedded temperature detectors may be used. For a slip-ring motor the brushes or slip-rings must be short-circuited for normal running tests.
(iii) Tests to check the motor has been constructed correctly. High-voltage insulation tests are always carried out. The windings may be checked by measurement of the d.c. resistance; for slip-ring motors the rotor winding is usually checked by measuring the open-circuit slip-ring voltage with locked rotor.

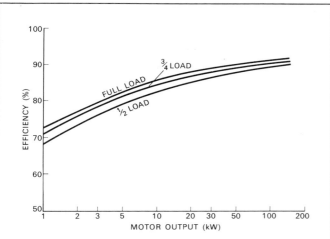

Fig. 12. Effect of degree of loading on motor efficiency of 3-phase TEFC cage induction motors (design-B types, BS2613: 1970).

Special tests
The equivalent-circuit parameters can be determined from locked-rotor and no-load tests, and measurement of the d.c. stator winding resistance.[1,7] The segregation of no-load losses, and stator and rotor copper losses is fairly straightforward.[7] Accurate separate measurement of the stray load loss component is, however, rather difficult. For slip-ring motors it is possible to inject d.c. into the rotor and determine the loss in a similar way as for a synchronous motor.[6] American standards specify the reverse-rotation test[6] to be used for measuring stray-load losses in cage motors, but, due to justifiable disputes over its accuracy, it is not at present included in British Standards.

Thermal protection in service
The life of motor insulation decreases rapidly with increasing temperature. Insulation is graded into classes in terms of maximum hot-spot temperatures which can be withstood for long periods without significant deterioration. Maximum winding temperature rises (such as in BS2613:1970) are based on these maximum hot-spot temperatures, on an agreed maximum ambient temperature (usually 40°C), and on typical differences between measured temperatures and hot-spot temperatures, which depend on the method of measurement. If the life of a motor is not to be impaired it is important that these limits are not exceeded under any operating conditions, *ie* during starting (for which accelerating time is important—see below), duty cycles, or normal running.

Several methods of providing on-site over-temperature protection are available which rely on detection of either high currents for a long period, or a high temperature in the windings. The latter is more precise for stator windings, but the necessary thermocouples, thermistors or thermal trips must be inserted into the winding during manufacture, and extra terminals provided. For over-current protection the trip time must be related to the degree of overload, and must be long enough to prevent false tripping during starting. Complete protection in this way is impossible unless the overload devices have a similar thermal time constant to that of the motor they are protecting.

These methods do not protect the rotor winding which, except in small motors, is likely to overheat quicker than the stator (most larger induction motors are thermally rotor critical).

Accelerating time
During motor acceleration up to speed the rotor and stator heat up very rapidly due to the higher currents and poorer ventilation (unless forced ventilated) than at full speed. To prevent overheating the rotor must therefore be close to normal running speed before the maximum temperature is reached (this is slightly longer than the maximum allowable time with locked rotor).

The motor accelerating time depends on: (i) the total moment of inertia (in $kgf - m^2$) of the motor rotor (J_R) and of the load referred to the motor shaft speed (J_L); (ii) the torque-speed characteristics of the motor and load, taking into account any supply voltage-dip due to the high starting currents. The speed increase over successive time intervals, Δt (chosen for desired accuracy) is given by:

$$\frac{\text{av. motor torque} - \text{av. load torque}}{J_R + J_L} \times \Delta t$$

where speed is in rad/s; torque in Nm; t in seconds. This gives the relationship between time and speed and hence the acceleration time. The motor losses and temperature rise can then be calculated during this time.

Clearly if the difference between motor torque and load torque is small for any region of the speed range, the time required for a given speed increase in that region can be abnormally long. This sometimes causes difficulties in large cage induction motors (*eg* from speed N_1 to N_2 in Fig. 8).

APPLICATIONS
The 3-phase cage induction motor is by far the most widely used type of industrial motor, for four main reasons. (i) *Cheapness*: due to simplicity of windings (in particular the rotor winding) and general construction; (ii) *ease* of starting, stopping and braking, etc; (iii) *reliability* due to inherent robustness of rotor and absence of commutator and slip-rings; (iv) *minimal maintenance* requirements due to absence of brushgear.

The last two factors make the cage induction motor, with suitable casing, ideal where a motor is required to be completely isolated or encapsulated (*eg* submersed in water) or for duties in an atmosphere which would cause commutator or slip-ring corrosion.

The cage induction motor does, however, have three major limitations. Firstly, it is essentially a constant-speed motor and cannot be run for long periods at a slip of more than a few per cent without overheating. Secondly, it has a poor starting torque and takes a very high current during acceleration, almost at starting value up to about three-quarters speed, as shown in Fig. 6. It is, therefore, not inherently suitable for accelerating high-inertia loads or for duty cycles involving rapidly repeating stops and starts. Thirdly, it has a poor power factor, particularly under light loading. The first two of these limitations can be largely overcome by using a slip-ring induction motor, and controlling the rotor current or power to give the desired torque/current-speed relationship.

Constant-speed drives

The induction motor is ideal for general-purpose constant-speed drives. It has a shunt-type speed-torque characteristic and a substantial short-time overload capacity without significant loss of speed (*ie* up to pull-out torque).

If precise speed holding is required under varying loads, however, a synchronous-type motor may be necessary. Considerable development work is being carried out on small/medium sized synchronous reluctance motors which are fairly simple and robust like the induction motor. But present pull-up characteristics are inferior to induction motors of the same size.

A system of synchronised drives using slip-ring induction motors, known as 'power selsyns', can be provided if two or more motors have their stators fed from the same supply and rotors electrically connected but not mechanically coupled. Such a system then consists of one 'transmitter selsyn', which must be driven at the required speed by another motor or some existing equipment, and one or more remote 'receiver selsyn', which exactly follows the speed and position of the transmitter. There is however a tendency for 'hunting' to occur around harmonic speeds, and for the receivers to lose synchronism and run up towards their respective synchronous speeds, but these can be overcome by the use of damping resistors and special stabiliser circuitry in the supply lines.[8]

For many years induction motors have been available with a choice of two synchronous speeds (speed-change types), all having certain limitations. The recent development of the theory of pole amplitude modulation (PAM) has enabled a wide range of speed ratios to be achieved, with little cost over that of a single-speed motor and very similar performance. The most common types of speed-change induction motors are described below.

Dual-wound motors

Dual wound units have two completely separate windings in the stator. Each winding can be designed and used separately, although care has to be taken to prevent circulating currents in the idle winding due to voltages in the active winding. The size of stator slot, and hence frame size and cost, must be greater than for an equivalent-rated single-speed motor. For a slip-ring motor two separate windings also have to be provided on the rotor.

Pole-change motors

A simple method of 2:1 pole changing (*ie* 1:2 speed changing) is possible using only one stator winding. This is

Connections		Relative air-gap flux density on a fixed-voltage supply		Suitable type of loading
p-poles 60° spread winding	2p-poles 120° spread winding	p-poles	2p-poles	
parallel star	series delta	0.707	1.00	constant output power
parallel star	series star	1.22	1.00	constant torque

Table 1. Pole-change motor arrangements and properties.

achieved by connecting all the coils of each phase in series for the one pole number, and series-parallel for the other pole number. The two most common arrangements and some of their properties are shown in Table 1. Only six motor terminals have to be provided. For the parallel-star connexion, the three terminals not connected to the power supply must be short-circuited. For a slip-ring motor the rotor winding has to be brought out to six slip-rings with six resistors, but no switching is required. The coil pitch is necessarily a compromise between the ideals for each pole number, and the performance is generally inferior to that of a single winding designed for either speed.

Pole amplitude modulation

Since the original publication of the principles of PAM,[10] a large number of two-speed PAM cage motors up to 2000 kW have been built, tested and put into service. The most common pole ratios are 4/6, 6/8, 8/10, 10/12 poles. The performance of these motors has been proved to be insignificantly different, at either speed, from that of normal single-speed motors, despite the presence of extra mmf harmonics. The theory has recently been extended to cover slip-ring motors.[11]

The basic construction of PAM motors is the same as that of single-speed motors, but the stator coils are inter-connected differently. In a single-speed motor the pattern of coil groupings repeats itself every few pole pitches around the stator periphery, and the number of adjacent coils connected in series (*ie* to the same phase) does not change very much, if at all. In a PAM motor, however, the pattern of coil groupings and the number of adjacent coils connected in series vary considerably around the stator periphery. The way in which this is done is determined by the type of modulation required.

The switching arrangement between speeds is the same as for 2:1 pole changing, which has been found to be a special case of the general theory of PAM.

Variable-speed drives

Because of the cheapness and other inherent advantages of induction motors over motors which lend themselves more readily to continuous speed control, *ie* d.c. motors and a.c. commutator motors, there has been a considerable amount of work in recent years on development of efficient methods of continuously varying their speed. The various methods of speed control give rise to high losses or low losses at low speeds.

High-loss systems. The simplest control methods suffer high losses at high slip and are only really suitable for fan-type loadings or for restricted low-speed running.

(a) *Supply-voltage control.* Motor torque is roughly proportional to the square of the fundamental component of the supply voltage. Hence variation of the effective voltage, using thyristors, saturable reactors, induction regulators, etc, gives a form of load-dependent speed control applicable to small cage motors.

(b) *Slip-ring motors with rotor resistance control.* Speed control in this way is very inefficient at low speeds, the efficiency always being less than $(1 - s) \times 100$ per cent. It is often used, however, for drives of fan-type loads (torque $\propto$ speed^2) with low output at low speed, or for drives requiring only occasional short-duration low-speed running.

Low-loss systems. By the use of sophisticated auxiliary equipment it is possible to overcome or to avoid the high losses at low speeds. The former involves slip-energy recovery from the rotor circuit of a slip-ring motor. The latter requires control of the motor synchronous speed, *ie* supply frequency control (attempts to continuously vary the number of poles have met with limited success),[12] and is usually applied to cage motors.

(a) *Slip-energy recovery.* Depending on the loading characteristic (constant torque, constant power, etc) the slip energy can be converted to mechanical shaft power via a d.c. motor, or to electrical power back to the supply from an induction generator. Typical schemes are the modified Kramer system (with solid-state rectifiers) and Scherbius system. Some recent work has been done on static invertors.

(b) *Supply-frequency control.* A considerable amount of work is being done on the use of variable-frequency supplies derived from the mains using thyristors, either d.c.-link invertors or cycloconverters. By suitable control of the frequency (synchronous speed) and terminal voltage, in conjunction with monitoring of the motor slip (load dependent), it is possible to achieve virtually any torque-speed characteristics, *ie* shunt-type or series-type (traction purpose).[11] The control equipment is very expensive, but is likely to become cheaper in time.

REFERENCES

1. P L Alger. The nature of induction machines. *Gordon and Breach* 1965.
2. M G Say. The performance and design of alternating current machines. *Pitman Press* 1958.
3. A E Fitzgerald and C Kingsley. Electric machinery. *McGraw Hill* 1961.
4. D G Fink and J M Carroll. Standard handbook for electrical engineers. Tenth edition, *McGraw Hill* 1968.
5. F T Bartho. Industrial electric motors and control gear. *Macdonald & Co Ltd* 1965.
6. B J Chalmers. Electromagnetic problems of a.c. machines. *Butler & Tanner Ltd* 1965.
7. R Bourne. Electrical rotating machine testing. *Iliffe Books Ltd* 1969.
8. P F Harrison. Power Selsyns. *Electrical Review*, 25 Oct 1968, pp 602–605.
9. B V Jayawant. Induction machines. *McGraw-Hill* 1968.
10. G H Rawcliffe, R F Burbridge and W Fong. Induction-motor speed changing by pole amplitude modulation. *Proc. IEE* 1958 vol. 105A, pp 411–419.
11. G H Rawcliffe and W Fong. Slip-ring PAM induction motors for two synchronous speeds. *IEEE Transactions*, Power Apparatus & Systems, Paper No. 70 TP 520-PWR. 1971.
12. F C Williams, E R Laithwaite and J F Eastham. Development and design of spherical induction motors. *Proc. IEE* 1959, vol. 106(A). p.471.
13. R W Johnston. Modulating invertor system for variable-speed induction-motor drive (GM Electrovair II). *IEEE Transactions* 1969, vol. PAS-88 no. 2, pp 81–85.

Chapter 8

Synchronous Motors

H T Price *DLC CEng MIEE*
Brush Electrical Engineering Co Ltd

In common with direct current machines, the action of synchronous machines is reversible. When supplied with a.c. power having the same number of phases as when operating as an alternator, it runs as a motor at synchronous speed, the speed being determined by the frequency of supply and the number of poles:

$$N_s = \text{(synchronous speed)} = \frac{60 \times \text{frequency}}{\text{no. of pole pairs}}$$

The name of this type of motor is derived from the fact that it has to run at the synchronous speed, which keeps it in step with the supply frequency.

There are three types of synchronous motor which may be defined by their rotor construction: (i) Salient pole synchronous motors with squirrel cage type pole face winding. (ii) Synchronous motor with a distributed wound rotor as for a slipring induction motor—synchronous induction motor. (iii) Synchronous motor basically of salient pole construction but with a distributed winding in the pole face—salient pole synchronous induction.

MECHANICAL CONSTRUCTION

In general the mechanical construction of a synchronous motor resembles that of an alternator or induction motor. The stator constructions are identical mechanically and magnetically and also with respect to stator windings. The stator consists essentially of two parts; the inner circuit or stator core which carries the magnetic flux; and the stator winding. The stator core is built up from low-loss silicon-steel stampings to reduce iron losses (due to eddy and hysteresis currents) to an acceptable value. The stator core is built up under pressure and securely held in position by end plates. For smaller stator cores, circular stampings are employed whereas for the larger stator cores the stampings are segmental. The stator winding is housed in slots punched around the inner periphery of the stator stampings. The type of stator winding is dictated by the power output rating, number of poles, and system voltage.

For smaller power ratings and low voltage supply, mush type stator windings are employed. For medium power ratings which are normally low voltage, bar wave type windings are used. For the larger power, the system voltage is high tension (2.2 kV to 13.8 kV) and diamond type windings carried in open type slots are fitted. The stator windings are suitably reinforced to prevent movement during starting, as for the induction motor.

The rotor mechanical construction is dictated by the mechanical stresses, which are determined largely by the peripheral stress. For low peripheral speeds, the poles may be secured to the rotor hub by bolts whereas for higher values of peripheral speed, some form of dovetail construction must secure the poles to the rotor hub. For 4- or 6-pole designs, it is the practice to make the poles integral with the body and to have bolted-on pole tips. For the largest motor, a three-part rotor construction is employed with a cast steel rotor body and forged steel stub end shaft. The poles are commonly steel laminations which are assembled under pressure onto a steel building bar and then firmly secured by through bolts or rivets. The rotor shaft is a steel forging. The excitation coils may be wound with strip on edge copper or with insulated rectangular copper conductors, depending on motor size and speed.

Enclosures may be screen protected/drip-proof, pipe/duct ventilated, NEAMA type II or III, filtered-air intake, or totally-enclosed water-cooled. The totally-enclosed fan-cooled construction is rarely used for synchronous motors. Selection of enclosure is dictated by site conditions, for example NEAMA II construction is employed for compressor drives or power station auxiliary drives on open sites.

In the lower power ranges, the machines are self-contained, of end-shield frame construction, with ball/roller bearings and overhung exciters. For larger power ratings, the pedestal bearing bedplate type construction is usual. Disc type or plain sleeve bearings with oil ring lubrication are used, but large high-speed machines require forced lubricated bearings. Where the drive is through a gear box, a single pedestal bearing construction, with part of the rotor weight being carried by the gear box bearing, is common practice. For certain applications such as compressor drives, it is usual to employ a bearingless construction with the motor rotor overhung on an extension of the shaft.

Synchronous induction motors are built as for an induction motor but with the addition of a d.c. exciter; a compromise rotor design is needed to suit both starting and running conditions. The open circuit rotor volts at start are considerably higher than for an induction motor, and wider spaced sliprings with flash barriers are required.

In the salient pole synchronous induction motor, the squirrel cage starting winding of the conventional salient pole synchronous motor is replaced by a distributed three-phase winding similar to that of a slipring induction motor rotor, necessitating pole stampings with deeper and wider tips. This winding is brought out to three sliprings which are additional to the two sliprings for the d.c. excitation winding.

CHARACTERISTICS

When the stator winding is connected to the supply, the polyphase currents flowing in the stator winding set up a magnetic field which rotates at synchronous speed. When the rotor is excited from a direct current source, a magnetic field is established which remains stationary relative to the rotor. Reversal of the direct current excitation source reverses the polarity of the poles, but does not alter the fact that the field is stationary relative to the rotor.

A characteristic of rotating electric machines is that the rotor field must have the same speed in space as the stator field. To produce a steady torque there must be a constant angle of displacement (α) between the stator and rotor fields. A synchronous motor must therefore run at synchronous speed to exert a uniform torque, and at standstill exerts no torque. The synchronous motor is not self-starting, and a starting winding of the induction motor type must be provided in the pole face. The variations in air gap reluctance due to the salient pole construction produce a reluctance torque which is superimposed on the torque due to the excitation winding.

Rotating electric motors have another feature in common, namely that the armature windings develop a back-emf (E_B) against which the applied voltage (V) has to circulate the armature current (I_a). In the d.c. motor, E_B is directly opposed to applied voltage and the armature current flowing is limited by the armature resistance (R_a). $I_a = (V - E_B)/R_a$.

With increasing load, the armature current must increase to exert the torque required, and the rotor adjusts itself to the increasing load by slowing down in speed, thereby generating a lower back-emf. But, in the case of the

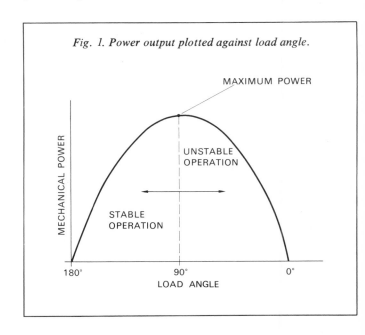

Fig. 1. Power output plotted against load angle.

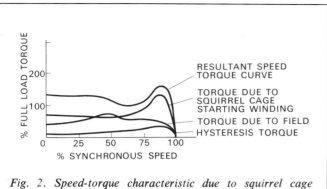

Fig. 2. Speed-torque characteristic due to squirrel cage winding of salient pole rotor.

synchronous motor, the rotor speed is constant and E_B cannot decrease automatically as for the d.c. motors. The armature current has to be circulated against the armature resistance and reactance in the case of the synchronous motor. The increased load torque required is produced by increasing the angle of displacement between the stator and rotor mmf.

In the case of a synchronous generator and motor running unloaded on a single phase system, assuming no losses in the machines, the circulating current is zero and (for the condition of equal excitation) the generated voltage (V) equals the motor back-emf (E_B). With increased load the angle of displacement between the stator and rotor mmf must increase. This is represented by the applied voltage (V) advancing on the back-emf (E_B). The resultant voltage E_R is the vector sum of V and E_B and circulates the armature current (I_a) against the armature impedance (Z_a).

Overexcitation: power factor correction

The back-emf (E_B) of the synchronous motor can only be increased or decreased by adjusting the rotor excitation. With under-excitation, the resultant voltage (E_R) causes an armature current (I_a) to flow lagging some 90° behind the applied volts. For overexcitation, the armature current leads relative to the applied voltage. By adjustment of the excitation, the synchronous motor can therefore be made to take lagging current or to deliver leading current, operating as a synchronous condenser. It is usual to design for a power factor between unity and 0.8 leading but, in applications where considerable power factor correction is required, the motor may operate at a much lower leading power factor.

Synchronous pull-out torque

If a synchronous motor is running under steady load, the characteristic vectors rotate at synchronous speed with no change. The effect of increased load is to slow down the rotor temporarily so that the applied voltage vector advances on the armature back-emf, and the motor settles down to run with the vectors in a new relationship. This means that with increase of armature current the angle (α) increases. Ultimately, the two effects counter-balance and the driving power ceases to increase. Beyond this point the mechanical power output decreases. If the drive load is so great that this point is passed, the motor will pull out of step and then stall. (Fig. 1)

For the salient pole synchronous motor, the synchronous pull-out torque is usually 150% of full load torque for unity power factor design and 200–225% of full load torque for 0.80 leading power factor design. The pull-out torque can be increased by increasing the field excitation. By employing current compounding, the excitation can be automatically increased with load. The following simple formula assists when considering the relationship between excitation, power factor, and pull out torque.

$$\% \text{ synchronous pull out torque} = \frac{AT_F}{AT_A} \times \frac{100}{\cos \phi}$$

where AT_F = Field ampere turns per pole; AT_A = Armature ampere turns per pole. Expressed in words, a synchronous motor pulls out of step when the watt component of the armature turns equals the excitation ampere turns.

With a sudden increase in load, the reaction due to transformer action results in transient increase in excitation and pull-out torque. This effect disappears in a few cycles and can be effective only on instantaneous peaks, the value depending on the armature reactance.

Starting and acceleration torque

The salient pole synchronous motor is not self-starting, and a squirrel cage winding must be provided in the pole face. This squirrel cage winding produces most of the torque during starting, but the field winding and hysteresis effects in the pole face also make a contribution. For large 4- and 6-pole salient pole synchronous motors with bolted-on alloy steel tips, the eddy currents in the pole tip produce most of the torque.

The field winding develops a torque similar to that of a slipring induction motor with a single-phase secondary. The mmf due to the single-phase secondary may be analysed into two components; a positive sequence component rotating in the same direction as the stator mmf, and a negative sequence component rotating in the opposite direction. During starting and acceleration the field winding is normally short-circuited through a field discharge resistance. The value of this resistance has a significance on the torque characteristic. It is customary to use a field discharge resistance 6 to 10 times the field resistance to ensure suppression of the negative sequence component, and to obtain maximum assistance from the positive sequence component at the load point.

The torque available for acceleration is the difference at any speed interval between the motor speed torque and the load characteristic. Step by step integration from standstill to synchronous speed gives the acceleration time. During acceleration the squirrel cage winding absorbs the energy required to overcome the load torque and also that stored in the system inertia during acceleration. It is important to ensure that the squirrel cage is not overheated; possibly by introducing a protection system.

Pull-in torque

Pull-in (synchronising) torque is that required to accelerate the motor and driven load from the maximum induction motor speed to synchronous speed. It is the critical stage in the starting of a synchronous motor rotor. Whether or not the rotor pulls into step depends on the position of the rotor poles relative to the poles of the rotating stator field. If like poles are aligned, they tend to repel each other, so that the pull-in torque is at a minimum and fails to pull the rotor into synchronism. But, if unlike poles are aligned, they attract each other and the pull-in torque is a maximum.

When the direct current excitation is applied to the field system and because the rotation is running below synchronous speed, the torque is a combination of the pulsating torque due to the d.c. excitation of the field system and the induction motor torque which also varies with speed. If the slip is low enough, the attraction between the unlike poles in the stator and rotor will pull the rotor into step with the rotating stator field. For given excitation, system inertia and load, there is a slip below which the synchronous motor will not pull into step. This is termed the critical slip.

The transient angular variations of the rotor of a synchronous motor, neglecting the effect of reluctance torque, following the sudden application of direct current excitation is characterised by the differential equation:

$$P_I \frac{d^2\theta}{dt} + P_d \frac{d\theta}{dt} + P_M \sin\theta = P_1$$

The first term represents the torque required to accelerate the combined inertia of the system; the second represents the torque that results from the squirrel cage winding for slip torque characteristics approaching synchronous speed. The third term represents the torque due to the synchronising action of the motor after direct current excitation has been applied. Solving for slip gives:

$$S_c(\text{critical slip}) = \frac{620}{\text{rev/min}} \times \frac{P_{ms}}{fWR^2}$$

where P_{MS} = maximum synchronising power

$$= \frac{3VE_B}{1000X_s}$$

V = applied volts; E_B = motor back-emf; X_s = synchronous reactance per phase. This gives the critical value of slip from which a synchronous motor always pulls into step when switching conditions are the worst. Besides the synchronising torque, the total system inertia is an important factor in determining the critical slip.

The maximum synchronising torque available can be increased by forcing the excitation. To achieve maximum pull in torque, the d.c. excitation must be applied at the right instant to ensure alignment of the north and south poles of the rotor and the revolving field of the stator. This is termed correct angle synchronising. When the d.c. excitation is suddenly applied, there is an associated current surge on the supply system. Correct angle synchronising gives the minimum current surge.

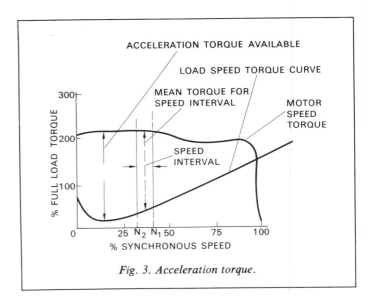

Fig. 3. Acceleration torque.

Hunting

Hunting may be defined as oscillatory motion of the rotor relative to the stationary stator field, which results in alternate motor and generator action. It is simple harmonic motion since the synchronising torque, the force accelerating or retarding motion is proportional to displacement. The frequency of oscillation for small angles of displacement is given by:

$$F = \frac{1}{2\pi}\sqrt{\frac{\text{displacement}}{\text{acceleration}}} = 3220\sqrt{\frac{P_s p}{N_s WR^2}}$$

where p = number of poles; P_s = synchronising power per electrical radian of displacement; N_s = Synchronous speed, rev/min; WR^2 = Total system inertia in lb ft^2.

Unless adequately damped a synchronous motor will continue to oscillate at its natural frequency once disturbed by load changes. This movement may be represented by:

$$WR^2 \frac{d^2\theta}{dt^2} + K\frac{d\theta}{dt} + P_s = 0$$

in which the three terms, from left, are respectively: inertia component; damping component; synchronising torque component. A positive damping component ensures stable operation, whereas with negative damping the magnitude of the oscillations becomes progressively larger until the power swing trips the circuit breaker. Damping is normally provided by the squirrel cage starting winding.

When a synchronous motor is subjected to a pulsating torque the angle of lag of the rotor behind the stator field follows closely the pulsating load imposed, and the current variation is in sympathy. Current pulsation, usually expressed as a percentage of full load current, is the difference between maximum and minimum values. Current pulsation in itself is not objectionable but the consequences are: such as voltage dip and light flicker. Some present standards permit a current variation of 66% of full load current; this current variation corresponds to an angular deviation of approximately ± 5 electrical degrees from constant speed of rotation.

If a synchronous motor is subject to a periodic disturbing force, the resultant oscillation is said to be forced and can be represented by the differential equation:

$$WR^2 \frac{d^2\theta}{dt^2} + K \frac{d\theta}{dt} + P_s = Q_d \sin\theta \, dt$$

where the resultant, right hand, term is the forced torque component.

The total inertia of the system for a given synchronous motor design must be proportioned to match two conditions: (i) The natural frequency must differ from any forced frequency by at least 20% to avoid resonance. (ii) The current pulsations must not exceed 66% of full load current, as defined above.

Methods of starting

When considering characteristics of synchronous motors, it was stressed that the synchronous motor is not self-starting and a starting winding as for an induction motor must be provided in the pole face. The method of starting selected depends on starting torque and pull in torque requirements and supply limitation relative to current. Direct on-line starting is the simplest and, when conditions permit, is first choice. But, when starting kVA overloads the supply system or generator capacity, some form of reduced voltage starting has to be employed.

Field excitation

There are three recognised methods of direct current excitation for the main field: (i) d.c. exciter; (ii) static exciter; (iii) a.c. exciter.

D.C. exciter: the circuit employed with a d.c. shunt exciter generator is out of circuit at starting, during which time the voltage builds up. The field application switch is single-pole make before break; it switches the main rotor field from the field discharge resistor to the exciter during field application. Automatic field application may be incorporated by employing a polarised frequency relay. This senses the slip frequency (induced in the field) from the voltage across a reactor in series with the field and provides automatic correct angle field application plus automatic application of the field discharge resistance in the event of pole slipping.

Static exciter: the simplest form of static excitation is via a tapped transformer and silicon diode rectifier bridge. The disadvantage of this simple form of excitation is that the motor is somewhat over-excited on light loads and operates at a low leading power factor. When using a static exciter it is important that the exciter cannot reverse current or produce a reverse voltage. It is therefore essential to isolate the exciter from the field during starting. Furthermore it is prudent to connect a surge suppressor across the motor field to absorb any reverse field current which may try to flow through the exciter during synchronising or pole slipping. If field control is required, the output of the field supply transformer may be controlled by means of tappings, transductors or thyristors. Automatic field application may be employed as for a direct-current shunt exciter.

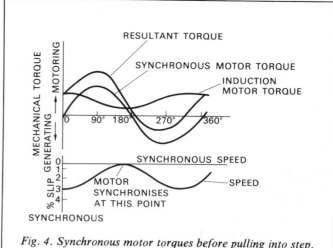

Fig. 4. Synchronous motor torques before pulling into step.

A.C. exciter: By mounting the diodes, field resistor, and field application thyristors on the motor, a brushless synchronous motor is obtained. Thyristors replace the function of the field application switch. They are switched by a control unit which senses the slip voltage in the field to trigger the thyristors. One is triggered at the most favourable instant in the slip cycle after the slip frequency has dropped to a predetermined level, which can be adjusted by means of a rheostat mounted on the control unit. Provision is included for field application in the event of the motor pulling into step by virtue of its own reluctance before the excitation is applied, as frequently occurs when starting on light load. The control unit is only energised during the starting period.

Control and protection

Automatic control of excitation is frequently required for certain applications — for example, power factor control. A simple open loop compounding scheme may be employed or an error-actuated automatic voltage regulator. Protection incorporated in the control scheme is dictated by the power rating, the synchronous motor design and the type of load. Protection may include: (1) Overcurrent and instantaneous earth fault release, all acting on the main circuit breaker. For large powers, a differential protection scheme may also be included. (2) Stator winding overheating by means of temperature sensitive detectors embedded in the winding. (3) Protection to prevent overheating of the squirrel cage winding. This is normally achieved indirectly by means of a relay sensitive to slip frequency voltage with a time delay action matched to the thermal time constant of the squirrel cage winding.

(4) *Field failure relay*. This is basically a static power factor relay responsive to lagging power factor but with a time delay to allow for the currents during starting and synchronising. This relay may be arranged to suppress the field immediately and thus permits the motor to attempt to re-synchronise. If the excitation is not restored or the motor field fails to re-synchronise, the motor will trip out on load. Alternatively the field failure relay may be arranged to trip the circuit breaker immediately without attempting to re-synchronise.

APPLICATIONS

In general, synchronous motors can handle any load which can be driven by a squirrel cage induction motor design B to BS2613:1970. Another rough guide is that synchronous motors are cheaper than squirrel cage induction motors if the rating exceeds 750 W (1 hp) per rev/min. Whether or not to use a synchronous motor for a drive requires sound application engineering. Due consideration must be given to matching the characteristics of a synchronous motor to the driven unit and also to the economics of the drive inclusive of control gear, together with capitalisation of the higher efficiency and power factor available. The factors which have to be considered are: machine characteristics; control and protection; economics.

Machine characteristics: starting torque/kVA inrush; accelerating torque available; pull-in torque; pull-out torque. *Control and protection*: d.c. excitation source; protection; automatic excitation control. *Economic considerations*: capitalisation of higher efficiency and power factor; cost of control gear; motor capital cost.

At speeds below 500 rev/min and powers above 150 kW (200 hp) where induction motor performance leaves something to be desired, the synchronous motor is often a good application on the basis of higher efficiency and power factor. For 1000/1800 rev/min drives, it is doubtful whether or not the synchronous motor shows an economic advantage below 750 kW (1000 hp). If low starting kVA associated with high starting torques is required consideration may be given to the synchronous induction motor or the salient pole synchronous induction motor.

Synchronous motors are inherently constant speed machines. However, operation of centrifugal fans, compressors and pumps under part-load conditions at reduced speed is a frequent requirement. The combination of synchronous motor and slip coupling is a worthwhile consideration for such applications, particularly for a limited speed range not exceeding 1.5/1.0 and where the torque falls with speed.

Compressor drives

NEAMA recommended torques for reciprocating compressors are expressed as % of full load. Typical applications are: air and gas (starting unloaded); ammonia (starting unloaded or bypassed); Freon (starting unloaded); vacuum pumps (starting unloaded). For such applications the recommended torques are: starting 40%; pull-in 30%; pull-out 150%. The salient pole synchronous motor is ideal for this application, particularly the low speed type.

Magnitude of current pulsation and resonance with a forced frequency from the compressor have to be watched. The magnitude of the current pulsation depends upon the torque characteristic of the compressor, the power load, synchronising torque of the motor, and the total inertia of the system. In general, the problem is how much inertia (WR^2) is required to limit the current pulsation. This may be determined from the 'XY' curve for the compressor motor combination (Fig. 5).

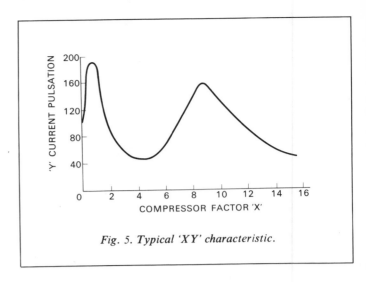

Fig. 5. Typical 'XY' characteristic.

X is termed the compressor factor which for a given compressor torque angle diagram and motor characteristic enables the inertia (WR^2) required for the selected current pulsation to be determined. Current pulsation is proportional to Y. The compressor factor may be determined from:

$$WR^2 = XfP_sG$$

where WR^2 = total system inertia in lb ft^2; X = compressor factor; f = frequency of supply in Hz; P_s = synchronising power per electrical radian of displacement; $G = 1.34(100/\text{rev-per-min})^4$. The compressor factor X has been established by NEAMA motor and generator standards MG10-45 for over 75 different types of compressor and operating conditions.

Turbo type compressor are usually multistage, designed for large capacity air or gas flow and geared to run at speeds well above normal motor speeds. They are generally started with the by-pass valve open or discharge valve closed and torque requirements are then: starting 20%; pull-in (unloaded) 40–55%; pull-out 135%.

Centrifugal pumps

Centrifugal pumps are normally of the radial flow type. However, mixed flow and axial flow pumps have also to be considered. In the case of radial flow pumps, the pull-in torque requirement against an open discharge valve must be watched. With 'mixed' flow and 'axial' flow centrifugal pumps the percentage shut off torque must be watched. For an axial flow pump with completely closed discharge valve the pull-in torque may reach 220%.

NEAMA recommended torques for centrifugal pumps. For various types of pump, torques (as % of full-load torque) are quoted for starting(s), pull-in (p-i), pull-out (p-o): *Horizontal* discharge valve closed (s)40 (p-i)60 (p-o)150; discharge valve open (s)40 (p-i)100 (p-o)150. *Vertical* discharge valve closed (s)50 (p-i)60 (p-o)150; discharge valve open (s)50 (p-i)100 (p-o)150. *Adjustable blade* vertical unloaded (s)50 (p-i)40 (p-o)150; *screw type* started dry primed (s)40 (p-i)(30) (p-o)150; discharge open (s)40 (p-i)100 (p-o)150.

Rubber mixers and mills

Crude rubber is mixed and kneaded with other materials in a (Banbury) mixer which consists of slow speed rolls through which the mixture is squeezed. The mixer is usually started unloaded but, in the case of an emergency shut down, it is necessary to clear the mill and therefore high torques are desirable. Torques are: starting 140%; pull-in 120%; pull-out 250%. This is an application for which a compound excitation system is needed.

Rubber mills: Following mixing, the rubber is milled between slow moving rollers to get required softness and thickness. If processing is interrupted the material hardens, so torques must be adequate to restart and synchronise in case of mill shut down, due to power failure. Rubber mills must have emergency braking as a safety measure; *eg* should an operator come into contact with the rolls. Synchronous motors are usually dynamic-braked to rest. The torque requirements vary according to the type of mill: for *mixers* starting 125%, pull-in 80%, pull-out 250%; with comparable figures for *refiners* of 140%, 100%, 175%.

Paper industry

In *pulp grinders* wood logs are forced against a grindstone to shred the wood; the machines are started light, by relieving the feed on the logs. Torques are: starting 30%, pull-in 25%, pull-out 150%.

A *jordan* is used to separate the fibres of paper pulp and consists of a conical rotor or plug with knives, that is adjustable inside a close fitting housing. The motor is started light by backing out the plug. Torques are: starting unloaded 40%; pull-in unloaded 40%; pull-out 150%.

Mining

Ball mills used to grind ore consist of a long nearly-horizontal tube of large diameter, in which a charge of hardened steel balls abrade the ore as the tube rotates. The mills operate at very low speed so that the motor is usually coupled to a reduction gear mechanism. The ball mill is always started under load, since it contains the charge of balls and usually the charge of ore. When starting, high friction torque has to be overcome, then the shell has to be rotated through about 60° before the balls start moving, therefore a relatively high torque is required at starting. Rod mills are similar, rods being used instead of balls for grinding the ore. Other similar types are tube and compartment mills. Respective starting, pull-in, pull-out torques for the various types of mill are: ball 200–225%, 115–120%, 175%; rod 200%, 120%, 175%; tube and compartment mills 180%, 120%, 175%.

Fans and blowers

Torque requirements for induced or forced draught centrifugal fans are: starting 25%; pull-in, closed discharge, 60%; pull-out 125%. Large ventilation fans, particularly those used for mines and sintering processes, have considerable inertia in the system. Care must be taken not to overheat the squirrel cage starting winding during the acceleration period and that sufficient synchronising torque is available to pull the motor into step. Torques are: starting 60%; pull-in, open discharge 120%; pull-out 150%.

Metals industry

Metal-forming mills are started unloaded but, should material become jammed in the mill, sufficient torque for emergency reversing is required. Load torque increases rapidly if the temperature of the materials falls too low during rolling. High pull-out torques are therefore required.

NEAMA recommended torques for synchronous motors applied to steel mill drives are (torque as % full load torque; values respectively for starting, pull-in, pull-out): Structural and rail roughing mills 40, 30, 300; structural and rail finishing mills 40, 30, 250; plate mills 40, 30, 300; merchant mills 60, 40, 250; billet sheet or bar mills 60, 40, 250; hot strip mill 50, 40, 250; rod mills 100, 60, 250.
Tube piercing and expanding mill 60, 40, 300; tube rolling mill 60, 40, 300; sheet and tin mill (cold rolled) 200, 150, 250; brass and copper finishing mills 150, 125, 250; brass and copper roughing mills 50, 40, 250. A compounded excitation system can also be used in these applications.

Synchronous induction motor

The synchronous induction motor (s.i.m.) combines the starting performance of a slipring induction motor with the power factor correction capability of a synchronous motor; it is basically a slipring induction motor with a direct-coupled d.c. exciter. When it operates as an induction motor, it can develop any starting torque within the limits of its maximum pull-out torque. At unity power factor the s.i.m. produces full-load starting torque with 1.5–1.75 times full load current (*cf* about 1.25 times for a slipring induction motor). A motor designed to work at 0.8 power factor leading takes only 1.0–1.5 times full-load current, hence its starting performance is considerably better than that of the salient pole synchronous motor (s.p.s.m.). On the other hand, the efficiency of the s.i.m. is some 1–3% lower than the s.p.s.m. (depending on power factor and number of poles) because it uses a 3-phase distributed type field winding compared with a concentrated field winding. Efficiency may be improved by using a 2-phase rotor winding with asymmetrical wider and deeper slotting for the excitation winding, thus considerably increasing the copper cross-section in the design.

Reasonable excitation values must be obtained, but the rotor open circuit voltage must also be kept within acceptable limits (a generally recognised maximum of 2500 V). Starting is by an ordinary circuit breaker and rotor starter (resistance or liquid type). The exciter is usually connected in series with the rotor winding, without a changeover switch; however, in larger s.i.m. the rotor must be connected open-delta or parallel-star, and a changeover switch must then be introduced, together with four or six sliprings.

Torque. Pull-in torque of the s.i.m. is higher than that of the salient pole s.m. because of the low inherent resistance of its distributed-field type of winding. Pull-out torque of the s.i.m. is rather less than that of a synchronous motor (1.35–1.50 times full-load torque, compared with 1.50–2.00 for the standard salient pole synchronous motor design). A compound excitation system may be incorporated if higher values of pull-out torque are required.

Salient pole synchronous induction motor

The most recent development in synchronous machines is the salient pole synchronous induction motor (s.p.s.i.m.): a combination which gives the high efficiency of the s.p.s.m. (with the concentrated field winding) and starting performance of the s.i.m. (with its distributed stator and rotor windings).

Construction differs from that of the s.p.s.m. only in that a 3-phase distributed winding (brought out to three sliprings) is used in place of the squirrel cage winding in the pole faces. A wider and deeper pole tip is needed to accommodate the distributed winding; hence there is less space for the concentrated excitation winding than in the s.p.s.m.

The s.p.s.i.m. is started (as a s.i.m.) by a liquid or grid type starter; a field discharge resistance shunts the excitation winding. The excitation field is applied, and discharge resistance open-circuited, when the maximum induction motor speed is reached. The 3-phase distributed starting winding is short-circuited by the starter, and acts as a damper when the unit runs in synchronism. Starting performance is similar to that of the s.i.m., but with a higher kVA input at starting; eg current of 1.75–2.25 times full-load current for full-load torque at start (for unity power factor design) compared with 1.5–1.75 times for a s.i.m.

Synchronising torque of the salient pole machine is also lower than that of the s.i.m. because restricted space in the pole tip gives a higher resistance starting winding, and hence a greater slip from which to synchronise. Because the pole tip encroaches on the field winding spaces, the copper content (and hence efficiency) is reduced slightly compared with the s.i.m.

The salient pole synchronous induction motor is more costly to build than the two types (salient pole synchronous and synchronous induction motors) between which it is a compromise. When considering the economics of a particular application, the higher efficiency must be capitalised to show a gain over the s.i.m. The s.p.s.i.m. is inherently suitable for large low-speed drives such as those used in cement mills.

Chapter 9
A.C. Commutator Motors

R Mederer CEng BSc MIEE
Laurence, Scott & Electromotors Ltd

If a d.c. armature is placed in a rotating magnetic field, the emf induced in the individual armature conductors is proportional to the relative speed of field and armature, both in magnitude and frequency. Due to the action of the commutator the voltage appearing between brushes placed on the commutator has the same frequency as the rotating field irrespective of the speed of the armature, but its magnitude depends on the relative speed and the spacing of the brush arms around the commutator. These characteristics have been used to design a number of polyphase commutator machines, but currently only two types are in general use and are made in relatively large numbers: stator-fed and rotor-fed motors with shunt characteristics. Motors with series characteristics are useful for some applications, and are also made, but in relatively small quantities.

Figure 1 shows the basic circuit and winding arrangements of the stator-fed (NS) shunt motor[4,5] which consists of a motor and some form of voltage regulator. The primary winding of the motor is situated on the stator and, being the same as that of a normal induction motor, produces a rotating magnetic field of constant speed in space. The secondary (armature) winding is connected to the commutator and is similar to a d.c. armature winding. The emf induced in the coils of this winding (and therefore appearing between commutator segments) is proportional to slip both in magnitude and frequency, but the voltage between brushes is of mains frequency irrespective of the speed of rotation of the armature. The brushgear is fixed in position and is connected to the voltage regulator, which is usually some form of induction regulator capable of producing a secondary emf whose magnitude can be varied from a positive to a negative maximum, and whose frequency and phase position are constant. If the emf per phase at the brushgear of the motor is the same in magnitude and phase as that of the regulator secondary, a state of equilibrium exists and no current circulates in the secondary circuit. If this equilibrium is disturbed, *eg* by altering the output from the regulator, current is caused to flow, whose

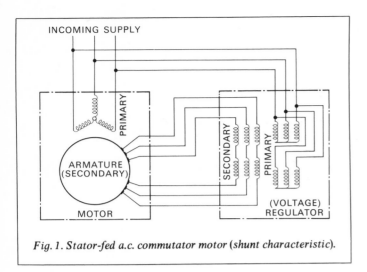

Fig. 1. Stator-fed a.c. commutator motor (shunt characteristic).

magnitude and power factor is determined by the impedance of the equipment, and the amount and direction of emf unbalance. The current flowing through the armature interacts with the main flux and produces a torque (proportional to flux × current × power factor) which causes the motor to accelerate or decelerate to a no-load speed at which the brushgear and regulator emf again balance, thus reducing the current and therefore the torque to zero. If the motor is to produce a mechanical torque to drive a load the motor speed has to drop slightly, thus causing an emf unbalance sufficient to circulate the required current to produce an electrical torque to balance the load torque. The power flow of this arrangement is shown in Fig. 2 for speeds at, below, and above synchronism. From this it can be seen that the motor is dimensioned to handle the synchronous power only and the regulator either returns slip power to, or draws slip power from the supply, depending on whether the motor is operating below or above synchronous speed. The size of the regulator is therefore determined by the maximum slip requirement.

The basic circuit and winding arrangement for the rotor-fed (Schrage) motor is shown in Fig. 3[6, 7]. The primary winding is a normal 3-phase winding, situated on the rotor, and connected to the a.c. supply by sliprings and brushes. It therefore produces a rotating field of constant speed relative to the rotor irrespective of the speed of the motor, but rotating at slip speed in space. The armature (regulating) winding connected to the commutator is in the same slots as the primary winding, and is usually nearest the airgap. The emf induced in the coils of this winding (and hence appearing between commutator segments) is therefore constant in magnitude and frequency, but the voltage appearing at the brushgear has a frequency proportional to slip, and its magnitude is independent of speed but depends on the separation of the brush arms. The stator winding is a polyphase winding with an induced emf proportional to slip both in frequency and magnitude. Both ends of each phase of this winding are connected to brush arms which are so arranged that they can be moved relative to one another along the commutator circumference and in fact cross over. The no-load speed of the motor always adjusts itself so that a state of equilibrium exists between the emf

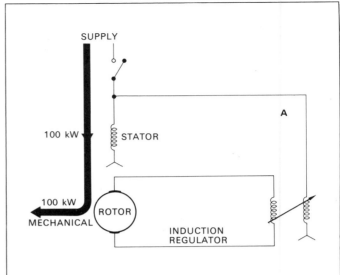

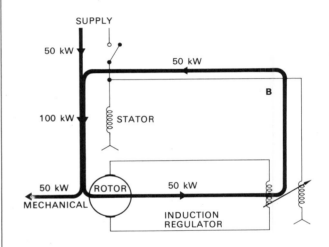

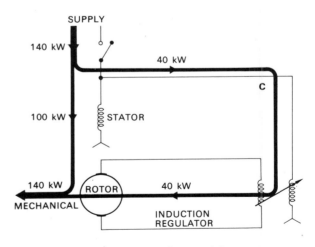

Fig. 2. Power flow diagram of stator-fed motor; constant torque, synchronised speed 1000 rev/min, motor rating 140/50 kW, 1400/500 rev/min. (A) 1000 rev/min, 100 kW delivered at shaft. (B) 500 rev/min, 50 kW delivered at shaft. (C) 1400 rev/min, 140 kW delivered at shaft.

induced in the stator winding and that appearing at the brushgear. As in the case of the stator-fed motor, torque can be produced either to alter the speed or to drive a load by disturbing this equilibrium; in this case, however, by altering the separation of the brush arms of one phase.

SHUNT MOTORS

In accordance with the simplified induction motor theory, it is assumed that the main flux and the magnetising impedance are constant irrespective of load and speed. Therefore, all the relative impedances, including that of the induction regulator if any, can be referred into the secondary circuit. More accurate vector diagrams can be constructed by dealing with the primary and secondary circuits separately, but normally the extra work involved is not justified. Figure 4A shows the secondary vector diagram below and above synchronous speeds, with the brush rocker set in such a position that the emf induced in the secondary winding is in phase with the regulating emf (either of the induction regulator or the regulating winding); ie the brush rocker is set in the neutral position. The difference between the secondary and regulating emf is made up by the impedance drops $I_2\bar{X}$ and $I_2\bar{R}$ where $\bar{X}$ and $\bar{R}$ are the reactances and resistances for the speed considered and referred to the secondary circuit. $\bar{R}$ includes an equivalent resistance to give the correct brush contact drop for the grade of brush used and the current flowing. Since some of the windings carry slip frequency, current $\bar{X}$ decreases and the power factor of I_2 improves as the speed increases,

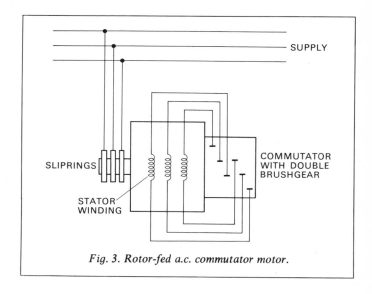

Fig. 3. Rotor-fed a.c. commutator motor.

but if it is desired to improve the power factor at, say, low speeds, the complete brushgear can be moved against the direction of rotation, therefore advancing the secondary emf in phase relative to the regulating emf as shown in Fig. 4B. From this it becomes clear that the power factor above synchronous speed is reduced, that brush shift has no effect in the region of synchronism, and that a forward shift of the brush rocker would have the opposite effect. Usually, it is necessary to compromise to get the best overall performances throughout the speed range. With the rotor-fed motor (where the brushgear has to be made adjustable

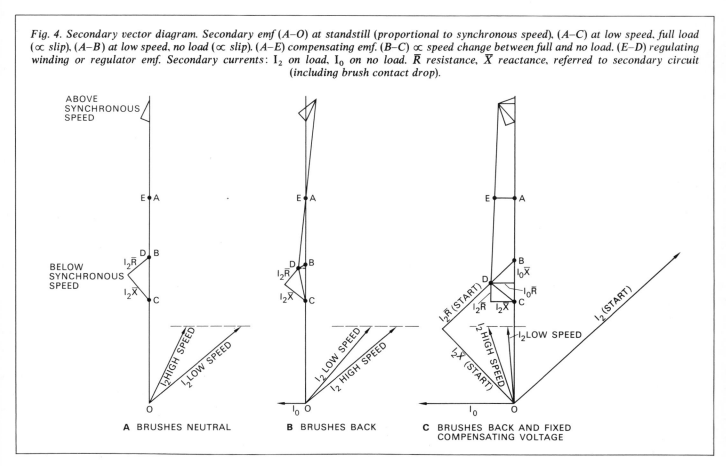

Fig. 4. Secondary vector diagram. Secondary emf (A–O) at standstill (proportional to synchronous speed), (A–C) at low speed, full load (∝ slip), (A–B) at low speed, no load (∝ slip). (A–E) compensating emf. (B–C) ∝ speed change between full and no load. (E–D) regulating winding or regulator emf. Secondary currents: I_2 on load, I_0 on no load. $\bar{R}$ resistance, $\bar{X}$ reactance, referred to secondary circuit (including brush contact drop).

during normal operation) it is possible, by suitable mechanical arrangements, to vary the effective brush rocker position as the speed is changed and therefore obtain some power factor correction, both at high and low speeds. In the case of the stator-fed motor it is normal further to improve the power factor throughout the speed range by introducing a leading emf of mains frequency into the secondary circuit. Figure 4C shows the effect of this additional compensation in conjunction with a fixed brush shift. The power factor of the rotor-fed motor cannot be improved by this method, since a slip frequency voltage would have to be used. There is, however, a limit to the amount of correction possible since excessive compensation (brush shift and/or fixed compensating voltage) causes large no-load currents to flow in the secondary circuit (Fig. 4B, 4C). The secondary no-load current, in order not to produce any torque, must have zero power factor, hence its magnitude is determined only by the amount of compensation and $\bar{R}$. Since the various resistances are usually fixed by other considerations, the total compensation must be kept sufficiently low to limit the no-load currents to safe values. It is occasionally advantageous to connect an additional permanent resistance in the secondary circuit to enable larger compensating voltages to be used, thus achieving a better phase position and smaller magnitude of the secondary current. The resulting reduction of the motor and regulator copper losses may well be more than the losses in the additional resistance, therefore keeping the motor as small as possible without reducing the overall efficiency.

The starting conditions can be obtained similarly, assuming zero speed as shown in Fig. 4C, the electrical starting torque being proportional to the in-phase component of the secondary current at start. Since it is usual to start these machines at low speed, the starting current taken from the supply is low, say, 1.5–2.5 times full load, while starting torques up to 2.5 × full load torque can be obtained at some extra cost where the application requires it. Typical speed

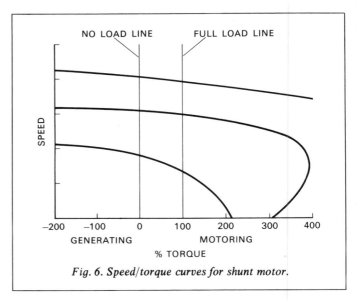

Fig. 6. Speed/torque curves for shunt motor.

torque curves (Fig. 6) can be obtained by using the secondary vector diagram. The energy flow in the whole equipment can be reversed, and this can provide regenerative braking or use the machines as generators.

The simplified primary current vector diagram is shown in Fig. 5, for constant torque condition below and above synchronous speed. In the case of the rotor-fed motor there is, of course, no induction regulator and therefore no regulator magnetising current. Various types of induction regulators are used in conjunction with the stator-fed motor, including the orthodox double induction regulator, and also single regulators in conjunction with transformers or special windings in the motor. The compensating voltage can be obtained from a separate transformer or a small additional winding in the motor stator, or from the regulator by means of special winding arrangements.[4,5]

SERIES MOTORS

For certain applications, *eg* fan and reeler drives, it is permissible or even advantageous to use a motor having a series characteristic.[4,5] The basic circuit diagram is shown in Fig. 7; the transformer between stator and brushgear is usually necessary for commutation, since it is generally not permissible to connect the brushgear directly in series with the stator, which in turn is connected to the supply voltage. To obtain speed variation it is necessary to displace in space the mmf produced by the currents flowing through the effectively series-connected stator and rotor. This can be achieved either by brush shifting or by replacing the series transformer with a phase-shifting single induction regulator, and keeping the motor brushgear fixed. In view of the effective series connection of stator and rotor, the vector sum of their voltages (allowing for the transformation ratio of the transformer or regulator) must equal the supply voltage, their phase position being determined by the stator/rotor phase shift (α) introduced by the brush or regulator position. Assuming a 1:1 turns ratio of the motor and transformer or regulator the primary vector diagram can be constructed as in Fig. 8. From this the slip can be

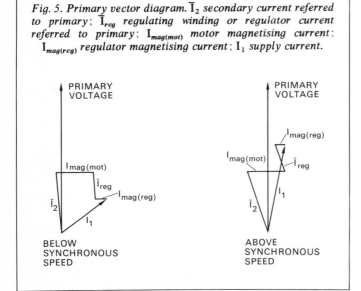

Fig. 5. Primary vector diagram. $\bar{I}_2$ secondary current referred to primary; $\bar{I}_{reg}$ regulating winding or regulator current referred to primary; $I_{mag(mot)}$ motor magnetising current; $I_{mag(reg)}$ regulator magnetising current; I_1 supply current.

obtained (E_R/E_S) and the torque is proportional to the product of the stator voltage and the component of the primary current in phase with it. In practice, the actual turns ratios of the motor and transformer or regulator are not 1:1, and must therefore be allowed for when constructing the vector diagram, and some allowance must also be made for the resistances and reactances of the equipment. The latter, however, have only a comparatively small effect on the performance of this type of machine. The motor and regulator magnetising currents are, of course, drawn 90° to the motor and regulator primary voltages respectively, and if no regulator or transformer is used its magnetising current must be omitted. Typical speed/torque curves for various values of α are shown in Fig. 9, on which is also shown the speed/torque curve of a fan, the intersections giving the operating points.

ARMATURE WINDINGS AND COMMUTATION

The design of these windings[8, 10, 11] is intimately linked with the problems of commutation and, unless this is carried out correctly, satisfactory brush and commutator wear is not likely to be obtained. In modern polyphase commutator motors there are usually two windings connected to the same commutator segments: the main winding to carry the load current, and some form of commutating winding designed to dissipate (or distribute around the armature) the energy contained in the leakage fluxes linking the main winding turns undergoing commutation. To minimise the commutator current, and therefore the number of brushes, it is usual to choose a winding that gives maximum voltage between segments (transformer voltage) consistent with good commutation and acceptable circulating currents in the turns short-circuited by the brushes. In small machines, series main windings are used where number of coils between adjacent segments is equal to the number of pole pairs. Series-

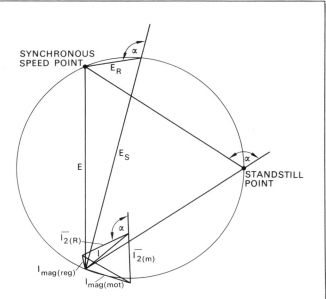

Fig. 8. Primary vector diagram of series motor. Voltages: E supply, E_s motor stator, E_R regulator primary. α stator/rotor mmf phase angle. Currents: $\bar{I}_{2(M)}$ secondary referred to motor primary; $\bar{I}_{2(R)}$ secondary referred to regulator primary; $I_{mag(mot)}$ motor magnetising; $I_{mag(reg)}$ regulator magnetising; I_1 supply.

parallel windings are also used where the effective number of coils between commutator segments is equal to half the number of pole pairs, thus halving the voltage between segments. Larger machines, with more flux per pole, normally have lap type windings, giving only one turn between segments irrespective of the number of poles; the effective number of turns and hence the voltage between segments can be still further reduced by utilising multiple lap windings (*eg* many stator-fed motors with quintuple lap armature windings have been in service for many years).

The quality of commutation depends on electrical and mechanical factors; the most important electrical ones are: (1) the 'transformer' voltage induced by the main flux in the short-circuited turns; (2) the reactance voltage due to the inductance of, and the change of current from one phase to the next in, the turns undergoing commutation.

Several methods have been suggested of dealing with the problems produced by these voltages; two of the most common ones take the form of a second winding connected to the commutator and located in the same slots as the main winding. A discharge winding of high resistance and low inductance (*ie* situated near the airgap) is often employed in the Schrage motor, and assists commutation by acting similarly to discharge resistances used in connection with the interruption of inductive circuits. Stator-fed motors usually use a transformer type of commutating winding in the bottom of the rotor slots and separated from the main winding by strips of magnetic steel; the commutating winding conductors are closely linked inductively (being surrounded by magnetic material). Coils of this winding (connected to the main winding turns being commutated)

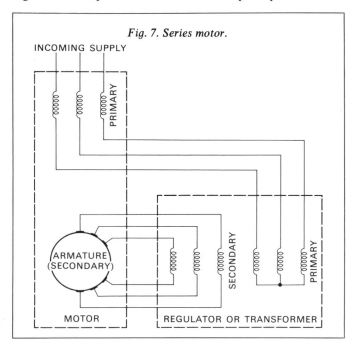

Fig. 7. Series motor.

are closely linked with other commutating winding turns in the same slot, which themselves are connected to main winding turns not undergoing commutation, which are closely linked with other main winding turns, which again are connected to other commutating winding turns, and so on. In this way the energy of the leakage flux of the turns that are being commutated is distributed by transformer action around the whole armature instead of causing sparking as a brush leaves a segment.

The discharge or commutating winding must be arranged so that the emf induced in it is substantially balanced by that induced in the main winding connected to the same segments. These auxiliary windings are also used to locate the potential of the separate chains of multiple lap windings relative to one another, and can be of the series or lap type. The auxiliary windings reduce the effective inductance of the coils being commutated, therefore reducing the reactance voltage. The total current change during commutation (and therefore the reactance voltage) can be reduced by increasing the number of phases. This also has the effect of reducing the current per secondary phase and therefore enables a shorter commutator to be used. On the other hand, the number of brush arms to be accommodated around the commutator is increased, which may present difficulties, particularly on smaller machines.

Sparking at the brushes is often attributed to difficult commutating conditions, but, in fact, is frequently due to unsatisfactory current collection. This is naturally influenced by atmospheric contaminants such as chemicals, oil vapour, *etc*, but also depends on mechanical stability of the brush. Care is therefore taken during design and manufacture to keep surface irregularities of the commutator to a minimum, and to ensure correct support of the brushes to avoid vibration and tilting.[9] This is particularly important with the movable brushgear of the Schrage motor.

APPLICATIONS AND LIMITATIONS

For a given rotor winding and flux the intersegmental voltage of the rotor-fed motor is constant irrespective of speed; it must also be kept low enough to ensure satisfactory commutation, so that the maximum flux per pole in such a machine is limited, which in turn limits the output per pole to about 30 kW. These factors, and the need to use a low-tension supply because of the sliprings, restricts this type of motor to small to medium sizes. It is frequently used in process lines, printing machines, etc, which require relatively wide speed ranges. In view of the high rotor losses, totally-enclosed machines of this type are not often used, since cooling of the rotor windings, sliprings and commutator can present problems. The rotor necessarily has a relatively large diameter to accommodate the windings, and therefore the inertia tends to be high. This can provide a useful flywheel effect to deal with cyclic load variations, but makes it more difficult to produce rapid speed changes on drives.

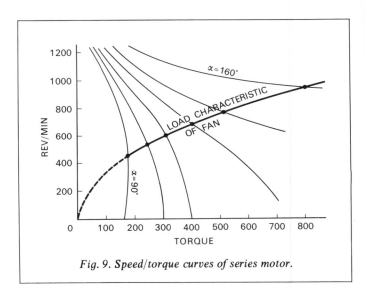

Fig. 9. Speed/torque curves of series motor.

The stator-fed motor has an intersegmental voltage that varies as the slip, so that very much higher transformer voltages can be used during the relatively short periods of starting than can be permitted for continuous operation. Depending on the speed range it is therefore possible to utilise very much higher values of flux per pole than with the rotor-fed motor, hence quite large equipments can be made. This type of machine is available and has been manufactured with shunt or series characteristics in sizes up to 1900 kW at 1000 rev/min, or 3000 kW at 750 rev/min, and there is no reason why even larger machines should not be made; on the other hand, many small machines with an output of only a few kW are in service. Large-diameter slow-speed machines with a high number of poles (up to 56) are also in commercial service. By choosing a suitable rotor winding to limit the transformer voltage, reasonably wide speed ranges can be obtained; naturally the smaller the machine the wider the speed range, but even on large machines 5:1 ranges are frequently possible. Since the stator winding is identical with that of an induction motor, it can be designed for supply voltages up to 11 000 V depending on size. The losses in the motor are only slightly higher than those of an induction motor and therefore the standard enclosures are available, *ie* ventilated, totally-enclosed fan-cooled, or totally-enclosed with air/air or air/water heat exchangers. The induction regulators can also be provided with similar enclosures if required, or can be oil-immersed in the smaller sizes.

In recent years static devices such as thyristors have been increasingly used to control d.c. and a.c. motors, and compete with the polyphase commutator motors, particularly in applications requiring very wide speed ranges, especially where the extremely high accuracy of speed holding and very rapid changes of motor speed obtainable with this type of equipment are required. However, for drives needing narrower speed ranges and only normal accuracies, the a.c. commutator motor is competitive with other forms of variable speed drives, taking all factors into account. Notably, for centrifugal pump drives needing speed ranges of only 2:1 or even less, the stator-fed commutator motor with its regulator (very small because it only handles slip power) achieves efficiencies, power factors, and first costs equal to or better than other low loss variable speed drives. The a.c. commutator motors can also accept short-time overloads which thyristors, because of their low thermal capacity, cannot tolerate. On occasion the shape of the inherent speed/torque curve is not acceptable and some form of automatic speed control is required. This is easily obtained by fitting electrical (or sometimes mechanical) actuators to the regulators or brushgear and controlling it (often by low power static switching devices) by means of a closed loop system from any of the normally available signals. If a failure occurs in the perhaps rather complicated control system it is usually possible to bypass completely all normal control circuits, and continue operation by means of manual control (push-button controlled servo-motor or handwheel) until it is convenient to have an orderly shutdown.

It is not sufficient only to consider first and running costs when deciding on the type of driving motor to be installed; the quality of maintenance personnel has an important bearing on the solution. In many instances the relative simplicity of the commutator motor is preferred to the complexity of modern static switching techniques, which makes diagnosing and curing of faults a highly skilled procedure. Another factor is whether ample warning is given of the onset of trouble (such as a commutator flat developing) therefore enabling the equipment to continue to operate for days or even weeks before remedial action is required, or whether failure is likely to be sudden and without warning (such as the failure of a semiconductor) leading to immediate shutdown. Apart from the cost and performance of the motor and its control gear, it is essential to consider their effects on the supply system when comparing different types of equipment.[12] In this connection there is one significant difference between the a.c. commutator motors and any type of drive relying on static devices such as rectifiers or thyristors, namely the harmful effects which the harmonic currents generated by the latter can have on the supply system and other equipments connected to it. The a.c. motors do not produce any significant harmonic distortion and are therefore often preferred for this reason alone, particularly if there is little difference in first cost or performance. Since supply authorities usually like to limit the amount of harmonic distortion on their system, the total rectifier and thyristor loads connected to it should be limited. Therefore, it is likely that users will reserve the capacity of their supply to accept harmonics, and will only install equipment that can cause distortion where significant economic and technical advantages are obtained from its use. In all other cases the a.c. polyphase commutator motor is likely to be the choice where stepless speed variation, together with low losses, is required.

REFERENCES

1. F J Teago. The Commutator Motor. *Methuen*.
2. E Openshaw Taylor. The Performance and Design of A.C. Commutator Motors. *Pitman*.
3. Adkins and Gibbs. Polyphase Commutator Machines. *Cambridge University Press*.
4. B Schwarz. The stator fed a.c. commutator machine with induction regulator control. *Proc. IEE*, vol. 96 (1949) part II.
5. J C H Bone. Recent developments in the field of N-S motors. Laurence, Scott & Electromotors, *Engineering Bulletin*, vol. 5, no. 2 1959.
6. A C Conrad, F Zweig, J G Clark. Theory of the brush-shifting a.c. motor. *Trans. AIEE* pt. 1 & 2, vol. 60 (1941) 824–836; pt. 3 & 4, vol. 61 (1942) 502–513.
7. O B Charlton. The Schrage type a.c. commutator motor. *BTH Technical Monograph* TMS 763.
8. B. Schwarz. The design of armature windings of a.c. commutator motors. *IEE Conference Publication* no. 11 (1964).
9. J C H Bone, K K Schwarz, R C Tessier. Electrical and mechanical requirements for satisfactory carbon brush operation. *IEE Conference Publication* no. 11 (1964).
10. H K Schrage. Multiple windings for a.c. commutator motors. *Bull. Assoc. Suisse des Electriciens* 1943 no. 6.
11. A C Lane. Recent developments in armature windings of polyphase commutator motors. *BTH Leaflet* AG 84J.
12. K K Schwarz. Effect of motor choice on supply system design. *IEE Conference Publication* no. 10 (1965) p. 28.

Chapter 10
Small/medium d.c. Motors

D Ramsden CEng MIEE
Bull Motors Ltd

The d.c. motor is inherently larger, more complex, and more costly than a conventional squirrel cage a.c. motor but for many areas of application it has distinct advantages which have ensured its continued use over the years. The advent of modern rectified power systems in the 1960s has enhanced its use quite considerably.

For applications where single speed operation is required with acceleration torque and control requirements within the capability of conventional a.c. motors there is no virtue in considering the d.c. motor with its increased cost and attendant problems of providing a source of d.c. power. D.C. motors become necessary, viable and advantageous where their characteristics fulfil the application's requirements more economically or more effectively than other simpler systems. These applications lie largely in the area of variable speed operation, precise speed control, and where specific control of acceleration, torque and speed are concerned.

Principles of operation

Fundamentally the d.c. motor consists of a stationary electrically excited magnetic field system inside which a rotating magnetic member or armature carries windings connected via a commutator to a source of d.c. power. The interaction of the field flux with the flux produced by the conductors in the armature produce a resultant force and hence a torque in the motor shaft. The commutator can be regarded as a switch which ensures that the direction of current flow in the armature conductors maintains the sense of the flux produced by the armature conductors and hence a continuous unidirectional torque at the motor shaft.

The designer can vary the strength and electrical and magnetic configuration of both the field system and the number and size of the conductors in the armature, and so has an extremely flexible piece of equipment from which he can achieve an enormously wide range of torque and speed characteristics. Unfortunately all these virtues are not without price.

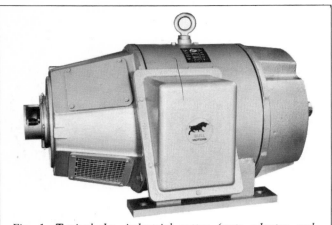

Fig. 1. Typical d.c. industrial motor (note adaptor and coupling at rear end for mounting tachogenerator).

While a current-carrying conductor moves in a magnetic field, there is generated within the conductor an emf proportional to the rate at which it intersects the field. Taking the total armature conductors there is generated a back emf in opposition to and somewhat lower than the applied armature voltage. The difference between the applied voltage and the back emf is that value which provides current flow in the armature according to simple Ohms law where the resistance to current flow is the total resistance of the armature and brush system. If the resultant current flow produces inadequate torque the motor speed will decrease, consequently the back emf will decrease and a greater difference between back emf and applied volts will give rise to an increase in current and hence torque. The system, therefore, in these simple terms achieves a self balancing condition.

Unfortunately though the above is quite true it is not, in practice, so simple. The effect of the armature field is to distort the main pole field and its effective axis is no longer at the centre of the main poles (see Fig. 9). During commutation of each armature coil it undergoes current reversal, and because each coil has self-inductance there is a self-induced emf generated at this time of flux reversal — this is normally referred to as reactance voltage. Combined with the effect of field distortion this reactance voltage tends to produce sparking under the carbon brushes since it is generated in coils momentarily short circuited by them.

To assist in this flux reversal it is normal to fit interpoles or compoles consisting of relatively narrow poles situated midway between each pair of main poles and carrying a winding connected in series with the armature circuit. Neglecting saturation effects, the interpole strength is thus proportional to the armature current.

CONSTRUCTION OF D.C. MOTORS

The basic components of a d.c. motor are: (a) A magnet frame or yoke. (b) Main poles and interpoles. (c) Armature core. (d) Commutator and brushgear. (e) Shaft, bearings, and end brackets. (f) Windings and insulation.

Magnet frame or yoke

Earliest machines frequently used cast iron but the magnetic limitations (low saturation levels *etc*) preclude its use currently. Fabricated or cast, low carbon steel has for many years been the generally accepted material for the magnet frame. Since the conventional form has been round or cylindrical in shape, apart from special cases, magnet frames have been rolled and welded forms.

More recently, in particular to meet the requirements of motors operating from thyristor rectified power supplies, has come the square frame motor in which the magnet frame is a fully laminated system; this gives optimum magnetic properties and highest performance, and uses what would otherwise be wasted space. The square form reduces the height and width of the space occupied by the motor, and this has become more important to meet the tendency to integrate the motor into the driven machinery or item of equipment.

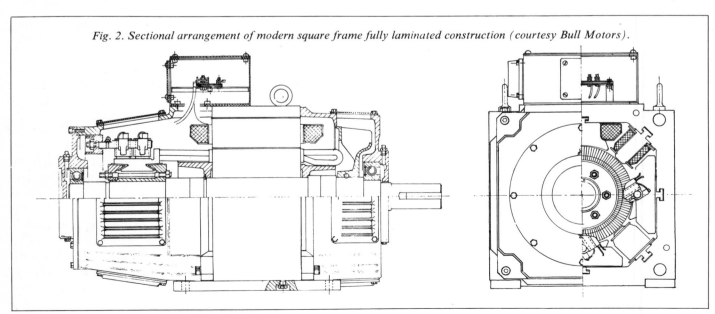

Fig. 2. Sectional arrangement of modern square frame fully laminated construction (courtesy Bull Motors).

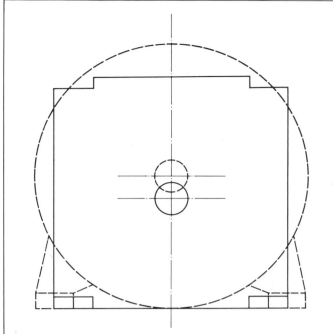

Fig. 3. Comparison of space occupied by conventional round frame motor and modern square frame motor of same rating.

Main poles and interpoles

Until relatively recently main poles were solid or laminated steel, and interpoles generally solid steel. Manufacturers have moved to laminated construction for both these items to produce the optimum shape, facilitate production, and meet the requirements of rectified power supplies where solid main and interpoles have undesirable characteristics. Pole face contours create a minimum air gap at the pole centre and thereafter gradually increase non-linearly — the purpose being to create sinusoidal flux distribution.

Armature cores

Armature cores are built from laminations. The choice of steel grade and thickness of laminations depends largely upon the number of poles and operating speed since (although these are d.c. machines) the magnetic frequency in the armature core can be quite high. Armature magnetic frequency = rev/min × $P/120$; where P = number of poles. Hence, for a 4-pole 2500 rev/min d.c. motor the magnetic frequency in the armature is 83 Hz and therefore armature iron losses can become quite significant. The armature core should be skewed by an amount equal to one slot pitch to avoid torque pulsation and noise.

Commutator and brushgear

Mica is still the generally accepted insulation material for the segments and vee rings for commutator construction, although modern techniques use mica splittings bonded with synthetic resins such as alkyds instead of the traditional shellac. Moulded commutators are available but their use is largely restricted to very small machines where large volume production exists. New synthetic commutator-insulating materials will be developed but the required characteristics of being partially compressible under heat and thereafter solid are not easily achievable by other means.

Brushgear is available in many forms and, according to the application, trailing, radial, or reaction forms may be suitable. The coil or clock spring type brushgear with pressure finger is still used extensively, but the constant pressure Negator/Tensater type of brush spring is becoming much more widely used because it can give constant brush pressure throughout the wearing life of the brush. Carbon brushes for general purposes use high speed electrographitic grade, but certain more difficult applications may demand a split or dual grade of brush.

Shafts, bearings, and end brackets

Shafts generally are of 35-t tensile carbon steel heat treated and normalised, such as BS grade EN8. Bearings are of the deep groove ball type at both ends of the machine with the inner races an interference fit on the shaft. The outer race, which is normally a transition fit, is secured at one end of the motor by the bearing caps and preloaded at the other end by a wavy spring washer to reduce bearing noise, since the resultant axial preloading eliminates ball/race clearances which largely contribute to bearing noise.

End brackets on industrial machines are cast iron, preferably with facings to accept fan blower units which are frequently required where motors have to operate for a significant time at low speed.

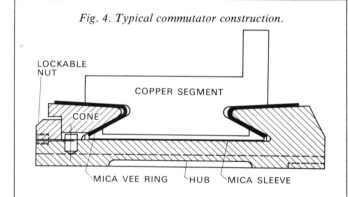

Fig. 4. Typical commutator construction.

Fig. 5. Brush and brushgear arrangements with constant force springs (Tensator type).

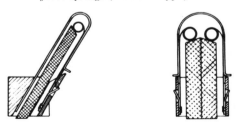

Fig. 6. Angle of brushes: (A) reaction; (B) radial (C) trailing.

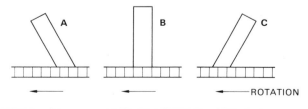

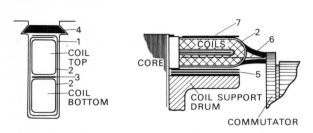

Fig. 7. Typical armature insulation system on small class B insulated motor. (1) Slot insulation; composite polyester-paper/polyester-film/polyester-paper. (2) Coil taping; polyester fibre or glass fibre tape. (3) Coil separators, as for (1). (4) Slot wedge; moulded polyester glass rod. (5) Drum insulation; multiple layers as (1). (6) Intercoil layer insulation; as for (1). (7) Banding; resinated glass fibre.

Windings and insulation

Class E insulation is little used today on d.c. motors — the high output per frame relationship demands Class B or F insulation using materials such as polyester enamelled conductors, with slot and coil insulations of synthetic composite materials incorporating, *eg* (with typical trade names in parentheses): polyethylene terephthalate film (Mylar or Melinex); polyester paper with epoxy or polyester isophthalate varnish impregnation: high temperature nylon paper (Nomex); polyimide film (Kapton); polyester fibre materials (Terylene or Dacron); glass fibre materials.

Although still extensively used, the traditional steel wire banding of the armature winding extensions is being replaced by resinated fibreglass tape banding (Resi-glass) which when cured is stronger and lighter than steel, is non-magnetic and an excellent electrical insulator.

Impregnation is normally carried out with modified phenolic, epoxy or polyester varnishes. Class H (silicone) insulations are employed on, *eg* traction motors where heavy overloads and abuse can create very high temperatures. Silicones have high temperature resistance but are not without disadvantages which have to be considered before employing them on d.c. machines since their mechanical strength is often low at high temperatures, and unless the motor is well ventilated silicone materials may contaminate the air inside the motor so as to deteriorate the commutation.

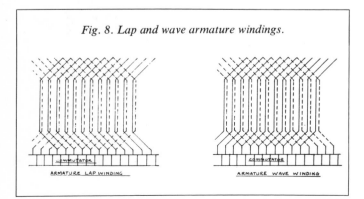

Fig. 8. Lap and wave armature windings.

Depending on the voltage, size, and speed of the motor the armature can be wave or lap wound. This is a difference in the coil connection form, as shown in Fig. 8. The significant difference between wave and lap windings is that the wave winding always has two parallel circuits whereas the lap winding has P parallel circuits where P is the number of poles. Whenever possible wave windings are used since the lap winding needs an equalising connection between points of equal potential in the armature winding to avoid circulating currents which would otherwise increase losses and heating.

CHARACTERISTICS OF D.C. MOTORS

The d.c. motor is the most versatile of all electrical prime movers in that by correct design and control it can operate at any speed and can have speed/torque/load characteristics of almost any desired form. The designer has an extremely wide choice of armature windings and arrangements of field excitation windings, so that the operational characteristics of most drives and driven equipment can be met.

There are three fundamental forms of field winding: shunt, series, and compound, each with its own characteristics but relatively simple changes to these basic forms can have a profound effect on the characteristics produced. The characteristics of a motor, so far as its operation as a prime mover and its relationship to the driven equipment are concerned, resolve themselves into torque and speed and the relationship between the two.

Speed of a d.c. motor

The speed of a d.c. motor obeys basic fundamental laws and has the relationship:

$$N = \frac{120 \cdot E}{\Phi \cdot P \cdot Z_a} \text{ for wave wound armatures}$$

or

$$N = \frac{60 \cdot E}{\Phi \cdot Z_a} \text{ for lap wound armatures}$$

where N = speed, in rev/min; E = internal armature volts (or back emf); Φ = flux per pole in webers; P = number of poles; Z_a = total armature conductors (Note: The internal armature volts (or back emf) is, in the case of a motor, the applied volts minus the IR volt drop in the armature plus interpole circuit together with the brush volt drop which can generally be taken as 2 volts irrespective of load.)

It will thus be seen that load current has a direct influence on speed, in that increase in load reduces E, thus tending to reduce speed. The current flowing in the armature produces a magnetic field with an axis displaced 90 electrical degrees, or in effect midway between the field main poles, this field produced by armature reaction has the effect of distorting and weakening the main pole field (see Fig. 9).

As a rough guide, it can be taken that the main pole flux is weakened by an amount equal to some 10% of the armature ampere-turns on normal machines under full load conditions. The precise degree of effect of armature reaction

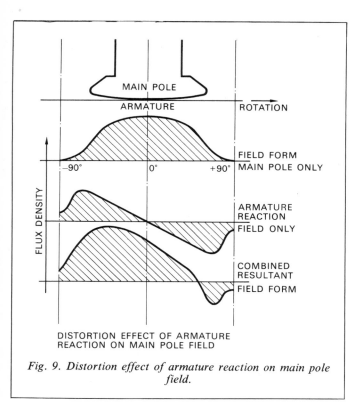

Fig. 9. Distortion effect of armature reaction on main pole field.

depends on the machine magnetic geometry, degree of magnetic saturation and the proportion of air gap and iron circuit ampere-turns to armature reaction ampere turns. Armature reaction therefore tends to cause reduction in main pole flux with a consequent tendency for speed increase with high loads.

Torque of a d.c. motor

For conventional forms of d.c. motor the motor torque at any speed can be defined as:

Motor torque = $0.07955 \cdot P \cdot \Phi \cdot I_a \cdot Z_a$ for wave wound armatures

or = $0.1591 \cdot \Phi \cdot I_a \cdot Z_a$ for lap wound armatures

where torque is expressed in Newton-metres; P = number of poles; Φ = useful flux per pole in webers; I_a = total armature current; Z_a = total armature conductors.

For any given machine then the torque can be varied by altering the armature current or the flux or both, but in considering machine characteristics as a whole it must be realised that changing flux and armature current also affect speed. In studying or designing a machine, no individual parameter can be taken in isolation but must be considered in the light of the total machine characteristics at all conditions of speed and load.

PLAIN SHUNT WOUND MOTOR

The simplest form of all d.c. motors is the plain shunt wound type which has, in addition to the armature and interpole windings, just one coil on each main pole — this coil or shunt winding can be excited by the same d.c. supply being fed to the armature or can be fed with d.c. from an independent source such as transformer rectifier unit. Generally speaking shunt coils consist of a relatively large number of turns of small gauge wire.

On single speed plain shunt motors where the motor is operating under full flux conditions the designer can achieve stability with relative ease; if necessary he can decrease the effect of armature reaction by using large air gaps. However, on variable speed shunt motors it is only at bottom speed that the motor is working on full flux and at progressively higher speeds the flux is correspondingly reduced. There is a limit to the degree of speed increase by flux reduction since one reaches a point where the ratio of main pole ampere turns to armature reaction ampere turns is such that increase in load decreases the main pole flux by armature reaction. This proceeds until the resultant tendency for speed increase is greater than the tendency for speed decrease due to armature circuit volt drop with the net result that an actual speed increase occurs (see Fig. 11) with resultant increasing load and instability.

Generally speed ratios of the order of 2:1 present little problem, but in modern compact machines speed ratios in excess of this may well involve de-rating the motor to overcome this limiting factor.

Because, roughly, at twice the base speed the flux level is halved then it will be seen (from the general expression for

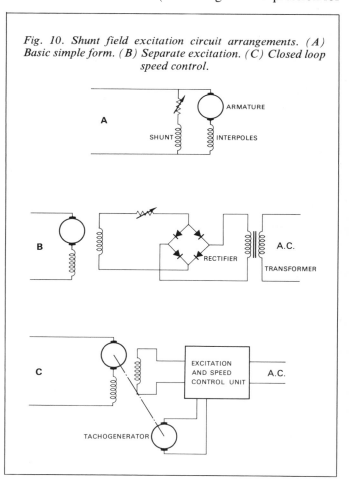

Fig. 10. Shunt field excitation circuit arrangements. (A) Basic simple form. (B) Separate excitation. (C) Closed loop speed control.

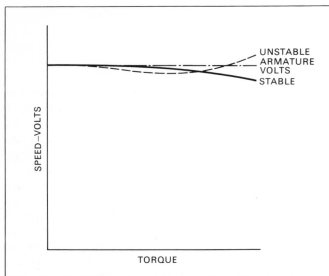

Fig. 11. Stable and unstable operation of shunt motors.

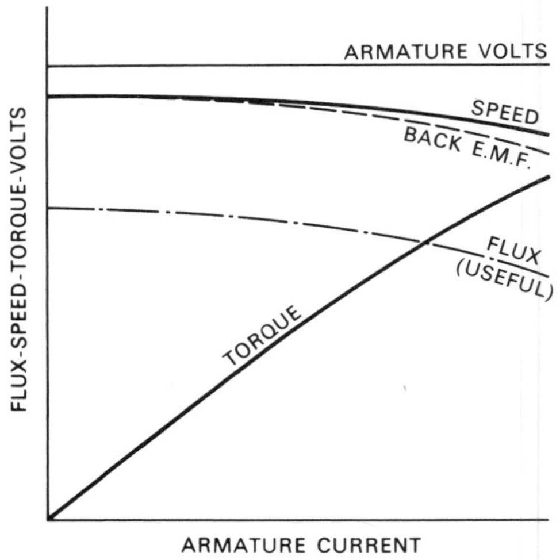

Fig. 12. Single speed characteristics of shunt motor on full field.

torque on any d.c. motor) that the torque also is halved; ie relative to base speed conditions the torque is inversely proportional to speed and the variable speed characteristics of a shunt motor are effectively 'constant power', that is, the maximum continuous rated power is the same at all speeds within the operating range. Naturally if the driven load demands a lesser output at some speeds than others the motor will deliver the reduced power. On a constant power basis the armature copper losses are roughly constant over the speed range but the armature iron loss increases with increasing speed due to the armature flux frequency increasing. (Note: 'constant power' is used in place of constant 'horsepower', since hp is an ambiguous and obsolescent term—Ed.)

The maximum armature specific electrical loading (or ampere conductors per unit of armature circumferential length) is constant over the speed range. Therefore, in determining the size of a motor for a given duty one must consider the maximum power demanded by load at any speed in the range. The maximum magnetic loading occurs at the lowest or base speed and therefore the aforementioned power value must be taken relative to the lowest operational speed; ie if the load characteristics are 6 kW (8 hp) at 1000 rev/min and 13.5 kW (18 hp) at 2000 rev/min the frame size of the motor will be that of a 13.5 kW 1000 rev/min machine.

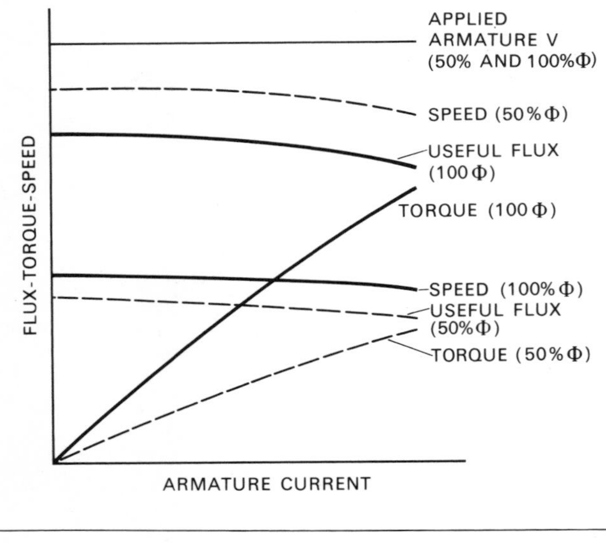

Hence, where wide speed ranges are required or where load varies appreciably with speed then the size of plain shunt motors with speed varied by shunt control tend to be disproportionately large and costly relative to output; this cost factor is however offset to some degree by the relative simplicity and low cost of the control gear. Where two or more motors are required to operate at identical or related speeds it is advantageous to operate with the motor armatures connected in parallel and with the fields of the motors connected in series with each other.

Fig. 13. Characteristics of shunt motor at 50% and 100% flux levels. In practice it is possible to get slightly higher power (at higher speeds than at base speed) with ventilated machines, because higher speed results in better cooling.

The properties of a plain shunt motor can be summarised as:
Speed control — Can be simple variable resistance in field circuit or can be variable voltage fed to field via integrated closed loop speed control using tachogenerator feedback; (can also be used on armature voltage control).
Speed range — This is limited by armature reaction stability and by economics of resultant machine size.
Output characteristics — Basically constant power over whole speed range, ie maximum torque capability varies inversely as speed.
Machine size — Economic where load has constant power characteristics, but where load is eg constant torque then machine size becomes large relative to output.
Advantages — Simple and easy to control; can be incorporated in multimotor co-ordinated speed control systems on complex machine drives. Speed regulation no-load to full-load is relatively small.
Disadvantages — Speed range can be limited by stability considerations. Machine size increases with speed range;

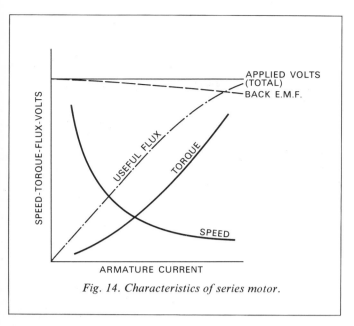

Fig. 14. Characteristics of series motor.

disproportionately so on wide speed ranges. Speed regulation cold to hot on constant load can be high; *ie* speed rises as machine warms up due to resistance/temperature coefficient in field windings.

PLAIN SERIES WOUND MOTOR

The plain series wound motor has one winding only on each main pole in the form of a relatively small number of turns of heavy section, usually rectangular strip copper. It is the most robust form of d.c. motor so far as windings are concerned, a useful characteristic since series wound motors are largely used on traction or crane applications where shock loading is frequent.

Due to the inherently low inductance of series field windings current builds up rapidly at switch-on, and consequently the torque build-up is also rapid. Under starting or accelerating conditions, where the current is perhaps two or more times full load current depending on design and operation of control gear, this high current also flows in the series field winding. Thus one has a field forcing condition resulting in a much higher field flux, and consequently achieving a high value of torque per ampere of armature current. High starting torques can be achieved with a lower armature current on series motors than on any other form of d.c. motor. The torque of a d.c. series motor is approximately proportional to current squared.

Speed can be controlled by variable series resistances which are also used for starting purposes, but for speed control these have to be heavier and of time rating to match the duty, and would therefore be much larger and more costly than normal starting resistances. Where more than one motor is used, a sequence of series and parallel field and armature connections can produce finite steps in speed; this form of control is largely confined to traction applications. Additional speed control to a limited degree can be achieved by using series field diverter resistance.

With the series motor the severely drooping speed curve makes it ideal for load-sharing on multi-motor drives, since if the load tries to increase on one motor the speed falls and the motor tends to take a smaller share of the load; *ie* it is a largely self-balancing load sharing system.

The most serious problem with the speed/load characteristics of a plain series motor is that as load decreases, so do the main field ampere turns, and the resultant decrease in flux causes speed increase. In the extreme condition of no-load the motor would theoretically achieve infinite speed but in practice the speed rises until centrifugal stresses burst the armature windings or lift the commutator bars. Plain series wound motors must therefore never be used on drives where the load can fall below a value that corresponds to a safe armature speed.

Where there is any risk of the load dropping to a dangerously low value, it is usual to fit an additional small shunt winding to maintain a minimum safe field strength and thus avoid dangerous speeds; this is known as a shunt limiting winding.

Characteristics of plain series motors are summarised as:
Speed control — by variable or tapped series resistance; by diverters across series winding.
Speed range — Wide speed range possible with series resistance control zero to maximum speed.
Output characteristic — Substantially constant torque, except for necessity to reduce output at lower speeds due to reduced cooling effect of fan on continuous rated machines.

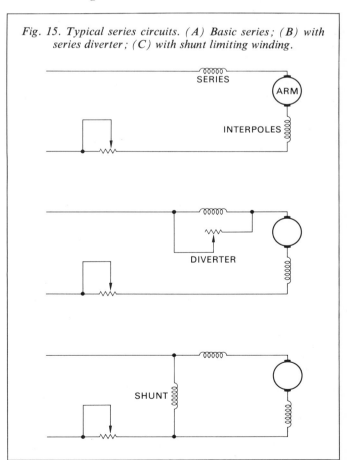

Fig. 15. Typical series circuits. (A) Basic series; (B) with series diverter; (C) with shunt limiting winding.

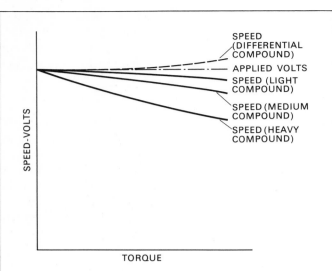

Fig. 16. Compound motor characteristics.

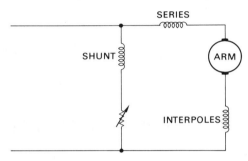

Fig. 17. Compound motor circuit.

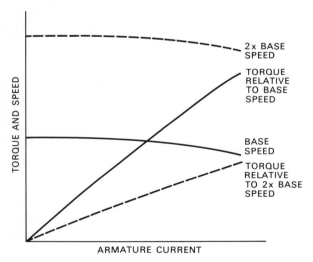

Fig. 18. Characteristics of compound motor with field weakening from base speed.

Fig. 19. Typical Ward Leonard circuit.

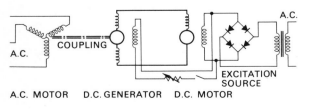

(Continuous rated series machines are unusual.) On short time or duty cycle ratings substantially constant torque over entire speed range.

Machine size — Economic; except on continuous rating duty, the variable speed does not increase frame size.

Advantages — Control is simple and robust, inherently high starting and accelerating torque capability. High torque/current ratio. High degree of stability.

Disadvantages — Small changes in load result in large changes in speed. Unless shunt limiting winding or other precaution is fitted then motor is totally unsuitable for applications where load can fall to zero or low value.

COMPOUND WOUND MOTORS

Where speed stability is required the compound wound motor is preferred. The proportions of series and shunt excitation can be varied in the design to produce a wide range of speed/load characteristics from a near shunt characteristic to a near series form. In the normal form, considered here, the series and shunt windings are connected in the same magnetic sense. Differential compounding on motors is rarely used — it increases motor size, is unsuitable for variable speed or where heavy overload conditions can occur. The lightest degree of compounding used is literally where the amount of series ampere turns are just sufficient to avoid the instability problems referred to under plain shunt motors. This form is often referred to as stabilised shunt motor.

Where two or more motors share a load, a much more severely drooping speed load characteristic is desirable. On a transient fluctuating load, such as power press drives, the use of a medium or heavy degree of compounding, and the use of a flywheel in the mechanical drive, can largely damp out what would otherwise be a corresponding heavy variation in the demand on the power supply.

Variable speed operation of compound motors fed from a fixed d.c. source is by variable resistance in the shunt circuit and the general principles and limitations are the same as for plain shunt machines except that instability problems under weak field conditions are reduced or eliminated. The maximum output/speed relationship is again one of constant power. The subject of speed control with variable armature voltage is discussed in a separate section.

Characteristics of compound wound motors may be summarised as:

Speed control — Simple variable shunt resistance or can be variable voltage fed to field via integrated closed loop speed control using tachogenerator feedback; (see later section on armature voltage speed control).

Speed range — Generally 3:1 or 4:1 speed range by shunt control is practical, but economics of machine size may be a limitation.

Output characteristics — Basically constant power over speed range with limited increase in power at higher speeds due to improved cooling.

Machine size — Economic on single speed. On variable speed by shunt control, economic if load characteristics are constant power but not economic if load is constant torque.
Advantages — Simple and easy to control with ability to incorporate into complex multi-motor co-ordinated systems. Speed regulation no-load to full-load can be small or large as required. Relatively the easiest of all d.c. machines to apply with flexibility in choice of characteristics.
Disadvantages — Machine size increases with speed range. Speed regulation cold to hot can be significant.

ARMATURE VOLTAGE CONTROL

Before the 1960s the most sophisticated and superior form of variable speed d.c. motor operation was the Ward Leonard system in which the motor is directly fed from a generator and is in effect operating on armature voltage control. During the next ten years or so (to the 1970s) major advances in rectified power systems have resulted in solid state thyristor controlled drives having the same performance capabilities for most applications. Since these types of drive employ many more d.c. motors than other systems, they are discussed in some detail, including their applications.

Ward Leonard control

The Ward Leonard control normally takes the form of an a.c. squirrel cage motor driving a d.c. generator, of which the field excitation is varied. The result is a variable voltage fed into the armature circuit of the d.c. work motor, of which the shunt field is excited at constant potential, usually from a small transformer rectifier system. The d.c. work motor is

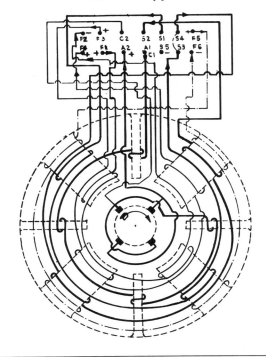

Fig. 20. Internal connection diagram of Ward Leonard generator (with series parallel field facilities and auxiliary shunt field). F1–F4 shunt (separately excited); A2,A1 armature; S1–S4 series; C2 potential lead; C1 diverter; F5,F6 auxiliary field.

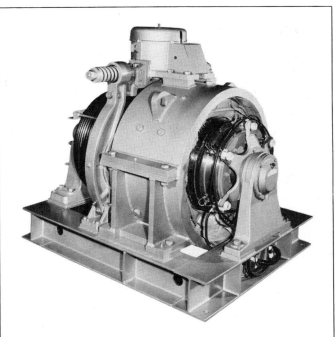

Fig. 21. Large low-speed d.c. motor for gearless lift drive; with integral brake and sheave (courtesy Bull Motors).

Fig. 22. Square frame motor with fully laminated construction specifically designed for operation on thyristor power supply.

usually a plain shunt motor, and instability under variable speed operation is eliminated for two reasons: firstly, the motor field is usually on full flux (although weakening of this field can also be used for speed trimming and additional speed increase); secondly, the generator voltage and load characteristics are designed to droop somewhat.

The excitation system of the Ward Leonard generator can be normal compound as in Fig. 19 or can be more sophisticated in that additional auxiliary bias, counter, or suicide fields can eaily be incorporated to match the requirements of the driven equipment.

Although still used for many other applications the greatest current use of the Ward Leonard system is on lift or elevator drives where the performance requirements are very stringent insofar as speed control down to zero is required, whilst precise control of speed and torque during starting,

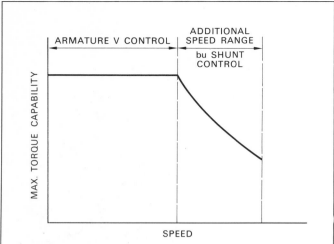

Fig. 23. Shunt or light compound motor; combination of armature voltage control with additional speed increase by shunt field weakening.

accelerating, running, decelerating and stopping are essential under conditions of motoring or regenerating with overhauling loads. Over 3000 drives of this type are produced in the United Kingdom each year. The motors generally range from about 5 kW to 37 kW (7.5–50 hp) with full load speeds of 75–1500 rev/min. The very low speed machines (say 75–250 rev/min) are of the gearless lift motor type and relatively small power ratings can result in very large heavy machines.

The Ward Leonard system can be designed to give literally any form of speed/load characteristics and, since the motor is always on full flux, then with constant maximum permissible armature current it inherently has a constant torque capability over the whole speed range.

Where the drive is of duty cycle form and operation at low speeds is intermittent or short time, then cooling of the windings is no problem. If long period operation at low speed is required, then forced air cooling by separately driven fan blower unit is usually applied to avoid increasing the motor size for purely thermal reasons.

THYRISTOR DRIVES

In simplified terms this is a solid state system of rectifiers, the thyristor being regarded as a high speed uni-directional switch allowing current to flow for a part or the whole of the half cycle of a.c. supply, thus providing an infinitely variable d.c. output voltage.

This form of armature voltage control of shunt or stabilised shunt motors has advanced extremely rapidly in the decade to 1971. Earlier equipments suffered from high cost, and were usually a six-thyristor fully-controlled system with additional smoothing chokes. With advances in motor design to improve commutating ability under the waveform conditions of this supply, the necessity for chokes has largely been eliminated and (except on large machine drives) it is usual to use a half-controlled system incorporating three thyristors and three diodes. This has now become an economic system and has replaced Ward Leonard systems on most industrial applications.

The inherent motor characteristics are generally as depicted in Fig. 23 for Ward Leonard control, but thyristor control lends itself readily to closed-loop tachogenerator feed-back speed control. High speed accuracy such as 0.1% over the full load range is achievable.

The 'wave form' of the d.c. supply to the motor has a superimposed ripple at three times or six times a.c. mains frequency in the respective cases of half and fully controlled systems. The form of the resultant current ripple depends on control equipment and motor inductance parameters, and is a complex variable over the speed range. This increases motor heating and commutation problems, and has to be especially catered for in the design of the motor.

The thyristor convertor has extremely low inherent losses operating with a theoretical efficiency in the order of 99%. Characteristics are:

Speed control — Can be simple; *ie* one rheostat in shunt of motor-generator set or thyristor unit, or can be complex dependent upon accuracy of speed or characteristic control required.
Speed range — Infinite from zero to maximum and stepless.
Output characteristics — Constant torque over range of armature voltage control but constant power thereafter where speed trim or increase by shunt control is applied.
Machine size — Economic — resulting in smallest motor relative to required output.

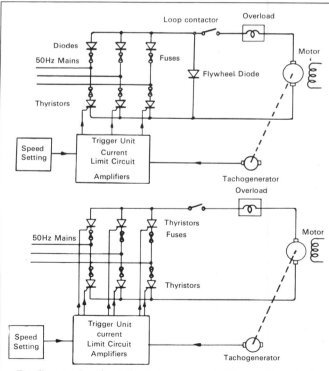

Fig. 24. (A) 3-pulse 3-thyristor system with feedback control; (B) 6-pulse 6-thyristor system with feedback control (courtesy LSE Ltd).

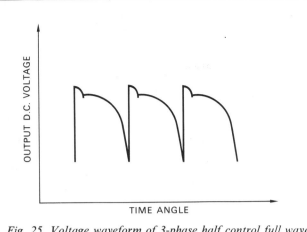

Fig. 25. Voltage waveform of 3-phase half control full wave bridge (3-pulse) system.

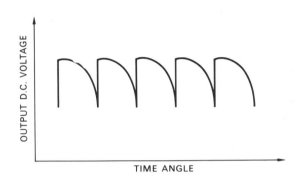

Fig. 26. Voltage waveform of 3-phase full control full wave bridge (6-pulse) system.

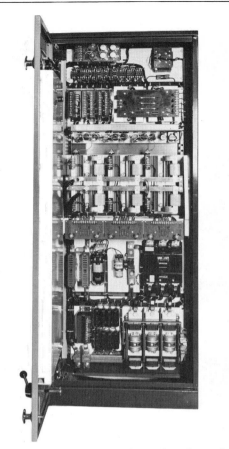

Fig. 27. Thyristor control unit for medium-large d.c. motors (courtesy K & N Electronics Ltd).

Advantages — Characteristics can be chosen by designer to suit application. Widest speed range of any system. Mixed characteristics achievable. Most versatile system. Completely stable.

Disadvantages — Few, apart from the relatively high cost of motor-generator sets or thyristor control units.

CONCLUDING NOTES

A newcomer to the field of rotating electrical machinery in the 1940s was advised to concentrate his efforts on a.c. machines, that d.c. was a dying subject, and that in any case all research in the subject had already been done. Far from being the case, d.c. rotating machine usage has gradually, and in recent years more rapidly, increased and new developments and techniques in respect to both motors and their associated equipments constantly appear and, what is more, will continue to appear.

The advent of more sophisticated machinery and equipment demand more and more sophisticated drive motor characteristics, and the d.c. motor is in many cases better capable of meeting these requirements than the alternative a.c. types. The reader should use all available comparative data to draw his own conclusions about particular applications of electrical motors.

AEI Semiconductors push the right buttons for solid state power control

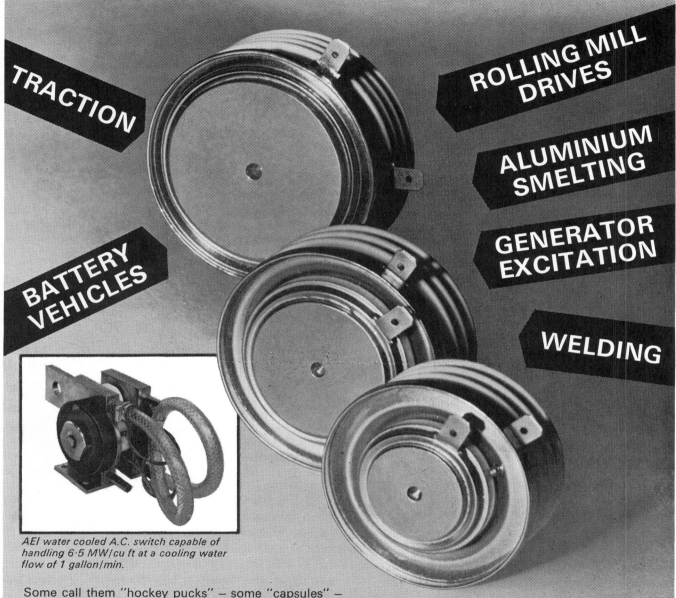

AEI water cooled A.C. switch capable of handling 6·5 MW/cu ft at a cooling water flow of 1 gallon/min.

Some call them "hockey pucks" — some "capsules" — to AEI they are "buttons" — the best in the business. Use them double-side cooled to get maximum power from the minimum space. A careful selection of silicon-wafer diameters enable surge current performance to be optimised in applications such as ac and dc motor control, turbo-generator excitation, ac power regulation for welding and furnace temperature control, dc "megapower" for electro-chemical duty.

A wide range of water-cooled assemblies are available with tested and guaranteed module performance.

There's a bright, bright future for these new components and they will help button up your high power solid state control programme. So contact AEI Semiconductors, now; their close association with every aspect of industry enables them to talk your language.

AEI SEMICONDUCTORS are in control

AEI Semiconductors Ltd., Lincoln, Tel: Lincoln 29992.

Chapter 11

Large d.c. Motors

B Skenfield *BSc CEng MIEE*
English Electric—AEI Machines Ltd

D.C. motors with armature diameters greater than about 600 mm are usually designed for specific duties. For metal-rolling mills, mine winders (hoists), wind tunnels, ship propulsion, excavators and many other specialised applications, detailed knowledge of loading conditions, type of power supply, mechanical conditions, environment and required accuracy of control is necessary to enable motors to be designed for optimum performance and reliability.

They are now usually associated with closed loop control systems, often with thyristor supplies for both armature and field circuits, since the common applications of large d.c. motors require accurate speed and torque control. It is uncommon nowadays to incorporate multiple field windings in large motors to give special speed/torque characteristics, and the normal methods of operation require separately excited motors with no series compounding, appropriate control being applied by speed, current or voltage feedback to the field current, armature voltage, and armature current to obtain the required characteristics.

When large overloads or a wide speed range by field control are required or when design considerations necessitate high segment-to-segment voltage, large machines have pole face compensating windings which oppose the mmf of armature reaction, and hence minimise distortion of the magnetic field in the airgap. The armature circuit inductance is also reduced, improving the accuracy of control.

Since one of the main reasons for using a d.c. motor is that it is capable of operating over a wide speed range, it is often necessary to provide cooling air from a separately-driven fan, as a shaft-driven fan would be ineffective at low speed, or lead to excessive losses at high speed. Various configurations of enclosure (to be described later) are used to ensure effective cooling and protection against dust, fumes and moisture.

PRINCIPLES OF OPERATION

Commutation and action of commutating poles. To provide a unidirectional torque, all armature conductors under N poles must carry current in one direction throughout a revolution of the armature, and those under S poles in the opposite direction. This is achieved by connecting the coil ends to the segments of a commutator and using carbon brushes rubbing on the commutator periphery to carry current to and from the armature. Then, as adjacent commutator segments pass under a brush, the current flow in each successive coil is reversed. Since the coil is embedded in the iron of the armature core it is highly inductive and the rapid current reversal produces an emf known as the reactance voltage, which would lead to unacceptable sparking at the brushes were it not for the effect of the commutating poles (compoles). These are excited by the armature current, and the rotation of the armature in the resultant magnetic field generates an emf in the coils being commutated, opposite in polarity and of approximately the same value as the reactance voltage. Ideally, therefore, the reactance voltage is completely cancelled, but in practice it is impossible to make the flux waveform in the commutating pole airgap of a shape that will result in a voltage exactly opposing the reactance voltage, and the commutating ability of the brush-to-commutator contact resistance then helps to eliminate sparking which would otherwise occur from the residual reactance voltage. Although the reactance voltage is largely neutralised as it arises, its theoretical value forms a useful yardstick of the commutating ability of a motor.

Fig. 1. Laminated frame with main poles, commutating poles, and compensating winding.

Numerous formulae exist for the calculation of reactance voltage, but these are mainly of interest to the specialist designer, who is able to assess the likely commutation performance in the context not only of reactance volts but of commutator speed, overloads, mechanical conditions, main pole flux waveform, maximum volts per segment, type of armature winding and other factors which have a direct or indirect influence on brush sparking. A vast literature is available on commutation, much of it academic, but nevertheless useful to the machine designer in understanding of this extremely complex process.[1-10]

The commutating poles are equidistant between the main poles and, for a motor, have the same polarity as the preceding main pole. Since the armature consists of a number of coils carrying current, an mmf (known as the armature reaction mmf) is produced, which has a maximum value between poles, *ie* on the compole axis, and zero value on the main pole axis. Thus it is necessary for the commutating poles to produce sufficient mmf to balance the armature reaction, and to provide an airgap flux sufficient to generate an emf in the coils being commutated to counterbalance the reactance voltage. Serious distortion can occur in the main pole airgap flux due to armature reaction, and high voltages are generated between certain commutator segments due to the peaky field form, with the possibility of causing flash-over between brush arms. This is particularly likely if a wide speed range by field weakening is required, as the armature reaction mmf then has a proportionately greater effect.

Compensating windings. It is beneficial to use a pole-face compensating winding. This winding carries the armature circuit current and has sufficient turns to neutralise (as nearly as possible) the armature reaction mmf over the pole-face. Field distortion is thus minimised, armature inductance is reduced and since turns are transferred from the compole to the compensating winding, the amount of compole mmf, and hence flux leakage, is reduced, permitting greatly increased overloads before the effect of saturation of compole iron destroys the linearity between load current and compole airgap strength.[11]

The armature reaction mmf per pole is given by

$$AT_A = \frac{ZI_a}{2ap} \qquad (1)$$

where Z = no. of conductors; I_a = armature current; p = no. of poles; a = no. of parallel circuits in armature.

Over the pole face, which commonly spans about 0.66 of the pole pitch, the mmf is $0.66ZI_a/2ap$ and to give perfect compensation the pole face winding mmf must be equal and opposite. To overcome armature reaction and provide the flux necessary to neutralise the reactance voltage, a total mmf for compensating windings and compole of between 1.15 and $1.3 \times AT_A$ is required, depending on the length of the compole airgap and other parameters. To give good compensation it therefore follows that, very approximately, the ratio (compensating amp-turns)/(compole amp-turns)

= 1.25, but unless average voltage per segment is high and the motor has a wide speed range by field control, satisfactory operation can be obtained with appreciably smaller ratios, provided that the sum of commutating pole and compensating turns is correct.

The commutating pole strength is adjusted on the test bed by connecting a d.c. supply across the commutating poles and compensating windings (which are usually connected in an interleaved arrangement), and at each of several loads increasing and decreasing the current flowing in the local circuit. The 'bucking' and 'boosting' currents at which sparking starts are noted, and plotted against load current, producing the well-known 'black band', the width of which is a measure of the commutating margin of the machine. The commutating pole strength is then modified to give a black band curve symmetrical about the axis by inserting or removing shims fitted between the frame and commutating poles. There is an optimum airgap between commutating pole tip and armature, and to keep the total reluctance correct, brass or other non-magnetic shims may also be fitted at the frame end of the commutating poles.[10]

Armature diameter and volts per segment. Although for many applications an armature of low inertia (and hence relatively small diameter) is required to permit rapid acceleration, from considerations of permissible armature loading and maximum volts per commutator segment the armature diameter for a given kW output cannot be reduced below a value given by

$$D_{min} = \frac{1000(kW) \times p}{V_b q \eta a} \text{ (mm)} \quad (2)$$

where (kW) = motor output; p = number of poles; V_b = average voltage per commutating segment; q = armature loading in ampere turns per mm; a = number of parallel paths in armature winding; η = motor efficiency per unit.

For a motor of 4000 kW output and 0.93 per-unit (p-u) efficiency having a single lap winding, (for which $p/a = 1$) and for which typical maximum values would be 19 for V_b and 33 for q, then D_{min} = 2200 mm. Using a double lap winding (which some designers try to avoid because commutation difficulties sometimes arise on site) then $p/a = 0.5$ and using a value for V_b of 16 (the tendency being to use a lower value for V_b for a duplex winding) D_{min} = 1300 mm for 4000 kW. A figure of 20 V is seldom exceeded for V_b, to ensure freedom from flashover, and q is limited by heating and commutation considerations. With a rectifier supply the average voltage per segment is kept lower than with a pure d.c. supply to allow for higher transient voltages.

Output coefficient and limiting core lengths. It can be shown that the output obtainable from a motor is given by

$$kW = 0.328 B_m q D^2 L N \times 10^{-9} \quad (3)$$

where B_m is the mean airgap flux density in teslas, q is as defined for equation (2), D is armature diameter and L core length, both in mm. B_m and q are parameters determined by experience for a particular duty, and for drives requiring minimum inertia D is obtained from equation (2), so a value can then be determined for L.[12]

There are however mechanical and heating limitations on the ratio of core length to armature diameter and it may be necessary to employ two or more mechanically coupled armatures to obtain the required output. Reactance voltage depends on core length, and this may also impose a restriction on output from a single armature, particularly if a wide speed range is required.

Rates of change of armature current and main pole flux. For most applications the control system must provide very fast changes of armature and field current to give rapid and accurate control and in a motor having a yoke of solid steel, eddy currents are set up which delay the build-up of the corresponding fluxes. Thus in the armature circuit the commutating pole flux will lag behind the armature current and sparking may occur at the brushes. Similarly when a rapid change of main field current occurs the flux change can only occur relatively slowly. It is now common practice to use a laminated yoke (Fig. 1) if the motor is associated with a sophisticated control system, and improvements are obtained in instantaneous correspondence between flux and current. The NEMA (USA) specification for electric motors gives the following formula for the maximum permissible rate of armature current change:

$$k = \frac{(\text{change in current/rated load current})^2}{\text{equivalent time (seconds) for change to occur}} \quad (4)$$

For a solid yoke machine $k = 15$. The 'equivalent time' is defined as the time which would be required for the change if the current increased or decreased at a uniform rate equal to the maximum rate at which it actually changes.[26] With laminated yokes a rate of change of at least 60 × full load current per second (*ie*, $k = 60$ in eqn 4) can be successfully commutated.

If the supply has a large ripple content (*eg* a 6-pulse thyristor supply operating with retarded firing) a fully laminated commutating pole flux circuit ensures that the effect on commutation is minimised. The coils being commutated, however, still have undesirable voltages induced in them from the ripple current in adjacent coils in the same slots.[13]

When reversal of a motor is obtained by reversing its field, (a system sometimes employed on rectifier-fed mine hoists and reversing mills to economise in the rectifier equipment) voltages are induced in the coils being commutated because they link the rapidly changing main pole flux. These voltages cannot be corrected by the commutating poles, and on a large machine may have a significant effect on commutation, so it is important that the machine designer is made aware when such a scheme is to be used.[23]

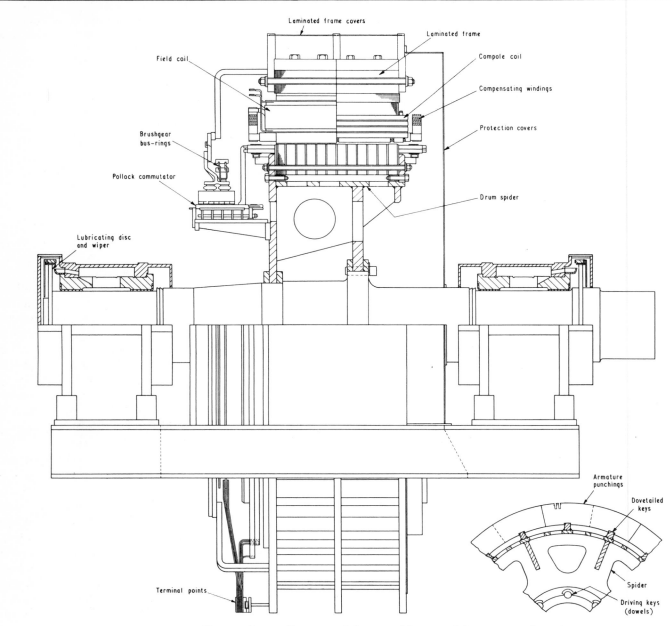

Fig. 2. Section of large rolling-mill motor with laminated frame and drum-type spider.

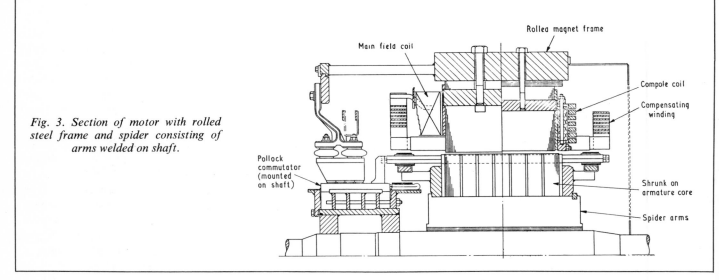

Fig. 3. Section of motor with rolled steel frame and spider consisting of arms welded on shaft.

CONSTRUCTION
Armatures

Mechanical construction. Available magnetic sheet steel width is limited to about 1050 mm. For smaller armature diameters, armature core plates are circular, and for larger diameters must be segmental. Material is usually 0.5 mm thick, varnished, and of a suitable grade to give acceptable core loss at the frequency and flux density of the motor.

For metal rolling mills in particular, very large transient torques can occur in operation,[14,15] and a robust spider construction is therefore essential. Fig. 2 shows a core and spider construction which has been very successful for armature diameters of 1500 mm and above. For smaller machines, cores may be built by pressing core plates on to a spider, with a key to provide the driving torque, or by shrinking circular armature punchings on to arms welded on the shaft (Fig. 3), the degree of shrink fit being such that no slipping will occur even at short circuit torque (approximately $10 \times$ full load).[16] Core ducts for ventilation may be either radial or axial. Endplates, which compress the core punchings axially, carry integral winding supports, the armature windings being held firmly to these by binding bands of steel wire or glass rovings bonded with synthetic resin. Glass binding bands eliminate heating and harmful effects on commutation which occur from eddy currents in wire bands, and should an armature winding fail, there is no consequential damage from unravelling of the binding wire.

It is usual to try to limit the peripheral speed of large armatures to about 55 m/s, since increasing problems arise with stresses in slot wedges, binding bands and core construction at higher speeds.

Armature windings. These are usually of the single turn double layer type. Simple wave windings, with only two parallel paths through the armature, are seldom used, since on multipolar machines poor current sharing between brush arms would be likely, and the armature currents are usually large enough to take advantage of the fact that simple lap windings have as many parallel circuits through the armature as there are poles on the motor. Equalising connectors are necessary with lap windings, to give good current sharing between the parallel circuits in the armature. A duplex lap winding may be either singly re-entrant, closing on itself after two tours round the armature, or doubly re-entrant consisting of two independent windings, each of which connects alternate commutator segments round the commutator. Duplex lap windings require particular attention to design of equalising connections, as explained below.

For windings of large d.c. motors there are typically between 1 and 5 commutator segments per slot. When speed and voltage are suitable for 1 segment per slot, the Whittaker winding, utilising a split conductor, gives excellent commutation. The inductance between adjacent commutator segments is lower than for an equivalent conventional lap winding, because there are two armature coils in parallel in two different slots, and in addition the coupling between two coils in a slot always permits a discharge path for the stored energy in one half of the coil when the segment leaves a brush, the other half coil in the same slot being still short circuited.

Class of insulation	Conductor insulation	Group (earth) insulation	Slot wedge
B	half lapped woven glass tape, impregnated with insulating varnish	Wrap of shellac bonded mica-folium; consolidated	Bakelised fabricboard
F	half lapped woven glass tape, impregnated with an isophthalate varnish	Wrap of glass-backed mica paper bonded with epoxy resin, and butted varnished glass tape; consolidated	Epoxy-glass laminate
H	half lapped woven glass tape, pre-treated with silicone varnish	Wrap of micaceous paper, reinforced with polyamide fibre and impregnated with high temperature epoxy resin. Butted glass tape; consolidated	Epoxy-glass laminate

Table 1. Typical slot insulation.

Coil insulation may be Class B, F, or H, permitting temperature rises in a 40°C ambient of 70°, 85° or 105°C (measured by thermometer).[25] Table 1 shows typical slot insulation for the different classes of insulation.

Connections to the commutator risers, and between top and bottom conductors at the other end are preferably brazed. The coils are usually retained in the slots by dovetail wedges, though glass binding bands along the core are sometimes used even on large motors.

Equalising connections. For simple lap windings, equalising connections, which join points in the armature winding two pole pitches apart, and therefore of very nearly equal potential, may be placed behind the commutator risers (Fig. 4), at the outside end of the commutator, or at the end of the armature opposite to the commutator. Their purpose is to ensure that circulating currents do not flow external to the armature between brush arms of the same polarity. Any circulating currents in the armature winding and equalisers arising from small differences in the voltages generated under different poles flow in such directions as to produce mmfs on the main pole axes which reduce the differences in the airgap fluxes. Also, if the commutator surface has slight irregularities, causing the brushes on one arm momentarily to make poor contact with the commutator, the current which that arm would have carried is transferred through the equalisers to other arms of the

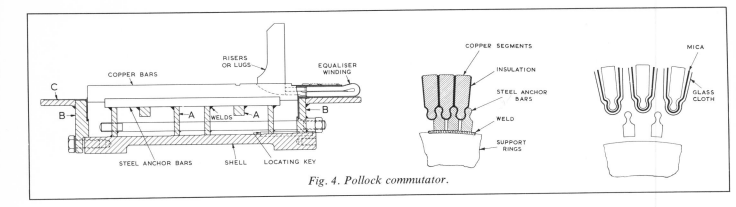

Fig. 4. Pollock commutator.

same polarity, minimising the sparking and consequent further deterioration of the commutator surface. With this condition, it is clearly advantageous to employ equalising paths having the lowest possible inductance, and thus equalisers at the commutator end are preferable in this respect. This arrangement has the additional advantage that it is often sufficiently compact to permit every commutator segment to be equalised.

Equalising rings mounted at the non-commutator end have connections from the rings to the coil noses, and are thus readily accessible, but have the disadvantages that less frequent equalising must be employed, and equalising currents arising from imperfect brush contact must flow through the inductive armature coils, possibly leading to increased sparking at the brushes.

Singly re-entrant duplex lap windings benefit considerably from having equalising connections at every commutator segment, as this gives the maximum possible interconnection between the two 'plexes', reducing the possibility of alternate commutator segment marking which can occur if the two plexes do not share current equally. For doubly re-entrant duplex lap windings, each 'plex' is separately equalised, usually by connections at the back of the commutator. In addition, to ensure that the potentials of segments of one plex lie midway between those of adjacent segments of the other plex, connections (bunched to minimise inductance) are taken under the core from coil noses to the intermediate segments.

With all duplex windings careful attention is necessary to the selection of numbers of slots, segments per slot and winding pitches to ensure that equalisers are in fact connecting points as nearly as possible truly equipotential.

Commutators

Commutators are built up of alternate segments of copper and insulation, usually mica about 1 mm thick. The most important requirement is that they must retain a truly cylindrical shape under all operating conditions of speed and temperature. The conventional V-ring commutator is still widely used although mica V-ring insulation is now sometimes replaced by an epoxy-glasscloth moulding. The Pollock commutator (Fig. 4) and the glass-bound commutator embody features which have led to their adoption by some manufacturers in recent years. Pollock commutators (BP602288) have all segments supported over almost their complete length, thus permitting high speeds and long face lengths with complete mechanical stability. The radial depth of the copper segments is much less than for a V-ring commutator, with consequent reduction in rotational stresses and inertia.

Glass-bound commutators have grooves turned in the periphery into which resin-bonded glass bands are wound under high tension, the segments being held firmly down to a sleeve insulated with epoxy-impregnated glasscloth or similar insulating material.

With all types of commutator, risers soldered or brazed into slots in the segments provide the connections to the armature windings. Various methods of bracing the risers are used to resist the severe stresses imposed by rolling mill and other drives. (Fig. 5.)

To ensure good current collection, commutator peripheral speeds are usually kept below about 35 m/s, but much higher speeds are possible if it is accepted that more commutator and brushgear maintenance may be required.

Fig. 5. Riser bracing on large rolling-mill motor.

Brushgear and brushes

Brushgear on large machines, where commutation frequently approaches safe limits, demands careful design to ensure maximum stability of brush contact. The use of aluminium for brush arms and supports permits rigidity with lightness, and high frequency vibrations are well damped, but steel castings or fabrications can also be designed to give satisfactory operation. Because maintenance labour costs are high, brush holders with constant-force springs (Fig. 6) which allow the use of long brushes, are increasingly incorporated in d.c. motors. Not only is the frequency of brush changing reduced, but complicated finger mechanisms which wear and require replacement, are eliminated. Constant-force springs are very reliable, and are simply and quickly replaced.

Brush holders may be radial to the commutator, or arranged to hold the brushes at an angle to the commutator surface. For unidirectional motors, it is common practice to run the brushes trailing since this is found to give very stable operation. Reversing motors may have radial brushes (which are preferably very short or with a cantilever top for stable operation) or have brushes which run reaction for one direction of rotation and trailing for the other. For the latter an angle of about 22° gives a good compromise; it is almost impossible to obtain quite such good brush stability on a reversing motor as on a unidirectional one, but careful attention to the geometry of the design enables satisfactory performance to be obtained. To allow more time for commutation and to minimise the effect of commutator surface irregularities, circumferential stagger (*ie* mounting half the brush boxes on each arm up to about 15 mm in advance of the remaining brush boxes) is often used. Axial stagger of brush holders helps to prevent circumferential ridges forming on the commutator face.

If constant-force springs are used, it is good practice to incorporate insulation pieces on the tops of the brushes, both to prevent passage of current through the springs and to provide some measure of heat insulation, since under high sustained overloads, brushes can become very hot, so that the temper of the springs might be affected.

The most commonly used forms of carbon for brushes are open texture electrographitic grades, which commutate well, give reasonable brush life and cause very little commutator wear. Special circumstances, such as long periods of light-running which tend to produce commutator skins that have high friction and high resistance, may necessitate use of sandwich brushes, one half being natural graphite (which has a slight abrasive action) and the other half electrographitic carbon.

To ensure the best possible contact with a commutator, which however well designed and made cannot be perfectly smooth and round, brushes are often made of two or more separate elements, joined at the top by a slightly resilient bridge which allows some independent movement between the several elements. This arrangement also increases the cross resistance of the brush, tending to suppress circulating currents across the face.

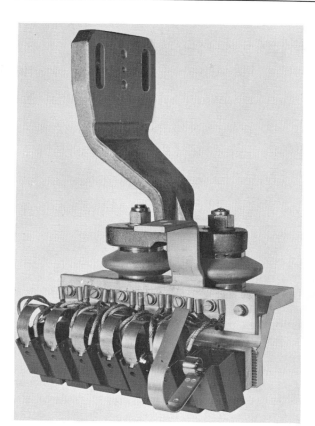

Fig. 6. Brush arm with constant pressure type brush holders.

Operating conditions and environment sometimes result in less satisfactory performance in service than on the test bed, and it may then be necessary to obtain the co-operation of a brush manufacturer in trials of various grades to find the most suitable type of brush.[17,18]

Frames and poles

Solid frames are formed from low carbon steel and except for very large machines have dimensions governed by magnetic densities. On machines with frames greater than about 3.75 m diameter, external ribs may be necessary to give rigidity. Laminated frames use external skeletons of fabricated steel to give support to the laminations (Fig. 1). Main poles and compoles are usually built up of punchings and may be riveted together or bolted with insulated bolts, depending on the rate of change of current to be met in service. Fig. 1 shows a typical laminated yoke assembled with poles and compensating winding. Field windings may be formed from strip on edge, or wound with insulated wire, depending on size and excitation voltage.

Bearings

The bearings in Fig. 2 are of the disc and wiper lubricated type, often used on large motors. If the duty entails frequent reversing, oil may be fed under pressure at the bottom of the bearing to ensure a continuous oil film. Heavy end thrusts can occur with some types of drive, and a thrust collar, with Michell pads, can then be incorporated in the housing.

Journal dia (mm)	125	150	175	200	250	300	350	400	450
Speed (rev/min)				Oil quantity (litres/min)					
300								0	20
350							0	18	25
400						0	14	20	30
450						11	16	23	35
500					0	14	18	27	40
600					9	16	20	32	50
700				0	0	11	18	25	36
800			0	4	7	14	20	30	
900		0	4	5	9	16	23		
1000			5	7	11	18	27		
1250	2	7	9						
1500	5	9							

Table 2. Cooling oil requirement for sleeve bearings.

Fig. 8. Dredger pump motor with closed air circuit and air-to-water cooler.

At journal speeds greater than about 7 m/s, cooling must be employed either by circulating oil or by use of cooler tubes in the bearing, through which cold water is passed. Table 2 gives approximate cooling oil quantities required for sleeve bearings at various speeds, assuming armature weights commensurate with the journal sizes. Grease lubricated ball and roller bearings are sometimes used but have the disadvantage that it may be necessary to remove a coupling or other shaft fitment having a large interference fit if a bearing requires replacement. Split roller bearings (Fig. 7) which avoid this difficulty, have been used up to about 750 mm dia, supporting an armature of up to 85 tonnes. They can be used with grease lubrication up to a value of $DN = 220000$ (D is shaft diameter, mm; N is rev/min). At this value of DN grease must be added at intervals of about 2000 hours with correspondingly longer intervals at lower values of DN.

Fig. 7. 'Cooper' split roller bearing on large rolling-mill motor, with cap and top outer race removed; an overspeed device and tachogenerator are shown at right.

The armature winding of a large machine has an appreciable capacitance to earth and, if fed from a supply containing a ripple voltage, a current will flow through the bearings, causing effects similar to spark erosion as the oil or grease film repeatedly breaks down and reforms. On motors larger than about 750 kW it is advisable therefore to fit earthing brushes on the shaft to short circuit the bearings.

Enclosure and ventilation

The simplest construction is an open machine, which relies for cooling on convection and on air being moved by rotation of the armature, possibly with the addition of a paddle fan. Many mine hoist motors are so constructed, one reason being to ensure maximum reliability by complete independence from any external source of cooling. A canopy may be fitted over the commutator to prevent damage from falling objects and water.

Forced ventilation is more commonly used for other duties. Cool air is blown in at the non-commutator end and may either be discharged over the commutator or be prevented from reaching the commutator by a baffle of insulating material fitted in front of the commutator risers, the air then either being discharged into the motor room or cooled and recirculated. The commutator, which is excluded from the main air circuit to prevent the ingress of carbon dust from the brushes, may be open or included in a separate air circuit, sometimes using air bled from the main system through local trunking.

Downdraught ventilation entails drawing air from the motor room into the non-commutator end, and discharging to free air through ducts below the motor at the commutator end. More complex systems sometimes use baffles to share the cooling air appropriately between armature and poles. Marine motors commonly have separately driven fan units at the drive end, the air being discharged through coolers into the motor compartment, but sometimes a local recirculating system with fans, coolers and filters all mounted above the motor frame, is used. (Fig. 8.)

Special mechanical arrangements. Where very large powers are required (*eg* at the rolls of metal-rolling mills) the twin drive arrangement, in which the upper and lower rolls are separately driven by identical motors, is now common. To minimise the distance between the drive shaft centres, one motor is mounted behind the other, with a jackshaft passing close to the forward motor frame. (Fig. 9.) Other arrangements of two armatures include both armatures mounted on a common shaft between two bearings, and tandem units using coupled armatures and three bearings.

CHARACTERISTICS AND PERFORMANCE

A d.c. motor drive is selected in preference to others when all or some of the following requirements apply: (1) Variable speed with stepless control over a wide speed range. (2) High overload torque. (3) Accurate speed and torque control. (4) Controlled reversibility. (5) Full overload torque down to creep speed.

The two essential equations which specify the performance of d.c. motors are the speed and torque equations:

$$N = \frac{60aE_B}{p\phi Z} \text{rev/min} \quad (E_B \text{ back-emf}) \quad (5)$$

and $$T = 0.158 \frac{p}{a} Z \phi I_a \text{ newton-metre} \quad (6)$$

(where ϕ is flux per pole in webers, and other symbols as equation 1) which for a given machine become

$$N = K_1 \frac{E_B}{\phi} \quad (7)$$

and $$T = K_2 \phi I_a \quad (8)$$

It is apparent that manipulation of ϕ, I, and E enables speed and torque to be controlled in any required manner. In the 'full field' region, ϕ is kept constant, and the speed can be adjusted from zero to base speed by increasing the armature voltage up to the full design voltage. In this region torque is proportional to armature current, and for most types of drive, short time (say 15 s) overload torques of at least twice full load are available.

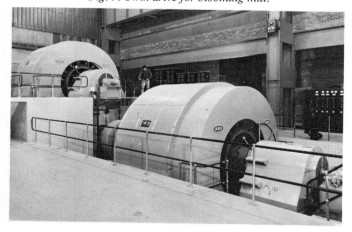

Fig. 9. Twin drive for blooming mill.

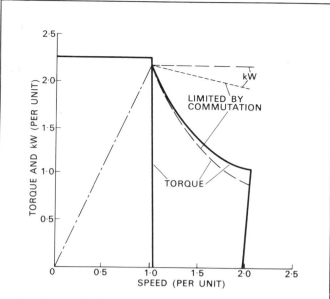

Fig. 10. Overload characteristics for large rolling-mill motor.

The region above base speed is obtained by adjustment of field current, and if required by the drive it is possible to design machines for up to about 5 × base speed. Since the flux changes in inverse proportion to speed, the torque for a given armature current is reduced accordingly. Fig. 10 shows typical short time overload capabilities of a large reversing rolling mill motor.

Current limit in the supply gives control of maximum torque, and various ramp functions incorporated in the control enable rates of acceleration, rise of armature current, etc, to be kept to values compatible with the motor capabilities and drive requirements.[23,24]

Fig. 11 gives an approximate guide to the output for which it is possible to design single-armature machines at various speeds and at speed ranges of 2:1, 2.5:1, and 3:1.

A complication which arises in variable speed motors is the effect of eddy currents in the conductors, which circulate within the depth of the conductors and produce cross-slot mmfs opposing those due to the normal conductor currents. Since these become greater as the speed increases, the effective inductance of the conductors is reduced and the commutating pole strength is consequently slightly too great. It is therefore sometimes necessary to modify the compole strength as speed increases, either by means of auxiliary coils on the commutating poles or by using diverters automatically introduced at higher speeds. The effect on commutation during transients must be taken into account, and inductive diverters are sometimes used to ensure that the proportion of diversion remains constant at high rates of change of armature current.[8]

Motors may be reversed by reversing the field or armature supply. As the main field inductance is high, a field voltage much greater than for full-load excitation is required to

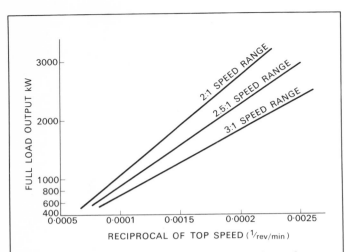

Fig. 11. Maximum outputs at various speeds with speed range by field weakening; temperature rises to BS2613 class B; overload 2 × full-load, 15 seconds (single lap armature winding). (Example: maximum top and base speeds for 2000-kW 2:1 speed range—from curve, reciprocal of rev/min = 0.00146: hence top speed = 684 rev/min, base speed = 342 rev/min.)

reverse the field in an acceptable time. During the field reversal an appropriate control of armature current is maintained by phasing back the rectifier firing pulses. In a scheme involving armature supply reversal, a separate bank of converters is required for each rotation and this scheme is therefore more expensive than field reversal, although it results in shorter reversal times. An alternative scheme employs an armature reversing switch.

Some drives particularly in metal rolling mills, must withstand suddenly-applied loads without a large drop in speed. For rod mills, in particular, if the impact speed drop is more than a few percent the gauge of rod already in the stand can be affected by the entry or exit of another rod. Modern control schemes can minimise this effect but it may still be necessary to design the motors with higher than normal inertia to give low inherent impact speed drop.

In calculations involving ripple currents from a rectifier supply and rates of rise of current in the armature supply, an approximate value of armature circuit inductance is often required. For compensated machines of modern design, Umansky's formula[22]

$$L_a = 0.2 \frac{E_a}{I_a} \cdot \frac{60}{p\pi N} \text{(henry, H)} \qquad (9)$$

gives reasonable accuracy and entails no more knowledge of the machine than is carried on the rating plate. In this equation E_a = rated voltage, I_a = rated armature current, p = number of poles, N = speed rev/min.

FUTURE DEVELOPMENTS

For conventional machines wider use of Class H insulation can be anticipated, together with the development of Class H materials that have better mechanical properties than were available up to the 1970s.

The homopolar d.c. motor with superconducting field winding developed in the UK under the auspices of the National Research and Development Council has aroused wide interest. It is however too early to predict what impact this will have on conventional machines.[27] Progress with motors incorporating thyristor assisted commutation has been slower than was anticipated in the mid-1960s, but motors of this type are being further developed and may well have an effect on future designs.

Acknowledgment. Illustrations Fig. 1–9 by courtesy of English Electric–AEI Machines Limited.

REFERENCES

Commutation (A brief selection of some important papers)
1. Ludwig A Dreyfus. Die Stromwendung grosser Gleichstrommaschinen. (Theory of commutation faults). Proceedings No. 212. Royal Swedish Academy of Engineering Sciences 1954.
2. A Tustin and H Ward. The emfs induced in the armature coils of d.c. machines during commutation. *Proc IEE* v109, 1962. Part C, pp456–474.
3. H Ward. The emfs induced in the end turns of armature coils during commutation. *ibid* pp475–487.
4. M Tarkanyi, H Ward, A Tustin. Electronic computers applied to commutation analysis. *ibid* pp488–507.
5. E I Shobert. Commutation. *AIEE Trans* (Power apparatus and systems) v81 1962 pp594–601.
6. R Holm. Theory of the sparking during commutation on dynamos. *ibid* pp588–594.
7. T M Linville, G M Rosenberry. Commutation of large d.c. motors and generators. *AIEE Trans.* 1952 ptIII v71 pp326–336.
8. A I Dvoracek. Eddy currents and their effect on commutation. *IEEE Trans* (Power apparatus and systems) 1964 v83 pp977–988.
9. B R G Swinnerton, J E Thompson, M J B Turner. Some properties of copper-graphite interfaces and their effects on the commutation of double lap wound d.c. machines. *Proc IEE* 1965 v112 no1 pp189–197.
10. J Hindmarsh, N K Ghai. Objective methods of assessing commutation performance. *Proc IEE* 1961 v108 pt1 pp55–58.

Compensation
11. Compensating d.c. motors for fast response. *Control Engineering.* 1960 Oct pp115–118.

Limitations in output
12. J Hindmarsh. Fundamental ideas on large d.c. machines. Volts per bar limits on large d.c. machines. *Electrical Times* series: 1958 7 Aug, 25 Aug, 11 Sep.

Rectifier supplies
13. R M Dunaiski. Effect of rectifier power supply on large d.c. motors. *AIEE Trans* 1960 v79 ptIII pp253–9.

Transient torques
14. E Schneider. Mechanical problems associated with rolling mill motors. *Brown Boveri Review* 1964 July.
15. Charles W Thomas, H Jewik, R P Stratford. Torque amplification and torsional vibration in large reversing mill drives. *Iron & Steel Engineer* 1969 May.

Short circuit torque
16. H J McLean, O C Coho. D.C. motor flash-over torque. *AIEE Trans* 1961 v80 ptIII pp850-3.

Brushes and brushgear
17. Carbon brushes and electrical machines. *Morganite Carbon Ltd*, London.
18. W C Kalb, F K Lutz. Carbon brushes for electrical equipment. *British Acheson Electrodes Ltd*, Sheffield.

Motor parameters and operation
19. H D Snively, P B Robinson. Measurement and calculation of d.c. machine armature circuit inductance. *AIEE Trans* v69 p11 pp1228-1237.
20. R M Saunders. Measurement of d.c. machine parameters. *ibid* 1951 v70 pt1 pp700-706.
21. R W Ahlquist. Equations depicting the operation of d.c. motors. *ibid* 1954 v73 ptIIIb pp1499-1505.
22. A G Darling, T M Linville. Rate of rise of short circuit current of d.c. motors and generators. *ibid* 1952 v71 ptIII pp314-325.

Control
23. L R Hulls, G H Samuel. Motor field control of large d.c. reversing mill motors. *ibid* 1958 v77 ptII pp579-585.
24. A Hansen, A W Wilkinson. Automatic speed regulation of d.c. motors using combined armature voltage and motor field control. *ibid* 1961 v80 ptII pp59-64.

Specifications
25. BS2613:1970.
26. NEMA publication MG1. 1967.

Superconducting motors
27. Duncan Peters. *Engineer* 1970 v230 no5967 p40.

EPE VARIABLE SPEED DC MOTORS 1-125hp

E.P.E. are specialists in supplying DC motors for thyristor drive applications. Our motors are used extensively in the rubber, plastics, paper, cable-making and metals industries and increasingly for machine tool drives. We have had over 50 years' experience in the design and manufacture of DC motors, so our standards of workmanship, materials and reliability can be taken for granted. Get in touch with the specialists at the address below, and find out all about E.P.E. service.

E·P·E

The Electrical Power Engineering (B'ham) Ltd., Mackadown Lane, Birmingham B33.OJQ. tel: 021-783 2261. Telex: 33-8589 Telegrams: Torque, Birmingham Telex.

30 hp T.E. variable speed DC motor

Materials in Engineering 1972

A complete appraisal of materials used for the construction of components across a wide range of industries, this publication will be the culmination of careful and accumulated research in the materials field. Essentially a book for the working engineer, all the information has a practical bias and all data shows an awareness of the materials' end use.

Five major sections result in a 500 pp. publication containing design ideas, materials origins, characteristics and application data, specific design area problems and materials for their solution, a survey of materials suppliers and a glossary of materials engineering terms.

The section on materials deals with thermoplastics and thermosets, metals and metallic alloys, non-metals, natural and synthetic rubbers, ceramics, glasses, sintered materials, composites, fibres, etc. Each material is given individual treatment and its characteristics and properties, usage and economics are discussed. Wherever possible comparative data are presented in tabular form.

Section 2 of the book appraises materials for specific design requirements — such as wear-resistance, flexibility, constant-dimension — and Section 3 covers specific design case histories where particular materials have fulfilled exacting requirements or shown considerable cost savings.

A fourth section provides a survey of engineering materials by type and supplier whilst Section 5 is a glossary of terms used in materials engineering.

EDITOR
The book's general editor is F C Cowlard BSc of the Plessey Company's Allen Clark Research Centre. Mr Cowlard has lectured on materials sciences in both this country and the United States and contributors to the book will include top UK, US and European materials engineers writing under his supervision.

FORMAT
The book will be published in two versions. The combined Vols. I & II will be A4 size, bound in a multi-ring binder and be approximately 500 pages. Containing all five sections it will sell at £10 with a pre-publication offer of £9 valid until publication date.

Volume I will contain the materials data only and will be approximately 330 pages. A4 size with illustrations and bound in a multi-ring binder, it will sell at £6 with a pre-publication offer of £5.

Both volumes will be published in January 1972. An order form is contained in this book.

Chapter 12

Stepper Motors

E H Werninck *AIEE MIMC*
Consultant

The ABC or dial telegraph invented by Wheatstone in 1840 was the first device which used impulses to rotate a remote motor driving a pointer. The impulses were generated by a specially-designed transmitter having a number of definite switching positions, thus enabling messages to be transmitted letter by letter. This procedure proved too slow and cumbersome and was soon abandoned for more suitable devices in the field of telegraphy. The principle was however used again in a device known as the Watkin dial, which gunners employed to transmit bearing and elevation until about 1925. Wheatstone's dial telegraph and the Watkin dial were based on small linear solenoids and mechanical linkages, and belong to the type of device which is sometimes referred to as a mechanical stepping motor. The first true stepping motor recorded appears to be the Vickers motor, which had a two-pole rotor and a three-phase stator. This motor was also d.c. operated through a special commutator transmitter which caused the motor to rotate in a number of discrete steps. Early designs had 12 steps, but on later models these were increased to 24.

Working in a naval equipment research laboratory, Hugh Clausen developed and perfected a new type of stepping motor, which he named the 'M' motor and which continued to be used in naval instruments for many decades. The usefulness of this motor, particularly in industry, was largely limited by the commutator switch, which had to deal with high current surges and mechanical wear. The application of stepping motors was therefore limited to low power and low speed applications, and the demand for higher resolution led them to be superseded by analogue links such as Magslips and torque synchros.

As digital computation and its application to control became increasingly used, the interest increased in output elements which could respond directly (*ie* without conversion into analogue form). The development of motors was further stimulated by the appearance on the market of devices which could rapidly and reliably perform the necessary switching sequences. At one stage thermionic devices were used, but their cost and slowness greatly restricted their applications. Modern semiconductor devices (*eg* transistors, thyristors or silicon-controlled rectifiers) will be difficult to improve upon for speed accuracy and reliability of switching even very large currents. The speed and accuracy of the control motor, and thus the position of the load it drives, depend on the number of steps per revolution, which for medium-sized units delivering some 0.5 Newton metres are typically 200.

Stator and rotor types

To cater for the resolutions demanded a great variety of stator and rotor types were developed:
(1) Single stator units with multiphase windings distributed in slots. These are very similar to induction motor stators, which were in fact used in some of the early stepping motors.
(2) Multiple stator units. To overcome the physical limitations of providing a sufficient number of steps (phases) in a given diameter, several stators displaced by an appropriate number of degrees are placed in line and made to act on the same rotor. The resolution increases in direct proportion to the number of stators used.
(3) Vernier type. In this type of construction a single or multiple type of stator is given a few more teeth than its rotor. The windings of the vernier motor are then energised so that each step causes the rotor to advance one rotor tooth pitch.
(4) Hybrid construction. This combines both the multiple stator and vernier slot arrangement in one unit.

Rotors: Three main types of rotor construction are generally considered:
(1) Variable reluctance rotors: usually made from magnetically 'soft' material such as one of the silicon irons and provided with teeth so that the rotor assumes definite positions when the stator windings are energised. This type of rotor is usually found in stepping motors capable of high stepping rates. It cannot provide any holding torque when the stator is not energised and provides less damping than a permanent magnet rotor.
(2) Permanent magnet rotors. Various arrangements exist to provide more than two poles even in very small motors but as a general rule fewer steps per revolution are possible than from the reluctance type. Against this the motor has an inherent holding or detent torque even when not energised. To combine the best features of both types of rotor the two types may be combined, thus providing a large number of steps, larger torques for a given excitation and a detent torque.
(3) Stereo-motor construction: could be described as an eccentric rotor construction of a permanent magnet rotor motor and was so named by the French inventors in about 1958. The rotor instead of rotating concentrically inside a stator is made to 'roll' around it, thus providing inherent hypocyclic gearing. In control systems the low effective inertia and small stepping angles which can be provided are particularly useful. Similar types of motor are made, *eg* in Japan and the USA.

For the sake of completeness, though not dealt with here in detail, mention is made of the rotary solenoid drives found in uniselectors and rotary switches. Some of these convert linear into rotary motion by means of a ratchet, others use inclined planes and steel balls. The nearest approach to a rotary solenoid is described in some detail in reference 3 of the bibliography. This device, like the others in this category, must be returned to the 'start' position before the next step. Comparatively slow stepping rates only are possible and the devices are inherently non-reversible. Their mode of operation and construction thus hardly qualifies them as prime movers in the same manner as motors.

PRINCIPLE OF OPERATION

The principle of operation of an 'M' type motor from a commutator transmitter is shown in Fig. 1, which illustrates how the motor rotor is made to step through 12 positions, in 30-degree steps. If the two-pole rotor is replaced by a four pole version there are twice as many positions, since the step angle is reduced to 15 degrees.

Two-pole and four-pole 'M' type reluctance rotors were rather difficult to manufacture and were, for many applications, replaced by permanent magnet rotors. The laminated construction of the poles was evolved to overcome the reluctance type rotor's lack of damping. The commutator type switch which was of barrel or face type construction, could nowadays be replaced by reed or semiconductor switching. Since the command signals mostly originate from a digital program or logic there is no need to rotate a shaft, so that the controller is usually a printed

Fig. 1. 'M' type step-by-step transmitting. Stepping sequence, starting with position shown:

Step	A_1	A_2	A_3	I_1	I_2	I_3
1	+	−	−	$\frac{2}{3}$	$\frac{1}{3}$	$\frac{1}{3}$
2	+	0	−	$\frac{1}{2}$	0	$\frac{1}{2}$
3	+	+	−	$\frac{1}{3}$	$\frac{1}{3}$	$\frac{2}{3}$
4	0	+	−	0	$\frac{1}{2}$	$\frac{1}{2}$
5	−	+	−	$\frac{1}{3}$	$\frac{2}{3}$	$\frac{1}{3}$
6	−	+	0	$\frac{1}{2}$	$\frac{1}{2}$	0
7	−	+	+	$\frac{2}{3}$	$\frac{1}{3}$	$\frac{1}{3}$
8	−	0	+	$\frac{1}{2}$	0	$\frac{1}{2}$
9	−	−	+	$\frac{1}{3}$	$\frac{1}{3}$	$\frac{2}{3}$
10	0	−	+	0	$\frac{1}{2}$	$\frac{1}{2}$
11	+	−	+	$\frac{1}{3}$	$\frac{2}{3}$	$\frac{1}{3}$
12	+	−	0	$\frac{1}{2}$	$\frac{1}{2}$	0

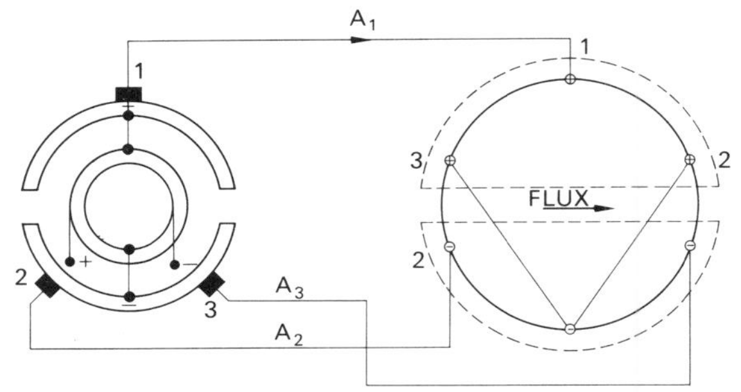

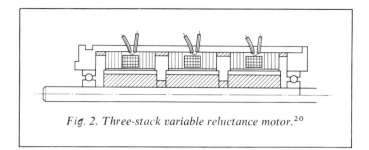

Fig. 2. Three-stack variable reluctance motor.[20]

circuit board with suitable semiconductors. Since highly inductive devices are switched 'commutation' must be effected by suitable resistances to avoid the destruction of the semiconductors by voltage surges.

Modern reluctance type stepping motors tend to use rotors. For very low power motors forms of construction similar to those of the synchronous clock motor are used. In larger motors use has been made of inductor type stators, thus greatly reducing winding costs. Another example of the great variety of configurations using this principle is the three-stack arrangement shown in Fig. 2. The exciting windings are three pancake coils and the stators are so arranged that the teeth are consecutively staggered by one-third tooth pitch. Sequential energisation will thus 'step' the motor. Where the windings are distributed in slots like those of an induction motor the switching of the windings plays the determining role. The single stator unit with three-phase star connected winding can be switched to provide 3, 6, or 12 steps for one complete revolution of the rotor when the star point is earthed and used as common return. If the phases are provided with a centre-tap (bi-filar winding) as is usual on two-phase windings, and these centre taps are earthed, the motor can be made into a 'four' or eight phase motor. In all cases the torque developed is determined by the vector sum of the fields created by the coils energised at the instant of stepping. The number of steps is thus also determined by the possible number of combinations. If both ends of the winding are connected to the control circuit and the current can not only be sequentially switched, but also reversed, the number of steps will be doubled. However, this greatly increases the number of wires to the motor, and hence wiring-up costs, to say nothing of the more complex switching logic; application is therefore influenced by economic considerations.

The Steromotor type stepping motor is a good example of the vernier type and is thus able to provide a large number of steps per revolution due to this inbuilt speed reduction. The principle of operation is shown graphically as in Fig. 3, which also shows how this is converted into conventional shaft rotation. Since this type, like many other stepping motors, can also be a.c. operated it can become a slow speed synchronous motor. Where absolute stepping or synchronism are required the tires, which allow the rotor to roll around inside the stator are replaced by toothed wheel and track. As in all stepping motors the direction of rotation depends on phase sequence and single phase a.c. must be supplemented by resistive or capacitive 'phase splitting'.

CONSTRUCTION

In the smallest sizes, intended for instruments, the construction as well as the magnetic circuit is the same as that of the synchronous clock motor. This makes possible mass production methods, and hence the construction of relatively cheap stepping motors. Since the stepping action puts greater strain on the bearings only the very smallest sizes are supplied with lubricant impregnated plain bearings.

The nature of stepping motors requires not only robust bearings but generally more rigid construction, particularly where several stators have to be accurately aligned. In this connection attention should also be drawn to certain types of permanent magnet motors which, to obtain maximum possible performance, are magnetised after assembly. These are very difficult to service and best returned to the manufacturer or authorised service agent.

As mentioned before, the steromotor rotor requires a special linkage to convert the gyrating motion of the rotor into conventional shaft rotation, which as shown in Fig. 3 the French inventors solved by what they describe as a combination of Oldham coupling and universal drive. In the 'Responsyn' actuator conventional bearings are used and the rotor, made of conducting material, is made to deflect and engage with teeth in the stator bore when energised.

Stepping motors are mostly flange-mounted to ensure rigidity and accurate alignment. As far as possible, stability against torsional oscillations when stepping, at any rate within its range, is built in and particularly good in permanent magnet types. Nevertheless special externally mounted damping devices have been designed and made.[10]

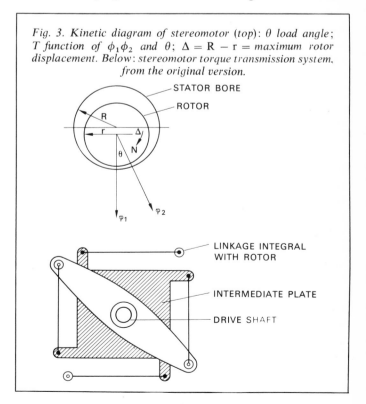

Fig. 3. Kinetic diagram of stereomotor (top): θ load angle; T function of $\phi_1 \phi_2$ and θ; $\Delta = R - r =$ maximum rotor displacement. Below: stereomotor torque transmission system, from the original version.

In general the mechanical construction of stepping motors is very similar to those used for induction type fractional-power (to 750 W) motors. For military and aircraft equipment, units are available in the NATO frame sizes. For use in process- and machine-tool controls rugged versions (usually totally-enclosed) are manufactured. As these motors are inherently reliable they find application in reactors and even space equipment, where their low power requirements are also a great asset. Stepping motors made for instruments, computer-peripherals, office machines, and so on can take advantage of the less hostile environment and larger quantities in which they are required. The Electrical Research Association's technical planning unit (UK) forecast[19] an annual growth rate of more than 20% in 1970 and established that one company were already making stepping motors at the rate of nearly 20 000 a year.

CHARACTERISTICS

Most stepping motor applications involve more or less frequent increases to fast stepping rates for quick return (slewing) in addition to the rapid and accurate initial response demanded of a servomotor. The stepping rate versus torque characteristic of the motor therefore plays a very important part in the selection process. Reluctance motors provide the fastest speeds, but their torques are lower than those of similar sized permanent magnet motors, which, as mentioned previously, also provide useful inherent damping. In some cases this detent torque of permanent magnet or hybrid motors must be traded off against the faster stepping rate of reluctance types. The stepping angle also plays an important part in the highest slewing speed which can be obtained, and here the compromise is usually the largest angle which will meet the specified resolution of the drive. Though the stepping motor is inherently a low-speed prime mover, gears may still be required to match inertias and supply large torques at very low speeds. If the inertia of the load is only marginally too large for the motor, play may be deliberately introduced to enable motor torque to build up.

Another problem to the purist servo-engineer is the open loop of the systems using stepping motors. Advances in digital systems have made it possible to control, for example, the position of a drilling machine table consistently to within $50 \mu m$ (0.002 in). Adequate mechanical construction and correct matching of the motor to the load are, however, still among the essential prerequisites of stability and consistent accuracy, which means that no steps must either be gained or lost. Thus the principle of matching motor and load are the same as in analogue servo systems; the speed-torque curve is now however replaced by the stepping-rate versus torque curves.

The discontinuous nature of the action of the motor, and therefore its describing or transfer function, as well as the variation of motor torque with rotor angle, make mathematical analysis of such systems rather complex. Nevertheless the advantages to be gained from an output element which is 'compatible' with the increasingly ubiquitous

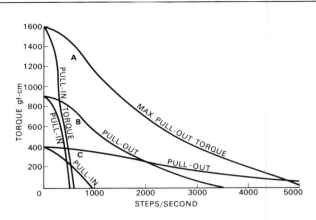

Fig. 4. Typical stepping motors (8-phase 2-pole). (A) NATO size 23 (57 × 78 mm), 96 steps. (B) NATO size 28 (70 × 108 mm), 48 steps. (C) size 28, 96 steps.

computer make the effort of programming the stepping rate to suit the torque characteristic of the motor well worthwhile.

It must also be pointed out that, for the motor to develop maximum torque for a given average power input, the current should reach and maintain its maximum value instantaneously not only when switching on, but also when stepping from phase to phase. Solid state logic units with semiconductors used in a switching mode have extremely small time constants so that it only remains to reduce inductances of motor windings. At higher frequencies or stepping rates series resistances are necessary to decrease the time constant (L/R) of the circuit. To maintain the correct voltage across the motor the d.c. supply voltage must now be increased.

PERFORMANCE

Since the exact method of control has considerable influence on the stepping motor, its performance is more difficult to specify than that of any other prime mover. The motor manufacturer generally tests his product with minimum inertia and optimum control which in some applications may be unnecessarily elaborate and costly. A considerable amount of information on drive circuits has been published by major manufacturers since about 1964. This gives individual characteristics for a wide range of motors, and tables in which the motors have been arranged in various orders with some of their main performance parameters:
(a) Stepper motors in order of increasing working torque ranging from 60 gf-cm to 1600 gf-cm (0.83 to 22.2 oz-in) with holding torque, pull-in and pull-out stepping rate with the corresponding speed (rev/min) and finally the step angle in degrees;
(b) stepper motors in order of increasing pull-in rate;
(c) stepper motors in order of increasing pull-out rate;
(d) list of motors according to type designation.

For more detailed study of the dynamic performance of the system characteristic curves such as shown in Fig. 4 are necessary. If load inertia and friction have been determined the torque available for acceleration as well as the permissible rate of increase in stepping frequency can be determined. The curves are typical of reluctance motors.

The characteristics of high resolution permanent magnet motors are also considered here, taking a very popular frame size, and combining data given by three manufacturers. Fig. 5 shows in order of magnitude the torque versus stepping rate curve from references 14, 16, 17. Curves from the last two sources indicate a depression in the torque at around 80 steps per second due to harmonics (not evident in reference 14's 3-in motor). The effect of series resistance on the time constant and therefore the stepping rate as well as maximum torque is also shown. The data-sheet points out that with suitable circuitry the stepping rate can be increased to about 4000, but omits to mention the price in terms of circuit losses.

Terminology

Detent torque: the torque an unexcited permanent magnet motor can resist before commencing to rotate.

Holding torque: the torque any excited stepping motor develops before being pulled out of step and starting to rotate. In taking the values given in any manufacturers literature careful note must be taken of the exact circuit conditions pertaining to them. In addition heating of the motor and supply variations must be taken into consideration. In some applications it may therefore be advantageous to reduce the holding current or use only half the exciting winding.

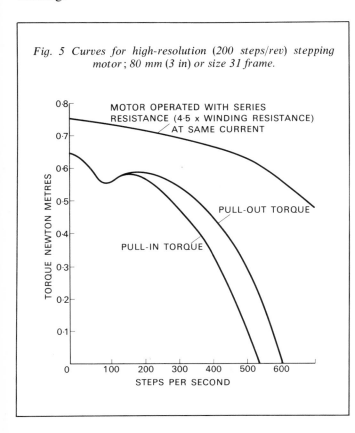

Fig. 5 Curves for high-resolution (200 steps/rev) stepping motor; 80 mm (3 in) or size 31 frame.

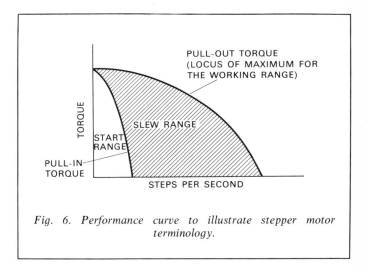

Fig. 6. Performance curve to illustrate stepper motor terminology.

Maximum pull-in or synchronising rate (speed): the maximum switching rate in steps per second [or its equivalent in speed (rev/min) = (stepping-rate/sec × 60) divided by the motor's number of steps per revolution] at which the motor will run up without losing a step.

Maximum pull-out or following rate: the switching rate at which an unloaded motor will begin to drop out of step as the stepping rate is gradually and smoothly increased.

Pull-in torque: the torque at which the motor can step absolutely synchronously when started from rest.

Pull-out torque: once a motor is running at a certain stepping rate the torque loading can be increased to the pull-out torque, when the motor will drop out of step and come to rest (or hunt). The implications of the parameters are shown in the general characteristic Fig. 6.

Stepping angle: the angle through which the rotor shaft turns when the stator winding is switched sequentially by one step. It is given in degrees as the reciprocal of the number of steps per revolution N; ie $360:N$.

Stepping accuracy: generally given as a percentage of the stepping angle and in practice is of the order of 5%. It should also be noted that for stepping motors the errors in individual steps are not cumulative.

Steps per revolution: depend on the number of stator poles and phases, how they are switched, and the rotor poles; can be defined as the number of discrete steps a fixed point on the output shaft makes in one direction of rotation before returning to its original position.

Slew range: the range of switching rates within which the motor can run in one direction, within a certain maximum rate of increase, and can follow the switching rate without losing a step.

BIBLIOGRAPHY

1. *IEE Conference on Servocomponents* November 1967: (a) L G Atkinson. Some rotary servo-components past, present and near future. (b) E H Werninck. Servomotors, actuators and stepping motors.
2. J Bell. Data-transmission systems. *JIEE* vol.94 pt.IIA 1947 pp222–235.
3. G W Cullen. The design and development of the 'Rotenoid'. *Component Technology* published by The Plessey Co vol.2 no.1 March 1966.
4. Naval Electrical Pocket Book. *HM Stationery Office* 1953.
5. R Bruce Kieburtz. The step motor—the next advance in control systems. *Trans IEEE* (automatic control) Jan 1964 pp98–104.
6. A G Thomas and J F Fleischauer. The power stepping motor—a new digital actuator. *Control Eng.* vol.4 pp74/81.
7. J G Truxall (editor). Control Engineers Handbook. *McGraw-Hill* 1958.
8. Factors to consider in choosing step servo-components. Selection guide feature *International Electronics* Dec 1965 pp24–28.
9. M Browning. DC stepper motors. *Product Design Engineering* Feb 1968 pp42–45.
10. Damper for step motor. BP 1 071 001 *ibid*.
11. K C Garner. Improving stepping motor performance. *Control* June 1968.
12. M J Edwards. Drive circuits for stepping motors. *Mullard Tech Comm* no72 Sep 1964.
13. D S Evans. Driving mode nomenclature for stepper motors. *Instrument and Control Engineering* Nov 1969.
14. Evershed and Vignoles Ltd. *Stepping motor and stereomotor catalogue.*
15. Impex Electrical Ltd/Polymotor International Brussels. *Stepper motor publication* April 1971.
16. Superior Electric Co, USA. *Slo-syn literature.*
17. Ing. S Auerhammer Groebenzell, Germany. Phytron Schrittmotoren.
18. Incremental servos; pt 1, 2, 3. *Control* Nov and Dec 1960; Jan 1961.
19. Stepper motors are making strides in the design field. *The Engineer* 13 Aug 1970.
20. D J Maxwell. Stepper motors and transmitters. Electronic Data Library, vol.2 Servosystems. *Morgan Grampian Publishers* 1969.

Chapter 13

Servomotors

P Vernon CEng MIEE
Muirhead Ltd

All types of asynchronous motors, and some types of synchronous motors (*eg* steppers), can be used in servo systems, but this chapter concerns small a.c. servomotors, as commonly used in closed-loop servo systems. Typically they consist of 2-phase induction motors and would fit into a servo system, as illustrated in Fig. 1.

Systems increasingly follow the trend of consisting of a group of components carefully integrated together and supplied as a complete package, such as that illustrated in Fig. 2. The output shaft of such a system will drive a load of 0.1 Nm (15 oz in) to follow a synchro, requiring only 0.0035 Nm (0.05 oz in) to rotate it, to an accuracy of $\frac{1}{3}°$ angular rotation. For substantially more accurate servo systems a coarse-fine system is usually used, when accuracies down to 2 min (angular) are often achieved. The limitation then becomes the accuracy to which the gears can be cut and mounted.

Induction motor characteristics

If a 2-phase alternating supply is applied to an iron stator, wound with two coil groups spaced 90° apart, a rotating field is set up. A suitable rotor inserted in the stator is dragged round in the same direction as the rotation of the field, and will eventually reach a speed slightly less than the 'synchronous speed' at which the field is rotating. This is the basis of the a.c. servomotor, and conforms to conventional induction motor theory. The advantage of using 2-phase system for servomotors arises because they are driven from a single-phase amplifier so as to provide a torque directly proportional to the (error) signal from this amplifier.

In the general induction motor, at any given speed, approximately

$$\text{torque} \propto (\text{rotor current})^2$$
$$\propto (\text{rotor voltage})^2$$
$$\propto (\text{supply voltage})^2$$

from which it can be shown that, in the case of a supply through two equal phases, V_1 and V_2: torque $\propto V_1 \cdot V_2 \propto V_2$ if V_1 is kept constant.

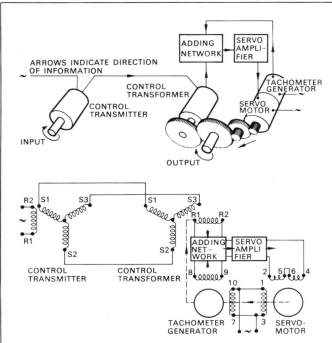

Fig. 1. Simple a.c. servo system using tachometer generator to supplement the damping.

Fig. 2. Miniature servo control (courtesy Thorn Automation Limited).

In conventional induction motors designed for 2- or 3-phase operation, the speed torque characteristic is usually in accordance with Fig. 3. A curve of this nature is usually measured by means of a dynamometer. The part of the curve where the speed is negative can be obtained by making the dynamometer drive the motor in the opposite direction to its desired rotation. Operation at negative speed occurs if the motor, when running normally, is suddenly switched so as to run in the reverse direction.

If no retarding torque is applied, the rotor speed is slightly less than synchronous speed. The progressive addition of load torque makes the speed drop until maximum torque is reached. The motor then stalls, unless the torque is reduced.

Such a motor will run quite satisfactorily on single phase, and this is usually explained by assuming that a single-phase supply creates an alternating field which (assuming a two pole device), acts along a straight line. This can be considered to be the vector sum of two rotating fields of constant amplitude rotating uniformly in opposite directions. Each rotating field has an amplitude equal to one half of the peak amplitude of the resultant field. If the clockwise field produces a speed torque curve as in Fig. 3, then the negative field will produce a similar field with the

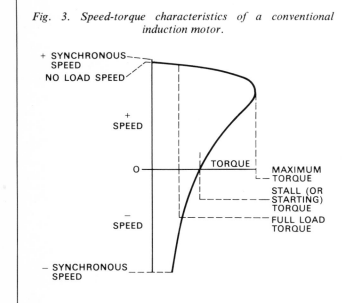

Fig. 3. Speed-torque characteristics of a conventional induction motor.

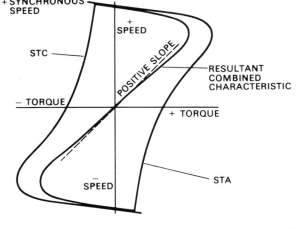

Fig. 4. Determination of single-phase speed-torque characteristics of a conventional induction motor. STC: speed-torque characteristic associated with clockwise rotating field. STA: ditto with anticlockwise rotating field.

normal operating range in the diagonally opposite quadrant of the diagram Fig. 4. The actual speed torque curve is the difference between the two.

Because no torque exists at zero speed, the motor is not self starting, but once started, the speed will increase because positive speed is associated with positive torque. (It will equally well rotate in the opposite direction as the same relative polarity of speed and torque still applies.)

THE SERVOMOTOR

The single-phasing characteristic of the ordinary induction motor is undesirable in a servomotor, in which close control of the speed by means of the control voltage is required. To achieve this the rotor is designed to have a large resistance/reactance ratio. This results in the speed–torque characteristic of Fig. 5. Single-phase operation may again be regarded as establishing two half strength rotating fields acting in opposite directions, and the characteristics arising from these are shown in Fig. 6.

If the rotor is already stationary, rotating it merely generates braking torque, which opposes the rotation, regardless of direction, because positive speed is associated with negative torque etc. As a result, the servomotor does not run on single phase, and actually develops useful braking torque in that condition.

In normal operation, one winding is permanently energised from an alternating supply, and the other, situated 90° from the first, is energised from a variable voltage supply in quadrature to it. The permanently energised winding is termed the 'reference' or 'fixed voltage' winding, and the variably energised winding is termed the 'control' winding, and is normally supplied from some form of servo amplifier.

By varying the supply to the control winding, a family of speed torque curves can be obtained. Fig. 7.

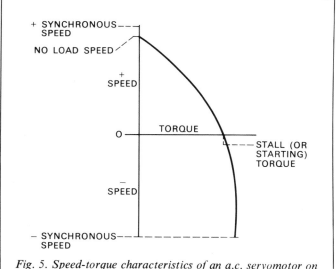

Fig. 5. Speed-torque characteristics of an a.c. servomotor on a 2-phase supply.

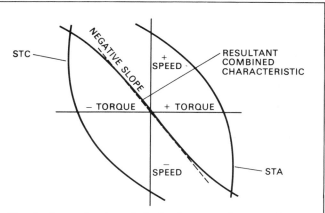

Fig. 6. Determination of single-phase speed-torque characteristics of an a.c. servomotor. STC: speed-torque characteristic associated with clockwise rotating field. STA: ditto with anticlockwise rotating field.

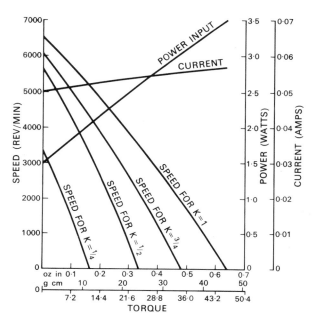

Fig. 7. Speed-torque curves for typical servomotor, showing effect of varying control volts. Size 11 400–Hz motor; values of current are for 115–V winding.

Inherent damping

The servomotor characteristics provide a form of damping which is analogous to viscous friction, with the torque varying linearly with speed. The presence of damping is evident if the motor is rotating, and the control voltage is reduced, or becomes zero. If damping were not present, the motor would continue to rotate at the original speed. The inherent viscous damping plays an important part in the operation of the servomotor in a servo system.

A.C. servomotor construction

The smaller servomotors, which were originally designed for military applications, are made to a high degree of standardisation, and are described by a frame size number. This number represents the outside diameter of the housing measured in tenths of an inch and rounded upwards to the nearest whole number.

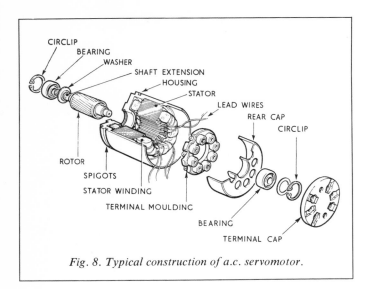

Fig. 8. Typical construction of a.c. servomotor.

To make the acceleration from standstill, *ie* the torque inertia ratio, as large as possible, the rotor tends to be made relatively small in diameter compared with length. Torque is proportional to diameter^2 × length, whereas inertia, *ie* mass × (radius of gyration)^2, is proportional to diameter^4 × length.

A cage-type rotor is used, generally with aluminium conductors. The rotor iron consists of laminations mounted on the shaft with the conductors and end rings cast in position. The rotor runs on ball bearings located in the housing and rear cap. Because of the small radial clearance between the rotor and stator (commonly in the order of 0.025 mm: 0.001 in), considerable precision in manufacture is necessary.

Servomotors in each frame size are available with various combinations of control and reference voltages. The control and reference windings normally have the same power inputs when fully energized. The control winding is usually wound in two separate halves which may be connected in series or parallel, giving a choice of two control voltages. Alternatively, the control winding can be connected directly to the output stage of a servo amplifier, joining the two halves to form a centre tap to which the output stage supply is connected. This avoids using an output transformer, so saving space and weight. The drawback is that d.c. current then passes through the windings from the supply to the output stages. This results in additional heating in the control winding and modifies the speed torque characteristics.

Choice of supply frequency

Standard a.c. servomotors are intended for use on one of two frequencies, namely 60 Hz and 400 Hz. There is little difference between the stalled torques in most 60 Hz and 400 Hz servomotors of similar frame sizes. 60 Hz servomotors have a synchronous speed of 3600, and a no-load speed of about 3200 rev/min; 400 Hz servomotors, however, usually have a six-pole or eight-pole construction giving synchronous speeds of 8000 and 6000 rev/min, and no-load speeds of about 6200 and 4800 rev/min. From this, it will be evident that a 400 Hz servomotor is more powerful than a 60 Hz servomotor of similar frame size.

A.C. servomotor temperature rise

The special torque-speed characteristics are achieved by using a relatively high rotor resistance. Considerable power is dissipated when running because of I^2R losses in the rotor conductors. Moreover, the motor may have to spend long periods at rest with zero control volts, and full volts on the reference winding. Again, I^2R losses will occur in the rotor. Consequently, the servomotor is inherently inefficient, and is subject to relatively large temperature rises. The relatively high losses make it impracticable to build an a.c. servomotor larger than about 750 W (1 hp), and production consists mainly of much smaller units. Large power outputs require d.c. motors or hydraulic drives.

Most of these small a.c. servomotors may be operated in an ambient temperature up to 125°C, though at these higher temperatures it is essential to mount them on an adequate heat sink. They are also often designed to operate at temperatures down to −55°C, though some warm-up time may be required after switch-on below −40°C.

Terms associated with servomotors

Slot effect (or cogging). Slot effect results in a reluctance of the rotor to move away from one of a number of positions. To overcome slot effect, a certain control voltage must be exceeded before the rotor starts to rotate. A common requirement is that the rotor must not remain stalled if 3% of the maximum control voltage is applied. The equivalent torque represented by this voltage may be found by assuming that stalled torque and control volts are proportional. This torque actually represents the torque associated with slot effect plus friction torque.

Friction. Friction may be separated from slot effect by supplying both windings with equal watts (ensuring that

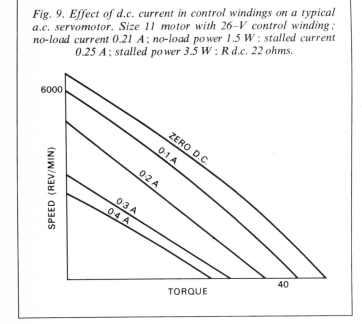

Fig. 9. Effect of d.c. current in control windings on a typical a.c. servomotor. Size 11 motor with 26-V control winding; no-load current 0.21 A; no-load power 1.5 W; stalled current 0.25 A; stalled power 3.5 W; R d.c. 22 ohms.

the voltages are in quadrature) so as to just keep the rotor turning. By assuming that stalled torque is proportional to control volts × reference volts, the friction torque can be determined. It is assumed that the slot effect can be ignored in these conditions. Specifications commonly call for the voltages to be not more than 15 volts (for 115 V windings) on this test, giving a friction torque of not more than $15^2/115^2 \times 100 = 1.7\%$ of maximum stalled torque.

The viscous friction at any point on any of the family of torque speed curves equals the slope of the curve at that point. The usual approximation for viscous friction (in Nm per rad/s) is: Viscous friction = (stall torque)/(no-load speed): where torque is expressed in Nm, speed in rad/s. In a practical servomotor, the damping is usually less at lower control voltages, and at zero control volts may be half the value found in the expression above.

Acceleration from standstill. The theoretical acceleration at stall of an unloaded servomotor is found by: acceleration (rad/s²) = maximum stall torque (Nm) divided by rotor inertia (kg m²). In practice, the torque falls off as the speed rises, and the acceleration diminishes.

Time constant. The time taken for an unloaded servomotor to reach 63.2% of its final speed is given by: time constant (s) = inertia/(viscous friction × g), where g is expressed in cm/s².

Torque constant. Stall torque and control voltage are approximately proportional: torque constant = stall torque (Nm) divided by control volts.

Velocity constant. To assess the ability of the motor to follow a uniformly moving command signal, we need the relation between speed and control volts. Velocity constant = no-load speed (rad/s) divided by control volts. The velocity constant is useful in determining the angular lag which occurs in a servo system between the command signal and the output shaft.

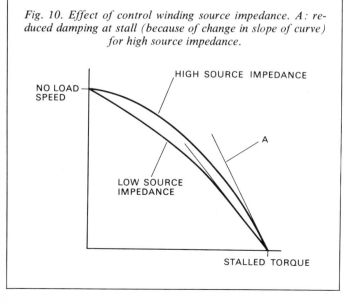

Fig. 10. Effect of control winding source impedance. A: reduced damping at stall (because of change in slope of curve) for high source impedance.

Transfer function of an a.c. servomotor

A servomotor may be represented by a second order differential equation:

$$\frac{I d^2\theta}{dt^2} + F\frac{d\theta}{dt} = KV$$

when K = torque constant (Nm/volt)
V = control volts
I = rotor inertia (kg m²)
F = viscous friction (Nm/rad/s)
θ = shaft angular position (rad).

An additional term, representing fixed or coulomb friction (independent of velocity) is sometimes added to the left-hand side of the equation. The Laplace transform of this equation is:

$$Is^2 + Fs = KV$$

and the transfer function

$$\frac{\theta(s)}{V(s)} = \frac{K}{s(sI + F)}$$

In terms of the time constant $\tau = I/F$

$$\frac{\theta(s)}{V(s)} = \frac{\frac{K}{F}}{s(s\tau + 1)}$$

This implies that the maximum phase shift of the motor, relative to the control voltage modulation, is 180°. In practice, it is usual to quote a more complicated transfer function, which includes an additional time constant.

$$c = \text{the ratio } L/R \text{ of the control winding}$$

$$\frac{\theta(s)}{V(s)} = \frac{\frac{K}{F}}{s(s\tau + 1)(cs + 1)}$$

Effects of control winding source impedance

Servomotor characteristics assume that the control phase is energised at the correct voltage for all conditions of load torque. In practice, the change in control winding impedance from no load to stall will result in an appreciable change in the phase of the current, and a lesser change in its magnitude. If connected to an amplifier with a relatively high source impedance, the terminal voltage on the winding will increase as the torque decreases. The no load speed will be relatively unaffected because it must always be less than synchronous speed. As a result, the speed/torque curve will become relatively humped. This condition may be exaggerated if a tuning condenser is shunted across the winding so as to make it a unity power factor load at stall. Fig. 10, showing the effect of varying the source impedance, assumes that the voltage is adjusted to give the correct stall torque.

The change in slope results in impaired damping at high torque, which may affect the stability of the system, and the general shape is much more conducive to single phasing, *ie* continuing to run when the control phase is not energised. (A standard test for single phasing is to disconnect the control winding completely when the motor is running, thereby introducing infinite source impedance. If the motor quickly comes to rest single phasing is not present.)

Trends in choice of servomotors

A.C. servomotors of the type described have been almost exclusively used in miniature high-speed servo control systems. Their robustness and low-inertia rotor are the main attributes responsible for this. However, as such systems are their only major application, production quantities are low and hence selling price relatively high. Automatic winding and precision injection-moulded plastic housings are likely to bring considerable cost reductions here.

Meanwhile d.c. motors employing permanent magnet (cylindrical) stators are becoming quite popular. As they have many applications other than in servo systems, production quantities are much larger and selling prices lower. Units of this conventional design have the advantages of higher efficiency and greater power-handling capacity for a given frame size. Their main limitation however still remains the comparatively short life of the commutator and brush gear. A brushless d.c. motor suitable for servomotor application may yet be the eventual answer, but some of those at present being publicised as having characteristics much superior to the a.c. servomotor unfortunately do not have the output-power proportional to control-voltage relationship required for servomotor applications. An iron-less rotor d.c. motor, in which a thin shell of armature windings, resin impregnated, revolves around a stationary inner core, is very suitable for light servo applications where extremely low inertia is the main consideration. Their main limitation is their relatively very small power handling capacities, even when compared with the a.c. servomotor.

VACTRIC

precision servo components

* **A.C. SERVO COMPONENTS**
  Servo motors
  Motor tachogenerators
  Synchronous motors
  Induction motors
  Rotary pick offs
  Synchros
  Resolvers

* **D.C. SERVO COMPONENTS**
  Servo motors
  Tachogenerators
  Torque motors
  Stepper motors
  Brushless motors

* **SERVO GEARHEADS**
  Precision gears

* **SHAFT ANGLE ENCODERS**
  Electronic amplifiers

* **ELECTRIC SERVO DRIVE SYSTEMS**
  for machine tools

* **DIGITAL READ-OUT SYSTEMS**
  for machine tools

VACTRIC CONTROL EQUIPMENT LIMITED

Garth Road, Morden, Surrey.
Telephone 01-337 6644
Telex 27796

A LUCAS COMPANY

Chapter 14
Linear Induction Motors

P J Markey BSc CEng MIEE
English Electric—AEI Machines Ltd

In the early years of the 20th century, when rotating electrical machines had become established and made to perform satisfactorily, it was realised that these machines could be made in a different form which would use or produce motion in a straight line, instead of the normal rotary motion.

The form of machine required to produce straight-line, or linear, motion, can be understood by imagining a rotating machine to be cut along a radius and then unrolled. Any type of rotating machine can be designed in this unrolled form: a generator or motor, synchronous or asynchronous, used with alternating or direct current, or whether its elements are continuous at the air gap or have salient poles. Relatively few types of rotating machines have been usefully developed in this manner, but the cage rotor induction motor has been outstandingly successful in a considerable number of different adaptations. The direct current motor is probably the next most successful type to be used as the basis for a linear motor.

Current world-wide interest in the design and use of linear motors is largely the result of the work done and described by Professor E R Laithwaite and his colleagues, first at Manchester University and later at Imperial College in London.

This chapter is limited to consideration of the cage rotor induction motor in its various linear forms, since this will include most types of linear machines that find useful applications.

LINEAR MOTOR FORMS

Two basic forms of the linear induction motor can be made. Each produces a driving force between the primary and secondary elements, acting in a straight line. The first form is produced, as mentioned above, as if by cutting the normal rotating machine along a radius and unrolling it. The second basic form can be considered as being made by rolling up the first form about another axis.

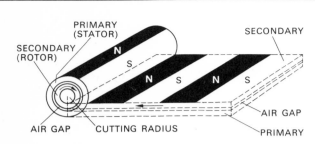

Fig. 1. One form of linear induction motor, represented by cutting a normal rotating machine along a radius and unrolling it.

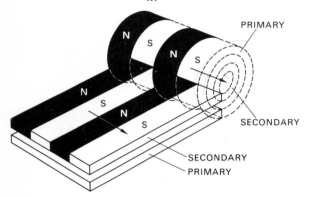

Fig. 2. Basic representation of linear induction motor produced by rolling up the form of Fig. 1 about another axis.

In both of the above types the term primary is used to describe the part of the machine carrying the windings, usually polyphase, energised from an external supply. In the conventional induction motor the primary winding is carried on the stator and the secondary is the cage rotor. For linear machines the terms primary and secondary are more appropriate since neither part normally rotates and either part may be the moving member.

The first basic form, in which both primary and secondary are in the shape of blocks or slabs, can be made in a variety of ways, since one of the fundamental restrictions imposed by the normal rotating machines (*ie* that the rotor must be contained within the stator) no longer applies. Either the primary or the secondary of the linear motor may be of any length and the two lengths are independent.

The primary is normally the more expensive element since it contains most, if not all, of the magnetic material and also the insulated winding. Therefore, the first arrangement has a short primary and a long secondary. With this arrangement either part may be the moving element, but if the primary is to move then it must have either long flexible leads or sliding or rolling electrical contacts. The secondary normally contains both nonmagnetic conducting material (sheet aluminium is commonly used) and magnetic material. This is the single-sided short primary arrangement.

The second arrangement is the single-sided short secondary type. This is obviously more expensive and normally consumes more power in producing a given thrust than the short primary type.

Either of these two arrangements may be made 'double-sided' and this produces two further varieties. The double-sided short primary type can be energised in two different ways. In the first, the two primaries are not co-operating, and some magnetic material is required in the secondary to carry the flux. The machine is thus the equivalent of two single-sided short primary machines. However, when the two primaries are energised as in Fig. 3, the magnetomotive forces (mmf) of both windings act together to drive flux through the secondary which may not now require any magnetic material. The double-sided short secondary type almost completes the varieties of the first basic form.

Further varieties of the flat block form of linear motor are possible by adding more primary and secondary parts but the mechanical complexity of the overall arrangement increases rapidly so that these types are much less commonly employed than the preceding ones. One example is effectively two double-sided short primary machines combined, which saves two portions of primary core steel, since the flux travels straight through the central primary.

The second basic form is the tubular motor. This form also has different varieties but since the secondary is bounded by the primary in two dimensions (a restriction similar to that found in rotating machines) these varieties are limited. The two normal variants are the long-secondary short-primary type, and the short-secondary long-primary type, of which the former is the most often employed.

DESIGN APPLICATIONS

The first important characteristic in most applications is the speed of the motion required. The rotating machine speed is given by the expression: revolutions per second equal supply frequency divided by number of pole pairs; or

$$\text{rev/s} = \text{Hz}/p$$

The equivalent expression for a linear induction motor is:

synchronous speed (m/s) =
  frequency (Hz) $\times$ 2 $\times$ pole pitch (m)

Therefore, whilst both the rotating and the linear motors have synchronous speeds dependent on supply frequency, the rotating machine speed depends also on the number of poles whereas the linear motor speed depends upon pole pitch. Thus a 2-pole 50-Hz rotating machine has a synchronous speed of 3 000 rev/min, whether it is a miniature motor or a turbogenerator, but a linear motor with a 150-mm pole pitch and 50-Hz supply has a synchronous speed of 15 m/s whether it has two or two-hundred poles. The result of this fundamental relationship is that many applications demand either the use of a frequency converter, or necessitate working at large slip values where the motor is relatively inefficient.

It is true that shortening the pole pitch reduces the speed of a linear motor, but there are practical difficulties in the way of designing polyphase motors with pole pitches less than say 25 mm for power applications, because insulation

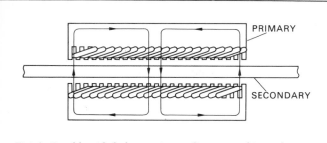

Fig. 3. Double-sided short-primary linear machine; showing the primary (in this case the moving part) energised so that the mmf of both windings act together to drive flux through the secondary.

takes a large share of the winding space, and magnetic leakage becomes a major part of the total flux. The motor thus becomes very inefficient. It is worth noting that these remarks also hold true for rotating machines whether the pole pitch is small because the machine is small or because a large machine has a very large number of poles.

The second characteristic to have an important bearing on the design of the linear motor is permissible length of the airgap. The length of the flux path in air or non-magnetic material in a slow speed single-sided linear motor with an aluminium or copper secondary may be of the same order as the pole pitch. If, in fact, the secondary has no magnetic material the effective air gap becomes pole-pitch/π. Even in good cases the linear motor is not likely to have a ratio of pole-pitch to air-gap of greater than 50, because whatever the motor designer might require is subject to the mechanical limitations of the application, whilst most rotary induction motors have a ratio greater than 100, since the whole of the motor is under the designer's control. The effect of this is to lower the power factor at which the linear motor can work, or for a given power output to increase the input of volt-amps from the supply.

A linear motor has the best chance of using a short air gap if the secondary is made of magnetic material, because the gap can then be as small as the application demands for mechanical clearance. The electrical conductivity of the secondary in this case is substantially less than that of the normal conductor materials such as copper and aluminium, so that the efficiency gained by having a short air gap is offset by the very high secondary resistivity.

Magnetic attraction

The third characteristic to require attention is that of the magnetic attraction. In a rotating machine the magnetic forces tending to pull the rotor towards the stator on one side are balanced by equal forces on the other side so long as the air gap is uniform all round and the mmfs are balanced. If the rotor is displaced so that the air gap becomes shorter on one side than on the other, then there is a stronger force exerted on the side with the shorter air gap than on the other and the resultant force is the well known unbalanced magnetic pull. The stator is, however, only reacted upon by the resultant or unbalance of two opposing forces. In the flat block types of linear motor, the whole of the force due to the total working flux acts in one direction. This force can be several times as large as the working thrust and is variable in amplitude for different values of slip. Further, the force is not uniformly exerted over the face of the motor and even this non-uniformity is variable with slip. Thus, firm mechanical restraint is necessary in order to maintain the desired air gap length, and in large motors this may become a major design restraint.

The magnetic attraction is not the only force operating if the secondary has a nonmagnetic conducting member. Between the primary and the secondary conductor there exists a force of repulsion which, depending on the type of motor, may be important or negligible. In a double-sided motor repulsion is exerted by each primary on the secondary, and if the secondary is equidistant from the primaries these forces will balance. If the secondary is offset then there is an unbalanced magnetic force on the secondary, and this resultant tends to increase the eccentricity. There is a strong attractive force between the two primary blocks regardless of the balance or otherwise of the forces on the secondary. In the case of a single-sided motor with a totally nonmagnetic secondary there is repulsion between the primary and secondary. If, on the other hand, the secondary contains magnetic material as well, then there is attraction between the primary and secondary magnetic material in addition to the repulsion existing between the primary and the secondary conductor. The magnitude of each of these forces changes with slip and, since the repulsion depends upon current and the attraction on air gap flux, it is quite possible for a motor to experience a net repulsion at standstill and slow speeds, changing as speed is increased to a net attraction at low values of slip. The problems of magnetic attraction and repulsion have to be resolved by suitable mechanical design of the motor parts and appropriate restraint between them. The tubular types of motor avoid most of these effects.

Mismatching

The next consideration is that of the basic mismatching of the primary and secondary lengths in the direction of motion. This results in an inefficiency since only a part of either the primary or the secondary can be working effectively at any given instant. Short primary motors normally are more efficient in this respect than short secondary machines, because in general the cost of the primary is higher than that of the secondary for a given length.

In addition to the cost implications, long-primary short-secondary machines have electrical inefficiency problems, since an appreciable length of the primary must be energised before the secondary reaches it. Only the portion of the primary opposite the secondary at a given instant produces useful work, the rest producing only losses.

The mismatching problem is only significant if the length of travel is great, but it can be a major design consideration in extreme cases, such as traction applications.

End effects

The fifth consideration, which is of particular significance for linear motors required to operate near to their synchronous speeds, is that of the end effects. The obvious physical difference between a rotating machine and a linear machine is that, in the direction of motion, the linear machine has a beginning and an end. This physical difference results in electromagnetic phenomena which are absent from the rotating machine.

Consider a portion of a linear motor secondary approaching an energised primary. As a portion of the secondary comes into the magnetic field, voltages are induced in it which set up currents attempting to stop the motion. When, with continuing motion, the current is fully established, it resists any attempt to slow down the motion; that is, it acts in the normal motoring mode. Later this same part of the secondary comes to the end of the primary and the magnetic field diminishes. The secondary current now attempts to continue in the previous state and maintains the field until the energy is exhausted. Therefore, there is an opposing force at entry and a removal of energy on leaving the primary, and both of these effects show up as losses which decrease the efficiency of the machine.

Both of these effects are in accordance with Lenz's law, which indicates that any electromagnetic circuit will resist any attempt to change its existing state, or put more conventionally: *the direction of the induced emf is such that the current set up by it tends to stop the motion or change in the flux producing it.*

The end effects are the result of motion and, therefore, are not evident at standstill but increase with speed until they determine the maximum speed at which a particular linear machine will motor. (Fig. 4.)

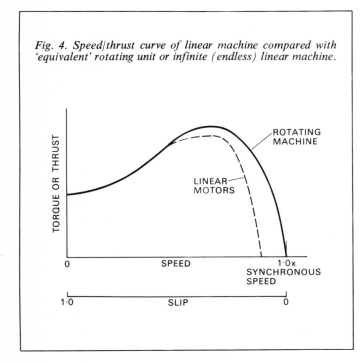

Fig. 4. Speed/thrust curve of linear machine compared with 'equivalent' rotating unit or infinite (endless) linear machine.

The smaller the number of poles on a linear motor primary block, the more severe are these end effects, so that they are very marked indeed for a two-pole primary, in which case the secondary might not achieve much more than half of the synchronous speed even on no-load. In general, if the primary is wound for six or more poles the maximum no-load speed will be within the range 5–10% of the synchronous speed for units with low resistivity secondary conductors. For pole numbers greater than ten or twelve the speed loss due to end effects for the same cases is very small. High resistivity secondaries produce higher end effect losses.

Efficiency

The normal definition of the efficiency of an electrical machine is given as the ratio of output power to input power for any particular working condition. It follows that both a synchronous condenser and a torque motor have an efficiency of zero because the former has no power component in its output and the latter, though it produces torque, does not move, and again no power is forthcoming. Therefore, two types of rotating machine, both exceedingly useful in their applications, have zero efficiency. These cases simply demonstrate the fact that considerations other than efficiency often determine the effectiveness or fitness-for-purpose of an electrical machine.

This is frequently the case with linear induction motors and so some of the other considerations are listed below to help in the evaluation of the effectiveness of a linear motor drive.

(1) The direct thrust which the application requires is available between the stationary and the moving parts without the need to employ any of usual intermediaries, *eg* ropes, belts, chains, gears, pulleys, couplings, wheels etc, and also without the restrictions which these items impose.
(2) There are no rolling or sliding parts in the motor. This implies no wearing parts and, therefore, no maintenance.
(3) In some applications the ability to transmit a force across a mechanical gap makes possible a process which was previously impossible. In other applications the operation can be performed with a degree of repetitive accuracy unobtainable by other methods of drive.
(4) The force may be applied at the point where it is most needed, thus removing the need for transmission shafts and linkages etc. The type of application will determine the benefit to be obtained from each of these factors.

To give an unbiased picture it is necessary to state the adverse factors which may also apply.

(1) Depending on the length of motion required the cost of a continuous primary or secondary may be very great. Normally if the length of motion is great then the primary is kept short, and the secondary is extended to keep the cost as low as possible, but this arrangement will normally require sliding contacts to bring the electrical power to the primary.

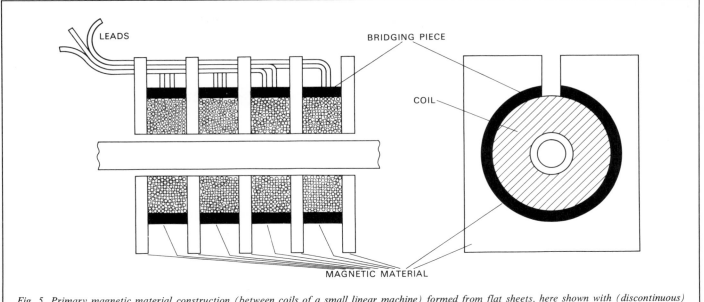

Fig. 5. *Primary magnetic material construction (between coils of a small linear machine) formed from flat sheets, here shown with (discontinuous) bridging pieces between the sheets.*

(2) the basic speed of a linear motor fed from a normal power supply is too high for many applications; hence frequency-conversion apparatus or voltage regulating means are required. Alternatively, if the drive permits, the motor may be fed at normal frequency and voltage but in controlled bursts of energy; *eg* five cycles in every twenty. Since this gives an impulsive drive it is usually suitable for drives which have a substantial load inertia to smooth out the resulting motion.

(3) The demand from the supply is likely to be considerably larger than for a rotating machine doing the same job because of the long air gap usually required, which results in low power factor and lowers the efficiency.

(4) The cost of linear motors is high unless standard units can be designed into a system, because usually a special design is required for each application. Often some experimental work is required in advance of the construction of the main motor, and this can be expensive with large machines.

DESIGN AND CONSTRUCTION

The mechanical design is often determined almost completely by the application, but a number of general comments can be made.

Flat block primaries

A flat block primary usually consists of a core made of rectangular electrical sheet steel laminations slotted along one edge, a mechanical arrangement for holding the core together, and an electrical winding which can be any of the types used for conventional induction motor stators. The windings can be insulated for any normal supply voltage and arranged for polyphase operation, usually three-phase or three-phase supplied from a capacitor split single-phase.

If the block is a small one then each lamination can extend to the full block length. This is convenient with blocks up to about 750 mm long, but longer ones are normally composed of interleaved laminations built up like a brick wall. Individual laminations must be designed so that, when built, continuous slots are formed. The core may be held together by bolts through holes in the laminations with side plates or angle pieces if necessary to stiffen the structure. The bolts must be electrically insulated if the motor is designed for continuous duty at power frequencies, but this is unnecessary for short time rated motors. Alternative means for holding the core together (such as welding at intervals across the back or bonding the laminations with a resin adhesive) may be employed, and the size of the core and the motor duty determine the most appropriate method.

The electrical winding can be of the single or two-layer conventional type or of the gramme ring type. The latter type is useful if a long polepitch has to be used and space for endwindings is limited. This winding is somewhat handicapped by extra magnetic leakage but has the advantages of space saving and the possibility of reconnection for changing the pole-pitch, a feature shared with the tubular motor. The reconnection arrangement is complex if several speeds are required from a long motor because two connections are required from almost every coil.

If the pole pitch is progressively increased along the motor and a short rotor employed, then the motor will act as a linear accelerator. Windings other than gramme ring types result in coil sides extending beyond the core at each end, or in some part-filled slots unless special measures are taken to avoid this. The winding is usually mechanically stronger if all the coil sides are in slots. In the case of small machines (where complete encapsulation is feasible and suitable) mechanical support for the windings is not a problem.

Flat block secondaries

The secondary may be constructed from slotted laminations, and carry a cage winding consisting of conducting bars and end connections. This is uneconomic for most purposes and the normal form of cage is a uniform flat sheet of conducting material. For a large number of applications, a single sheet of aluminium forms the entire secondary. For others, particularly when single sided motors are used, magnetic material is necessary on the side of the secondary conducting sheet remote from the primary in order to reduce the air-gap to a reasonable length.

Tubular primaries

The primary winding of the tubular motor is composed of ring shaped coils which can be connected in various ways like a gramme ring winding. Unlike the gramme ring winding, the coils of the tubular motor encircle the secondary. The primary magnetic material may be either slotted laminations used in groups around the coils or, for small machines, it may consist of flat sheets between the coils sometimes with bridging pieces between them. The bridging piece must not be a complete ring if the magnetic material is also conducting, otherwise it will act as a short-circuited secondary winding and produce losses and heat.

Tubular secondaries

A tubular secondary can be a rod of conducting magnetic material or a tube of similar material. In either case a thin non-magnetic conducting outer layer may be incorporated. If the conductor is liquid the motor becomes a pump in which the secondary has a fixed magnetic core and a liquid metal (sodium, potassium or alloys) moving conductor. In order to contain the conductor, a sheath made of non-magnetic non-conducting material is required. In practice, to meet other requirements such as mechanical strength and corrosion resistance at high temperatures, a thin-walled stainless steel tube has been employed.

ELECTRICAL DESIGN

No comprehensive method for predicting linear induction motor performance has yet been published, which deals adequately with the various types and sizes. The most usual approach to design is therefore to estimate the parameters of the machines equivalent circuit, and base the performance on this after making due allowance for the end effects.

The equivalent circuit uses phase values, and all the quantities are referred to the primary, as is normal for rotating machine calculations. The equivalent circuit assumes balanced values in the separate phases of a multiphase machine and this is not true for most linear motors. The degree of unbalance only becomes serious with small numbers of poles because the unbalance is caused by the end effects including any unbalance in the arrangement of the coils at the ends of the motor. The primary resistance per phase is readily calculable from the physical dimensions of the coils and conductors. The primary reactance has several components depending on the flux leakages in various areas. (Fig. 6.)

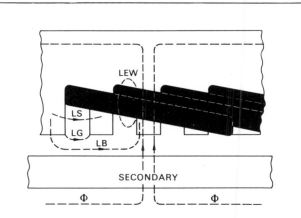

Fig. 6. Flux leakages in various parts of the primary of a linear machine: L_S slot leakage; L_{EW} end winding leakage; L_G gap leakage; L_B belt leakage; Φ useful flux.

Each component is calculable by normal design methods but the relative importances may differ from those of rotating machines due to the dimensions occurring in linear machines. Slot proportions and core length determine the slot leakage; coil-pitch, winding type, and the proximity of structural magnetic material determine the endwinding leakage; whilst the slot-pitch, core length, and air gap length determine the gap and belt leakages.

Secondary reactance may be calculated in a similar manner if the winding is in slotted laminations. If the secondary consists of a sheet of non-magnetic conductor then the reactance is more difficult to assess, but since it will be small it can usually be neglected with little error. A sheet of aluminium backed by iron is a more difficult case, and some allowance should be made even though this reactance is still small compared with that of a winding in slots. Estimation of the secondary resistance, variable with slip in the equivalent circuit, presents problems when the conductor is a non-magnetic sheet, and more if it is also magnetic. Both cases have been the subject of treatment for rotating machines, but for linear motors the problems are considerably increased by the fact that the end effect losses must be accounted for in a complex variation involving slip, number of poles, and the secondary time constant. In the absence of a fully satisfactory method for estimating R_2 a similar method neglecting end effects must be used and an allowance made for the losses caused by these effects.

Magnetising resistance is calculated, as for a rotating machine, from the iron losses and tends to be of rather small importance since the majority of applications demand long air-gaps and relatively unsaturated iron circuits. Accurate assessment is complex, again because of the end effects, since the core flux density must build up from zero, neglecting fringing at each end of the core, to densities at some points exceeding those that would be expected in the equivalent rotating machines.

The magnetising reactance calculation is, however, much more important. This is because an induction motor with a long air-gap has a low magnetizing reactance: thus a high magnetizing current producing a large voltage drop in the primary impedence, and therefore a low air-gap voltage E_A. It should be noted that excessive values of R_1 and X_1 produce a similar effect. A high R_1 value could come from an attempt to reduce the copper content of a winding having a very short time rating, the result being that increased power would be required by the motor only to be lost in R_1. High X_1 values can result from high values of the ratio air-gap-length to pole-pitch. Since E_A determines the secondary current and thus the output, an accurate assessment of X_M is vital to accurate output determination. The formulae used in rotating machine design and that given in reference 1 give good results for flat block motors.

As with all machine design problems, test evidence from a machine of similar proportions is the best check on a new design, but with linear machines tests on machines other than at standstill are difficult or expensive to obtain just because the motion is linear. Disc test rigs have been made to test the dynamic performance of linear motors, continuous motion being obtained without the need for an extended track. (Fig. 7.)

APPLICATIONS

A successful application using a double-sided short-primary motor is in operation at the Motor Industries' Research Association testing station in the UK. Car manufacturers carry out extensive testing of chassis and components under impact conditions in order to check the strength and safety of their vehicles. One test involves running a vehicle at a controlled speed into a large concrete block and recording, by extensive instrumentation and photography, the effect of the impact on the car body and its dummy occupants. The vehicle under test is accelerated along the test track by means of a quick release tow bar fixed to a trolley running in a duct below the track level. The linear motor stator is mounted on the trolley and propels both itself and the test vehicle by its reaction with the fixed aluminium alloy rotor plate. When, after reaching a steady speed of 13.4 m/s (30 miles/h), the vehicle is about 3 m from impact, the towing mechanism disengages, the trolley runs under the concrete block and is mechanically braked, whilst the vehicle crashes.

The linear motor has a maximum input of about 1800 kVA at 3.8 kV and produces an initial thrust of 1680 kgf. The rotor plate is 25 mm thick and the total effective air gap about 32 mm.

Conveyors

A number of conveyor systems are in operation using single-sided short-primary motors. One example concerns assembly jigs, which carry components undergoing successive operations at stations along an extensive track. When assembly is complete the component is removed from the jig, which must then be returned to the starting point to pick up the next item. This return motion is powered by the linear motors. The jigs each weigh about 500 kg, but a

Fig. 7. Part of disc test rig (during assembly) used to evaluate dynamic performance of linear motors.

relatively small force is required to move them along the conveyor track. The linear motor primaries are positioned at intervals along the track and, as a jig approaches, a dry reed proximity switch starts a timer set to the ideal passage time. If the jig is travelling too fast it will pass before the motor is energised, and if it is travelling too slowly the motor will give it an extra boost. Closely spaced primaries at the beginning of the travel and reversed primaries at the end complete this arrangement, which depends on the combination of high inertia and a constant low load to produce the correct speed. The motors work with a total air gap of 4.6 mm, the secondary consisting of 3 mm thick aluminium alloy sheets backed by steel. A standstill thrust of 34 kgf is produced and the rolling resistance of each jig is 14–18 kgf. The synchronous speed of the linear motor is considerably higher than the desired speed of travel.

A further application in this category is the long travel motion of cranes. In this case short primaries are mounted on the crane and the secondaries are the steel girders on which the crane travels. Again, relatively high synchronous speeds are used, and the speed is controlled by either voltage or on-off energy supply. In the latter case the crane inertia smoothes out the motion.

Impact testing

Another linear motor has been supplied to the National Experimental Laboratory, in Scotland. This motor, in which the stators are fixed, acts as an accelerator for a machine designed to test ropes under impact load. Aluminium rotor plates are fitted on two sides of a carriage which runs almost friction-free on air bearings. A pair of stator blocks on each side accelerate the carriage until the 3.66 m long rotor plates have passed through them. The carriage then runs free for a short distance before picking up the pretensioned rope, stretching it and breaking it. The carriage is then brought to rest mechanically. From the known energy of the carriage at the time of contacting the rope the dynamic load properties of the ropes can be checked. The ropes concerned are mainly of nylon or other man-made fibres.

There are now a large number of examples of applications of linear induction motors: ranging from aircraft launchers and artillery to small actuators and sliding door gear. A list of references follows to assist further study.

REFERENCES

1. E R Laithwaite. Induction machines for special purposes. *Newnes* 1966.
2. A wound rotor motor 1400 ft long. *Westinghouse Engineer* **6**, 160 (1946).
3. L A Byesti. An electromagnetic chute for molten metal. *Elektrichestro* no. 5, 74 (1962) (translated by *Pergamon Press*).
4. J C West, B V Jayawant. A new linear oscillating motor. *Proc. IEE* 109(A) 292 (1962).
5. W Johnson, E R Laithwaite, R A C Slater. An experimental impact extrusion machine driven by linear induction motor. *Proc. I Mech E* Part 1 179 15 (1964).
6. L S Blake. Conduction and induction pumps for liquid metals. *Proc. IEE* 104(A) 49 (1957).
7. R L Russell, K H Norsworthy. Eddy currents and wall losses in screened rotor induction motors. *Proc. IEE* 105(A) 163 (1958).
8. E R Laithwaite, D Tipping, D E Hesmondhalgh. The application of linear induction motors to conveyors. *Proc. IEE* 107(A) 290 (1960).
9. E M Freeman. Levitation or attraction due to a travelling field. *Proc. IEE* 1968 115(6) 894.
10. E M Freeman. Travelling waves in induction machines, input impedance and equivalent circuits. *Proc. IEE* vol. 115 no. 12 Dec 1968.
11. W J Adams, B A White. Applying linear induction motors. *Automation* June 1967 74.
12. Steps towards practical linear motor propulsion in France. *Electrical Review* 2 Jan 1970.

Chapter 15

Starting and Speed Control of d.c. Motors

R M Carter BSc
AEI Semiconductors Ltd

The subject of speed control is vast, and it is only necessary here to outline the possibilities. A large amount of space has been devoted to thyristor converter systems, which have been the main growth area in recent years. For large industrial drives the choice is now between a thyristor armature drive and a Ward Leonard system with a thyristor controlled field. For low power industrial drives, single-phase thyristor converters have become almost universal. Some transistorised systems are likely to attract greater interest in the future, as development of batteries extends the use of mobile and portable battery powered equipment.

STARTERS

Fig. 1 shows a conventional starter circuit for a shunt-wound motor. Interlocks, and protective devices have been omitted for simplicity. The operating sequence is as follows: The isolator IS1 is closed, the field contactor RL1 is closed, and the field rheostat RV2 is set to minimum resistance. The starting resistor RV1 is set to maximum resistance and the armature contactor RL2 is closed. The operator then observes the armature current and progressively reduces the setting of RV1 so as to maintain the armature current between specified limits. When all the resistance has been removed from the armature circuit the machine speed is adjusted to the required speed by means of the field resistor RV2. Interlocks are usually provided to prevent the closure of the armature contactor if either of the resistors is incorrectly set or if there is no field current. A third contactor is frequently used to short circuit the starting components RL2 and RV1 after the motor has run up. This contactor may be operated by a limit switch on RV1. Where a dynamic breaking resistor is used the dynamic breaking contactor is interlocked with the armature contactors. A further interlock is necessary to prevent the operator from leaving the machine running with part of the starting resistor still in circuit, as it is not usually continuously rated. This can be achieved by arranging the controls so that the start button has to be held in until the resistance has been reduced to zero. The machine is stopped by opening the armature and

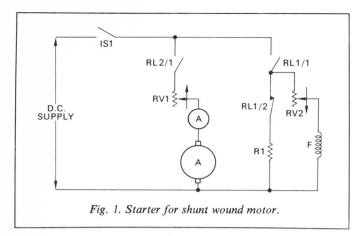

Fig. 1. Starter for shunt wound motor.

field contactors if no dynamic breaking is fitted but if there is dynamic breaking the field must be locked in until the motor has stopped. When the field contactor opens it connects a discharge resistor to prevent excessive voltage being developed by the field inductance. Both contactors must be of a type suitable for breaking inductive d.c. circuits. Arc chutes are necessary even at quite low currents and larger contactors also have magnetic blow out coils.

Starting resistors are usually constructed from grids of wire or strip connected to studs. The number of studs required depends on how much variation in armature current can be tolerated during starting. Fig. 2 illustrates the design of a starting resistor with n studs. The supply voltage V_S, motor back-emf E and armature circuit resistance R determines the armature current I_A

$$V_S = E + RI_A$$

If the armature current is maintained between a lower limit I_1 and upper limit I_2 then the total resistance including armature resistance at stall is

$$R_1 = \frac{V_S}{I_2}$$

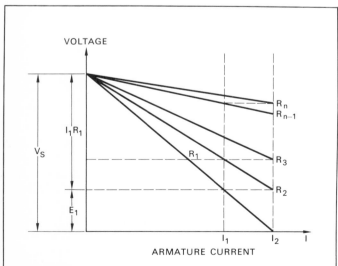

Fig. 2. Design characteristics of starter resistor used with field tappings to control a series wound motor.

This current allows the motor to accelerate to a speed at which the back-emf is E_1 and the current has fallen to I_1 at this point the current can be increased to I_2 by reducing the total resistance to R_2 so that $R_1 I_1 = R_2 I_2$ similarly the next resistance value R_3 is given by $R_2 I_1 = R_3 I_2$ etc until the final stud of the starter is reached and only the nth resistance value is that of the armature itself; $R_{n-1} I_1 = R_n I_2$.

Multiplying equations together

$$R_1 I_1^{n-1} = R_n I_2^{n-1}$$

ie

$$n - 1 = \frac{\log R_1/R_n}{\log I_2/I_1}.$$

SPEED CONTROLLERS

Electromechanical speed control systems are in most cases sufficiently simple for the diagrams to be self explanatory.

Control by series resistance. Fig. 1 includes two forms of speed control. The variable resistor in series with the armature is an adequate control for some very small drives which have fairly constant loads. It is frequently used with a permanent magnet motor to drive *eg* toy trains. Field weakening is an important method of control for larger motors as the power loss in the field resistor is relatively insignificant. This method cannot control down to zero speed and is usually limited to about a 4:1 speed ratio. The problem at weak field is that armature reaction and residual flux prevent the field current from adequately controlling the effective field flux.

Variable transformer and rectifier. Fig. 3 shows an armature drive in which a variable voltage is derived from the slider of a variable transformer and rectified by a bridge rectifier. The motor may have a separate rectifier for the field or may be series wound. Similar systems with tapped transformers are used for railway traction where the supply is a.c.

Tapped field series motor control. Fig. 4 shows a combination of a starting resistor and field weakening. Both the resistor taps and field taps can be on the same controller.

Series–parallel switching. In multimotor traction systems the motors can be connected in series initially and switched in parallel when sufficient speed is reached. Batteries may be switched in the same way.

Closed loop control

The basic element which makes closed loop control possible is the amplifier. The traditional devices such as rotating amplifiers, magnetic amplifiers, and valves have now been replaced almost completely by semiconductor devices. The introduction of transistors and thyristors has caused such a dramatic improvement in the cost, size, weight and reliability of electronic control equipment that high accuracy systems

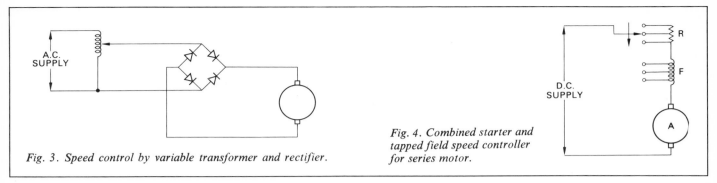

Fig. 3. Speed control by variable transformer and rectifier.

Fig. 4. Combined starter and tapped field speed controller for series motor.

can sometimes replace the crudest electromechanical systems at little or no extra capital cost.

Analogue computing techniques are usually used, and the low cost of transistor and integrated circuit operational amplifiers means that protective systems such as current limit and voltage limit are usually economically justified. There are two methods of providing a limit which over-rides the speed control: the 'spill-over' current limit system and the two-loop system. These systems are applicable in principle to all the power amplifier systems described later, but in the case of the Ward Leonard system it is usually necessary to limit both the generator field current and the armature current.

In Fig. 5 a stabilised d.c. supply provides a speed reference and a bias voltage for the current limit circuit. During normal operation the diode D1 is reverse biased and there is no current feedback. The system is then controlled by the speed reference through R1 and the speed feedback through R2. Stabilisation is by means of R4, C1, R5 and C2. When a predetermined armature current is exceeded the current feedback exceeds the bias boltage and the diode D1 conducts so that there are two feedback loops in operation. The current feedback must over-ride the speed control and so the steady state gain of the current loop must be an order of magnitude higher than that of the speed loop. The ripple filter and R6 and C3 determine the transient response in current limit.

The two-loop system requires two amplifiers and two sets of stabilising components. It is therefore usually more expensive, but for high power systems the difference is not significant. An advantage is that the current limit action is faster, and is achieved by limiting the output of the speed error amplifier which is the input of the current regulator. It is considered that the 2-loop system is both easier to design and to commission.

If the speed range of the motor is to be extended by field weakening, either of these armature drive systems may be used with the addition of another spill-over system which weakens the field when a certain armature voltage is reached. Armature voltage limit is also required and maximum and minimum limiters for the motor field current.

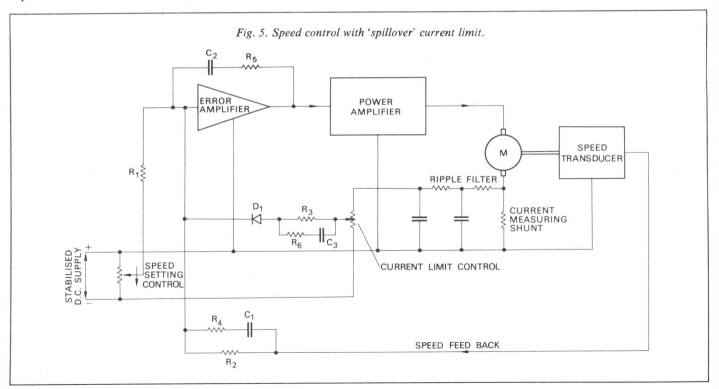

Fig. 5. Speed control with 'spillover' current limit.

Current feedback from a shunt is not possible if isolation is required between the motor and the control circuits. In this case a 'd.c. current transformer' (a type of magnetic amplifier) may be used. If the armature is driven from a fully controlled thyristor bridge then a conventional a.c. current transformer in the a.c. line can be used with a bridge rectifier.

There are several methods of obtaining a speed signal. A tachogenerator (which is usually a d.c. generator with a permanent magnet field) and devices to compensate for changes of voltage with temperature may be coupled to the motor shaft. A flexible coupling must be used, as the clearance in the motor bearings is much greater than in the tachogenerator. The coupling must allow axial movements and misalignment but must be torsionally rigid and have no backlash. An alternative device which is becoming more popular is a digital tachogenerator. This comprises a toothed wheel mounted on the motor shaft close to a magnetic or optical device which counts the teeth as they pass. The pulse repetition frequency is proportional to motor speed and can be easily converted into a d.c. signal for use in an analogue circuit. This is cheaper to manufacture and install and in some cases gives better performance.

A common method of deriving a speed signal from a motor with a fixed field is to compute the motor back-emf from the armature voltage and armature current. This technique can only give accuracies of about $\pm 2\%$ whereas tachogenerator systems may give accuracies of better than 0.1%.

Ward Leonard and similar systems

A multi-Ward Leonard system is shown in Fig. 6. A constant speed motor drives a d.c. generator, the output of which can be controlled by means of its field current from full positive to full negative armature voltage. With the generator coupled to the work motor a full range of positive and negative speeds is possible, and since armature current can flow in either direction the motor torque can be reversed and regeneration obtained; ie kinetic energy stored in the inertia of the work motor and load can be fed back through the generator and main motor into the supply. Where regeneration is required this system is still widely used but it is less efficient than a thyristor convertor drive as there are losses in three machines. The main advantage over a thyristor converter is the better power factor at low speed. For regenerative drives the cost of the Ward Leonard is similar to that of a thyristor convertor.

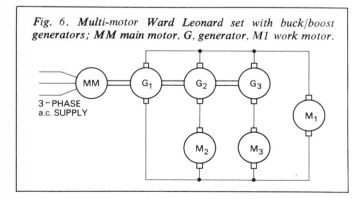

Fig. 6. Multi-motor Ward Leonard set with buck/boost generators; MM main motor, G, generator, M1 work motor.

In a multimotor system, such as a process line, a single motor generator set controls the speed of a group of motors by control of the main generator fields. Relative variations in speed between different motors in the system can be obtained by auxiliary generators used to buck or boost the voltage to individual motors. The system could be used for a reeling drive in which material is unwound from a reel driven by one motor, passed over rollers driven by a second motor, and wound up on another reel driven by a third motor. As reel diameters changed so the second and third motors would be required to change their speeds relative to the speed of the first. This speed change is achieved by varying the fields of related generators. (Fig. 6)

Systems in which the main generator is driven by a heat engine are similar to the Ward Leonard systems described above but regeneration is not possible. Such systems are used to provide electric transmission on vehicles such as cranes and excavators. In these systems the engine speed is variable, and is the main method of control.

Thyristor converter systems

Rectifier and converter supplies differ from d.c. supplies in that there is ripple in the output. This can be filtered by inductors and capacitors but it is usually uneconomic to do so, and the ripple has to be taken into account when designing the motor. The most usual modifications are lamination of the entire stator instead of just the polepieces and the use of larger brushes. When the converter is single-phase the ripple is frequently sufficient to necessitate using a larger motor frame size. There are many circuits in common use, and many more circuits exist for systems incorporating multisecondary transformers. In half-controlled systems some of the devices are rectifiers and some are controlled rectifiers. Such a system is cheaper than the fully controlled system but will only work in the rectifying mode and cannot invert or regenerate; ie power from the load cannot be transferred back into the supply.

In one circuit a half-wave fixed supply is applied to the field through a diode. Since the field is inductive a second diode known as a commutation diode (or freewheel or flywheel diode) is required. The armature also has a half-wave supply through a thyristor: the voltage is controlled by phase control of this thyristor. A second commutation diode, though not always used because the armature inductance is relatively small, improves the form factor at low speed. This is the cheapest type of drive and was widely used with thyratrons and selenium rectifiers before silicon devices became available. The bad form factor causes bad commutation and short brush life unless an inductor is used in series with the motor armature. Also the effect of half-wave loads on the a.c. supply can cause trouble.

Most drives up to about 7.5 kW (10 hp) now use a single-phase full wave bridge. The field is supplied by four diodes which rectify the a.c. supply voltage to give a field voltage. The field inductance smooths the current on one half cycle of the a.c. supply; two diodes conduct to give the first half-cycle current, and on the other half cycle the other diodes conduct to give the current. The armature has relatively low inductance and resistance as most of the impedance of an armature is provided by its back-emf. Fig. 7(4) shows the armature voltage when the motor is stationary and power is applied to start it. By firing the thyristors late in their respective half cycles only a small fraction of the available voltage is applied.

When the motor is running its back-emf modifies the wave forms. Fig. 7(6) shows the armature voltage when the motor is running at high speed on load. There are three sections of the waveform. At the beginning of each half cycle the armature current established in the previous half cycle has commutated to the commutation diode so that the armature voltage is the forward volt drop (about 1V) of the diode. When the current drops to zero the armature is on open circuit and the back-emf with its commutation ripple is seen. When a thyristor is triggered the armature is connected across the supply and a part of the sine wave is seen. The corresponding current is shown in Fig. 7(7). On light load the duration of conduction is less and the commutation diode may not come into use as in Fig. 7 (8 and 9).

The full wave supply to the field is usually used to synchronise the firing circuit and supply power to the control amplifiers through a dropping resistor and zener diode. Alternatively the field may be supplied with a half wave supply by connecting it across one of the bridge diodes. This may eliminate the other two diodes, and the lower voltage on the field winding usually reduces the cost of the winding.

In a circuit now very popular for fractional-power (to 750 W) motors, cost is reduced by using only one thyristor but the voltage characteristics applied to the thyristor and the armature leave only a very short time available for the thyristor to turn off. The field current is an important factor in determining the turn off conditions and so the circuit can never work reliably unless an inductive load is connected across the bridge. The thyristor must be selected for fast turn off. A variety of auxiliary circuits have been devised to assist reliable turn off. Since turn off is also affected by stray reactances, equipment of this type is difficult to design.

At powers above about 7.5 kW (10 hp) single-phase equipment becomes inconvenient from the supply point of view and the 3-phase system becomes more attractive. The lower ripple amplitude gives higher motor efficiency and the higher ripple frequency allows a faster response time. The field can be supplied half wave by connecting it across one of the bridge diodes as with the single-phase circuit. The field bridge may be replaced by a half controlled bridge if field weakening is required.

Commutation diodes in half controlled systems are a common cause of failure. The two main reasons are current overload and excessively fast turn off. When the motor is stalled and overload current is passed the proportion of armature current carried by the commutation diode is much higher than in normal conditions. After the overload,

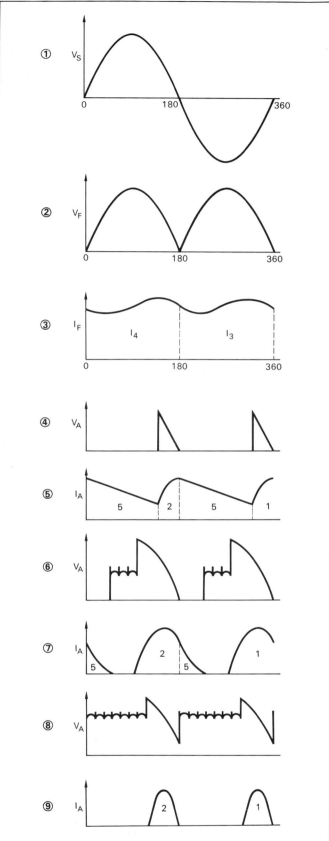

Fig. 7. Waveforms for half-controlled full-wave single-phase system: (1) supply voltage; (2) field voltage; (3) field current; (4, 5) armature at stall or on load at low speed; (6, 7) armature on load at high speed; (8, 9) armature on no-load at high speed.

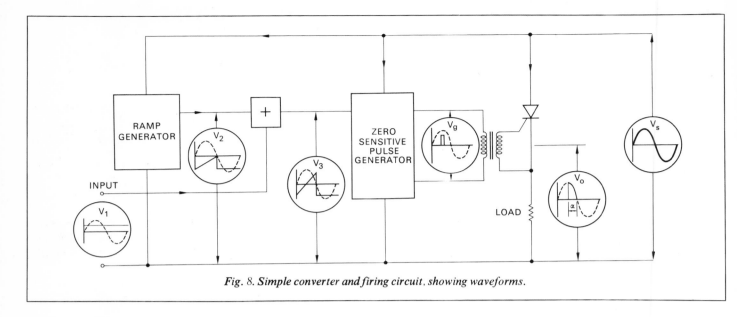

Fig. 8. Simple converter and firing circuit, showing waveforms.

when the thyristor output is reduced to zero, all the armature current is carried by the commutation diode. If the motor is caused to run in the reverse direction the commutation diode forms a short circuit. These conditions must be considered in order to select a diode of sufficient current rating. When the commutation diode is conducting and one of the thyristors is triggered, reverse current is passed through the diode until it turns off. The rate of rise of thyristor current is determined by the a.c. circuit inductance. This rate of rise together with the stored charge of the diode will determine the power dissipated during turn off. Since inductance has to be included in the a.c. circuit for other reasons this is usually the best solution as fast recovery diodes are expensive. However, if the supply frequency is 400 Hz or more a fast recovery diode will usually be necessary.

REGENERATION

Regeneration or inversion can be achieved either by reversing the voltage or the current. In the case of a field drive the energy stored in the magnetic field can be transferred back into the supply if the converter can be made to conduct during the negative part of the a.c. cycle instead of the positive. This is only possible if all the rectifying devices used are controlled rectifiers. In order to obtain positive current at negative voltage the load must generate a back-emf. In the case of a field this emf is determined by the rate of fall of current in the inductance. When the thyristors in a single-phase circuit are triggered early in their respective half cycles the average voltage is positive, and the current builds up to a positive value after a time depending on the inductive time constant of the field. When it is required to reduce the current to zero, the thyristor firing time is changed to a point just before the end of each half cycle. The average voltage is then negative until the current has dropped to zero. The thyristors then stop conducting except for a short time after they are triggered, and the current remains at a negligible level.

When a fully controlled bridge is used to drive a motor armature the back-emf determined by the speed and flux of the machine can be made negative by reversing the field or armature.

The 3-phase circuit functions in a similar way but the ripple frequency is 6-times instead of twice the supply frequency. In both systems it is necessary to trigger two thyristors simultaneously to initiate conduction.

For high power drives a series-connected multi-pulse converter circuit has advantages. By using a transformer with two secondaries having a 30° phase displacement between them it is possible to double the ripple frequency. Also the system simplifies the voltage sharing problem. An alternative arrangement is to connect the two bridges in parallel using a centre tapped inductor. These systems become economic when the drive is large enough to require its own supply transformer.

Fig. 8 shows a simple converter and firing circuit with the voltage waveforms occurring at half output. A d.c. input signal V_1 is added to a ramp waveform V_2, which is derived from and synchronised with the anode supply. The resultant waveform V_3 is used to control a pulse generator which produces a pulse when its input changes from negative to

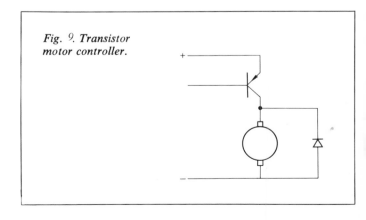

Fig. 9. Transistor motor controller.

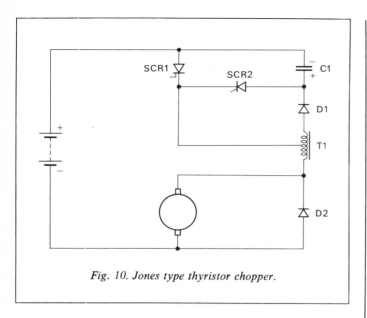

Fig. 10. Jones type thyristor chopper.

positive. The output pulse V_g triggers the thyristor through a pulse transformer. The load voltage V_0 is a part of the supply voltage V_s. As the input V_1 increases so the angle α for which the thyristor conducts increases and the mean load voltage increases. In the simpler firing circuits the ramp is linear. An improvement is to use a ramp which is generated by integrating the anode supply voltage so that for a sinusoidal supply the ramp is a half cosine. The firing angle α is a linear function of the input V_1 and the load voltage is a cosine function of the input. The incremental gain of the system at any one operating point is $A = dV_0/dV_1$. This gain drops to zero at zero input.

A closed loop system with a wide range of gains in it can only have fast response at the maximum gain condition and to obtain a good performance over the full range of firing angles the cosine ramp is preferable. The load voltage V_0 is a linear function of the input V_1 and so the gain is constant.

The gain of a practical converter depends on the nature of the load as the presence of inductance and back-emf affect the time for which the thyristor conducts after it is triggered. In a full wave system the current is sometimes discontinuous; ie it consists of a series of separate pulses or may be continuous. When the current becomes continuous there is an increase in gain.

Electronic control

A transistor operating as a linear amplifier is the simplest form of electronic control. Up to load powers of about 20 W this system is useful with a circuit as in Fig. 9. For higher powers the same circuit may be used but the transistor operated as a switch to reduce the dissipation in it. The average load voltage is then controlled by varying the duration and spacing of the on-periods. Transistors operating in this way are economic up to about 100 W. Above this power transistors are expensive and the more complicated circuits using thyristors are more often used.

Not only-
ELECTRIC MOTORS AND

GEARED MOTORS
$\frac{1}{100} - \frac{1}{3}$ h.p. from

 for $\frac{1}{25} - 15$ h.p.

but also-
SPEED CONTROLLERS

400 B Series
10–1 speed control by voltage regulation from A.C.-input

A.F. Series
20–1 speed control Closed loop Armature feedback

J.T. Series – from Tacho feedback

G.P. Series – similar to A.F. series for use with Fractional h.p. motors

A complete package deal

PERFECTLY MATCHED
manufactured and tested as a single unit

NORMAND ELECTRICAL
GROUP OF COMPANIES

Walton Road - Eastern Road
Cosham - Portsmouth - Hants
PO6 1SZ   Tel. Cosham 71711   Telex 86149

Fig. 10 shows the Jones chopper. When SCR1 is triggered the battery is coupled to the load through one winding of the auto transformer T1. The induced voltage in the other winding charges the capacitor C1 through the diode D1. The capacitor then remains charged until SCR2 is triggered and it then discharges through the load. In so doing it reverse biases SCR1 and turns it off. The current in the load then commutates to D2. The process can then be repeated by triggering SCR1. The voltage is controlled by varying the time intervals between the trigger pulses to the two thyristors.

BIBLIOGRAPHY

A Draper. Electrical machines. *Longmans*.

J J Distenfans, A R Stubberd, J J Williams. Theory and problems of feedback and control systems. *McGraw-Hill*.

F E Gentry, F W Gutzwiller, N Holonyak, E E von Zastrow. Semiconductor controlled rectifiers, principles and applications of p–n–p–n devices. *Prentice Hall*.

Power applications of controllable semiconductor devices. *IEE Conference publication* No 17.

A W J Griffin and R S Ranshaw. The thyristor and its application. *Chapman & Hall*.

Chapter 16
Starting and Speed Control of a.c. Motors

J H C Bone BSc MIMechE FIEE
Laurence, Scott & Electromotors Ltd

Comparisons are occasionally made between the starting currents taken by a.c. motors with cage-type rotors and those taken by d.c. machines, generally to the detriment of the a.c. machine. Although it is true that cage motors generally take starting currents in the range 4 to 7 times full load current whilst those of d.c. machines are normally much less, the comparison ignores one of the principal virtues of the a.c. machine, its very robust construction, which permits it to be started by a direct connection of the motor windings to the supply system; the lower starting currents of d.c. machines are almost always associated with multi-step resistance starters. Steps taken to reduce the starting currents of cage machines are generally due to the need to reduce these currents to values which the supply system can handle, rather than because of any limitations within the motors themselves.

A.C. STARTING

Starting current to full load current ratios of 7 or 8 are common on small machines having outputs of a few kilowatts; 4 or 5 are more typical ratios for machines with outputs of several thousand kilowatts. In spite of these high currents at start, the corresponding starting torques of a simple cage machine tend to be on the low side. Not more than full load torque is fairly typical of small and medium sized machines, falling to about 50% of full load torque for the largest sizes. Table 1 indicates how the starting currents and torques of simple cage machines vary with machine output and speed. Although it is generally convenient to quote starting currents in terms of full load currents, if accurate comparisons between machines are to be carried out, then the ratio of starting kVA to full load kilowatts provides a more satisfactory basis for comparison. This is because the machine with the poorer running performance, *ie* with the higher full load current (due to a lower efficiency or a worse power factor) could appear to have the better starting to full load current ratio. Multiplying by the full load efficiency and power factor converts from a ratio based on starting kVA to one in terms of current.

Similarly it is usual to use the terms 'starting torque' and 'starting current' to denote the torque and current at the instant of start whereas strictly speaking the terms refer to the torques and currents throughout the run up period. Fig. 1 shows typical torque/speed and current/speed curves for a cage machine, appropriately labelled with specific terms for the various significant portions of the curves. For the equipment to start to rotate when the machine is connected to the supply, the locked rotor torque must exceed the breakaway torque of the load. Thereafter the time taken for the equipment to run up to speed will depend on the inertia of the drive, and the average accelerating torque, ie the average difference between the motor torque and the load torque curves.

Various techniques are available to the machine designer to enable him to alter or improve the starting performance of the simple cage machine so as better to match the motor characteristic to the individual requirements of the driven machine. If, instead of using a rotor winding of circular cross-section bars, the windings are constructed from thin rectangular bars (Fig. 3), the so-called 'deep bar effect' can be exploited to improve the starting performance. Due to the self-inductance of the bars the rotor currents at start do not distribute themselves uniformly across the whole cross section of the bar, but concentrate on those parts nearest the bore where the reactance is lowest. Consequently, the effective resistance of the bar is higher than would be expected from simple considerations based on the whole cross-section of the conductor. As the change in effective resistance depends on the depth of the conductor, changing the shape of a conductor from a circular form to a deep thin bar provides a method of increasing the effective resistance (and hence torque) during the early part of the motor run up. The attraction of this method is that it is accomplished without penalty to the running performance, because, as the motor runs up to speed, the frequency of the rotor currents fall to such a low figure that the reactance of the rotor conductors can be neglected. The effective resistance therefore falls to the normal value corresponding to the condition when the current distributes itself uniformly across the whole cross-section to the rotor bar.

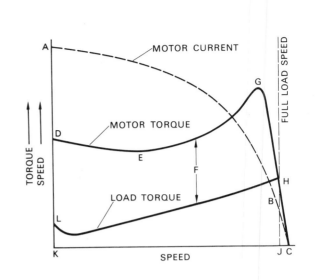

Fig. 1. Typical starting characteristic curves for an induction motor with cage type rotor. (A) Locked rotor current; (B) full load current; (C) no load (synchronous) speed; (D) locked rotor torque; (E) pull up torque; (F) accelerating torque; (G) pull out torque; (H) full load torque; (J) full load speed; (K) standstill; (L) breakaway torque of load.

Fig. 2. Torque/speed and current/speed curves showing improved starting performance of cage induction motors exploiting the deep bar effect; (A) 'deep' bar rotor; (B) equivalent round bar.

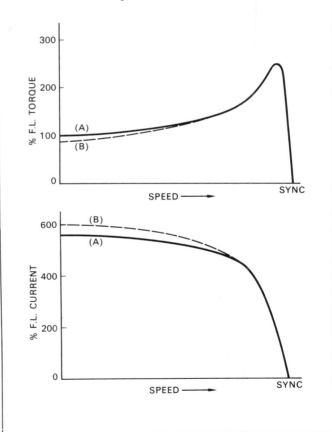

Table 1. Typical starting performance data for induction motors with simple cage rotors.

Rated output kW		Starting current ratio expressed as locked rotor kVA* / rated kW output	Locked rotor torque (per unit values)			
			2p	4p	6p	8p
1.0 up to	2.5	10.5	2	2	1.75	1.5
2.5	6.3	9.8	2	2	1.75	1.4
6.3	16	9.2	2	2	1.75	1.4
16	40	8.7	1.6	1.7	1.6	1.4
40	100	8.2	1.25	1.4	1.3	1.25
100	250	7.8	1.0	1.1	1.2	1.2
250	630	7.6	0.8	0.8	0.8	0.8
630	1600	7.4	0.6	0.6	0.6	0.6
1600	4000	7.2	0.5	0.5	0.5	0.5

* To obtain the starting current to full load current ratio, multiply by full load efficiency and power factor.

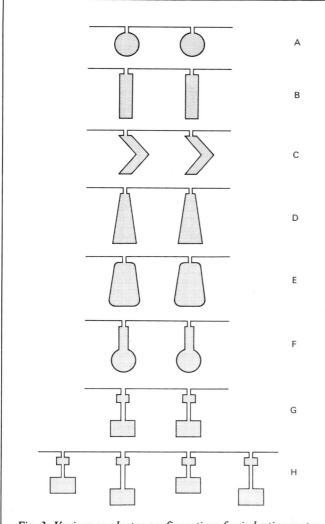

Fig. 3. Various conductor configurations for induction motor cage rotors; A to F single cage rotors; G to H double cage rotors.

Fig. 2 compares the starting performance of a deep bar machine with one with a circular cross-section bar of substantially the same cross-section. There is, of course, a limit to the depth of bar which can be accommodated on any particular machine. Purely mechanical considerations, such as strength, rigidity, and stiffness of the bars themselves set certain limits, as does the necessity to leave sufficient depth of iron below the slots to carry the magnetic flux. Fig. 3C shows one method of increasing the mechanical rigidity of a thin bar.

Additional improvements that can be made to the starting performance further exploit this effect by using various shaped bars. Fig 3(C–F) illustrates some typical bar shapes developed to enhance the effective resistance of the rotor bars at start. By distributing the greater part of the conductor cross-section at the bottom of the slots, in the high reactance region at start, the ratio of starting to running resistance can be considerably increased compared with that of a simple rectangular bar, although at the cost of some of the basic simplicity of the simple cage rotor. Shapes C, E, F are typical of those used on larger machines; D is more appropriate to smaller machines where the rotor windings are cast in aluminium alloy.

A logical extension to the use of these shaped bars is the use of a double cage design where there is no longer any direct electrical connection between the top and bottom of the conductor (Fig. 3G and H). With a double cage design the bottom cage lying in the high reactance region of the rotor carries little current at start, and hence has only a marginal influence on the starting performance of the machine. Conversely, the top cage, which almost entirely controls the starting performance, has correspondingly much less influence on the running performance. This is because, close to synchronous speed, the low-frequency load currents distribute themselves between the two cages inversely as the respective cage resistances. Hence, almost all the load current flows through the large cross-section bottom winding, and little through the high-resistance top winding. The ability to alter the characteristics of the two cages independently of each other offers considerable freedom to the designer, in that improvements to the starting performance can be obtained without a significant sacrifice to the running performance. Fig. 4 shows how the resultant speed/torque curve of the machine is built up from the individual curves of the two windings. Fig 5 indicates the range of starting performance characteristics which can be obtained from such machines, 5A representing a design in which the emphasis is on a low starting current, whereas 5B is for a machine in which a high starting torque has been considered to be of the greater importance. Table 2 lists some typical standard classifications for the starting performance of cage machines. Though indicative of the range of starting performance commonly available, they by no means represent the limits of what can be achieved.

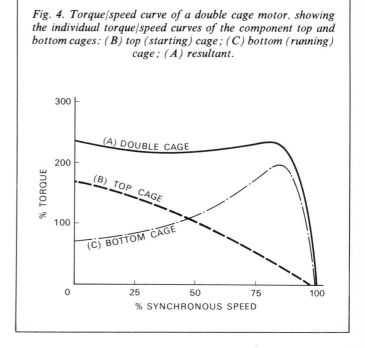

Fig. 4. Torque/speed curve of a double cage motor, showing the individual torque/speed curves of the component top and bottom cages: (B) top (starting) cage; (C) bottom (running) cage; (A) resultant.

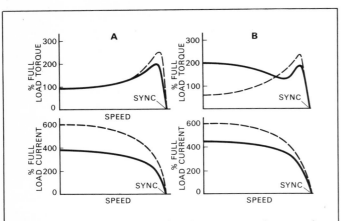

Fig. 5. Typical torque/speed and current/speed curves for double cage motors: (A) design emphasis on low starting current; (B) emphasis on high starting torque; broken lines show typical single cage characteristics for comparison.

Designs of double cage machines are readily available with significantly better starting performance for drives whose characteristics justify their use.

When a cage machine accelerates a load (for instance a fan) up to speed the heat generated in the rotor windings is approximately equal to the kinetic energy acquired by the drive in accelerating to full speed. Hence, with a high inertia load, the losses within the rotor windings are very high during the run up period, many times those which occur during normal running, and therefore much greater than can be dissipated by the normal cooling system of the motor. Consequently, the heat generated in the windings during this period is largely absorbed by the rotor windings and the surrounding iron core, so that the windings must have an adequate thermal capacity to ensure that they do not overheat. The top cage resistance of a double cage rotor needs to be relatively high, and if made of copper would require conductors with such small cross-sections that their thermal capacity would be very low in relation to their losses during acceleration. Therefore, where the drive inertia is substantial, the thermal capacity can be increased by using materials with higher specific resistances than copper (for instance cupro–nickel, which has substantially greater cross-sections for any required value of cage resistance than copper bars).

Table 2. Some standard classifications of induction motor starting performance.

Category	Starting current	Starting torque
Basic design	Table 1	Table 1
High torque with normal starting current	Table 1	Twice full load torque
Low starting current with normal starting torque	75% of values in Table 1	Table 1
Low starting current with reduced starting torque	75% of values in Table 1	75% of values in Table 1

When allocating a duty cycle rating involving acceleration, or considering the permissible frequency of starting, it is clearly necessary to define the capabilities of the machine in terms of the load inertias involved. This information can be presented in the form of the 'stored energy constant' of the motor and load, which is defined as the combined stored energy at rated speed divided by the motor rated output in kilowatts. Alternatively, a 'factor of inertia' can be used, this being the ratio of the stored energy constants of motor rotor and load divided by that of the motor rotor alone.

As already noted, one of the chief advantages of the cage machine is that it can be started direct-on line with the simplest of starters. Modifications to the rotor construction of the type described above enable a reduction in the starting currents of such machines down to around 3.5-times full load current. However, even such figures may be too high for some supply systems, in which case it is necessary to use more complicated starting arrangements. The function of most of these is to reduce the motor starting current by temporarily reducing the voltage to the motor stator windings. Compared with the results obtained by modifi-

Auto-transformer starter for a 225-kW (300-hp) motor, using the Korndorffer transitional step.

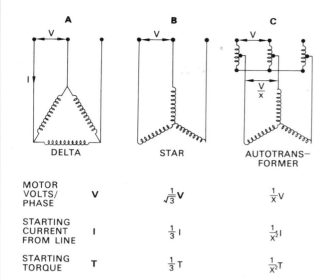

Fig. 6. Current, voltage, and torque ratios for various forms of starting: (A) direct-on; (B) star connection step of star/delta start (C) auto-transformer start with tapping to reduce voltage to $1/x$ of full value.

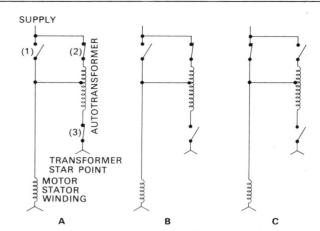

Fig. 7. Auto-transformer starting of induction motor, using the Korndorffer transitional step (only one phase shown): (A) start and star point switches (2) and (3) closed, feeding motor at reduced voltage from auto-transformer tapping; (B) switch (3) opens to give transitional step, feeding motor through part of transformer winding which acts as an inductance; (C) run switch (1) closes and then (2) opens to leave motor running at full supply volts.

cations to the rotor design, a reduction in supply voltage can produce much more drastic reductions to the starting current (which will fall approximately as the square of the voltage reduction), but this is accomplished at the cost of a correspondingly large fall in the developed starting torque. However, for drives where the starting torque requirements are low (eg fans or pumps where 30% full load torque is commonly sufficient to break the drive away) the low starting torques associated with all forms of reduced voltage starting are perfectly acceptable.

Star/delta starter

Multi-step primary resistance (or reactance) starters for cage-type motors are seldom employed because of their complication and expense, even though they offer the facility of breaking up the starting sequence into several discrete steps. A much more common form of reduced voltage starter is the well-known star/delta starter, which provides a single reduced voltage step by temporarily reconnecting a delta connected stator winding into star for the initial part of the starting period. Fig. 6 shows the two connections and the corresponding voltage, current, and torque ratios, from which it can be seen that this method is in fact a way of obtaining a voltage reduction of 1.73 to 1 on the starting step, with a consequent reduction in the current and starting torque to one-third of the values which would be obtained with a direct-on start with the winding in delta. Where star/delta starting is to be employed, it is of course essential that the stator winding be designed for a delta connection when running (not normally a problem unless one is considering applying the starter to an existing motor which happens to have been designed with a star-connected winding). In addition, all six ends of the motor stator windings need to be brought out to the starter; on small machines this is unlikely to present much of a problem, but on large medium-voltage machines the additional cable cost may be of significance if the starter is situated some way from the motor.

The use of a star/delta starter is sometimes advocated as a means of reducing the mechanical stresses imposed on the motor windings at start, but the changeover from star to delta connection involves disconnecting and then reconnecting the machines to the supply with the motor running fairly close to its synchronous speed. Consequently, heavy transient currents and torques are generated upon reconnection, which can well stress the machine as much or more than a direct-on start.

The star/delta starter has been the conventional method of obtaining a reduced voltage start on relatively small medium-tension machines up to, say, 150 kW, but modern supply systems feeding such motors are now much more likely to be able to withstand direct-on starting currents (using double cage machines if necessary) so that the use of star/delta starters is diminishing.

Auto-transformer starter

With a star/delta starter the voltage reduction ratio is fixed at 3:1. However, if an auto-transformer starter is used, the reduced voltage tapping can be arranged to give any required voltage reduction to suit the particular drive and supply requirements. Compared with direct-on, or even star/delta starters, the auto-transformer starter is expensive, particularly if, as is often the case, the Korndorffer transitional step is introduced (Fig. 7), which makes the transition from the reduced voltage tapping to full voltage operation without actually disconnecting the machine from the supply. This is accomplished by arranging the connections so that on the transitional step the machine is fed from the supply through a part of the auto-transformer winding which acts in this connection as a reactor. The Korndorffer connection, therefore, eliminates the possibility of transient currents and torques being generated during the transition from reduced to full voltage operation. There is a similar transitional connection available (although much less often used) for star/delta starters where the Wauchope connection introduces a transitional resistance step between the star and the delta connections.

With a direct-on full-voltage start magnetic saturation effects associated with the heavy starting currents of the machine reduce its reactance, giving rise to rather higher starting currents and torques than would be predicted from readings taken under unsaturated conditions. Hence, when considering the use of any type of reduced voltage starter, allowance must be made for the fact that the actual torque (and current) will be noticeably less on reduced voltage than would have been predicted from the theoretical reduction ratios.

Another phenomenon which is frequently overlooked in connection with auto-transformer starters is the high transient inrush magnetising current of the transformer itself. Because such transformers are designed to operate with flux densities very close to the saturation level, transient inrush currents may be produced, which persist for a few cycles, with peak values of ten or more times the full load current of the motor being started.

High starting torques

All the above methods reduce the starting currents by reducing the voltage to the stator windings, thus reducing the motor flux. Consequently the starting torques developed are low (reduced in a somewhat greater ratio than the square of the voltage reduction), so that these methods of obtaining low starting currents are inappropriate for drives requiring substantial starting torques. For such drives, it is necessary to sacrifice the simplicity and low cost of the cage rotor, and substitute for it a wound rotor whose winding ends are brought out to sliprings. This enables the resistance of the rotor circuit to be varied at will by the introduction of external resistances connected to the sliprings. By this means the effective resistance of the rotor circuit can be increased to its optimum value, giving a high starting torque with a relatively low starting current. As the machine runs up to speed the external resistances are progressively reduced until, by the time the machine is up to speed, the external resistances have been reduced to zero, the sliprings are short-circuited, and the rotor circuit resistance is reduced to that of the rotor windings alone.

SPEED CONTROL

Multi-speed machines

When fed from normal constant-frequency supplies the induction motor is essentially a constant speed machine, limited to one of a series of speeds corresponding to the number of poles for which the stator is wound. Two or more speeds can be obtained simply by altering the number of poles, either by the use of two independent stator windings wound for a differing number of poles, or by specially designed windings that can be reconnected in alternative ways to give differing pole numbers. Certain speed ratios, particularly 2 to 1, can be obtained by very simple reconnections of a suitably designed single winding. More recently other methods of reconnection, notably pole amplitude modulated windings (PAM), have become available, to give a much wider choice of pole ratios than was hitherto possible with single windings.

Wound rotor machines

Where two, or at the most four, fixed speeds are required, the pole change machine offers a cheap and simple solution. However, if a range of adjustable speeds are required, the wound rotor induction motor can be used to obtain speeds below synchronism using external speed-regulating resistances connected to the sliprings. The motor will run at a speed determined by the voltage applied to the rotor windings, *ie* the voltage at the sliprings. With external speed regulating resistances, this voltage will be that due to the passage of rotor load current through these external

Fig. 8. Speed control of a wound rotor induction motor with rotor circuit resistances, showing a family of speed/torque curves for various values of rotor resistance.

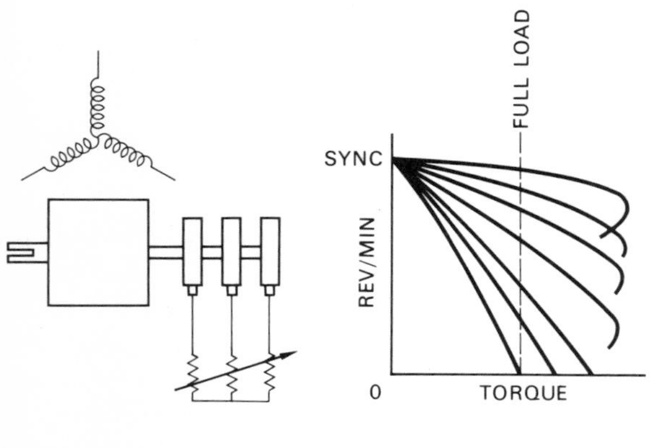

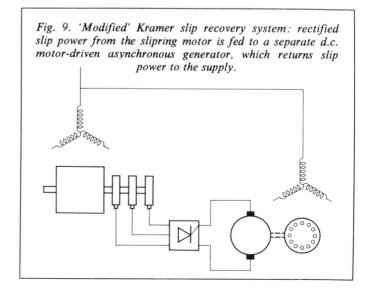

Fig. 9. 'Modified' Kramer slip recovery system: rectified slip power from the slipring motor is fed to a separate d.c. motor-driven asynchronous generator, which returns slip power to the supply.

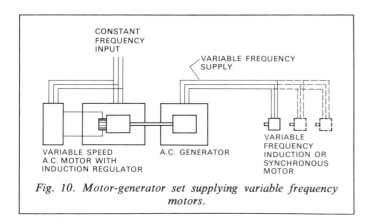

Fig. 10. Motor-generator set supplying variable frequency motors.

Rotating frequency converters: three 52-kVA 2.8/0.5-Hz pony motor driven frequency converters (right, near wall) and three 125-kVA 15/3-Hz self-driven converters (left) with their associated induction regulators in foreground.

resistances. As the slipring voltage is zero at synchronous speed and increases linearly to a maximum value at standstill, then for a given load current the motor speed can be reduced by the introduction of additional resistance. Similarly, for a fixed value of external resistance the motor speed will rise as the load on the motor is reduced, and thus, irrespective of the initial setting of the speed regulating resistances, on light loads the motor will tend to accelerate up to synchronous speed (Fig. 8). If a relatively wide speed range is required the lower speeds will be very load dependent, and the speed/torque characteristics in this area will be correspondingly steep. In addition, the fact that speed is controlled by the passage of load currents through resistances is indicative of major losses in the system. For example, if the speed is to be dropped to half synchronous speed, then the losses in the resistances will be equal to the output of the motor (and for a speed of a quarter of synchronous speed the resistance losses will be three times the motor output). The efficiency of the scheme is therefore low, particularly at low speeds, and its application tends to be limited to drives requiring relatively narrow speed ranges (eg power station boiler feed pumps) or to those drives where low speed running is required only intermittently, such as cranes or hoists.

Slip energy recovery schemes

If, instead of wasting the 'slip energy' in resistances, this power can be fed back into the supply system, then a fundamental improvement in the efficiency of the drive at reduced speeds is obtained. A direct connection between sliprings and supply is not possible because both the slipring voltage and frequency vary with the motor speed. Early attempts to utilise the slip power were based on the use of either rotary convertors (Kramer system) or low frequency exciters (Scherbius system). Although these are now rarely of more than academic interest, the availability of thyristors has led to the limited use of modified Kramer systems in which the rotary convertor is replaced by a thyristor bridge converting slip frequency power from the sliprings to d.c. The d.c. output from the thyristors may be handled in several ways. For instance, it can be fed to a d.c. motor coupled to the main slipring motor (so that the combined torques of the two motors provide a constant power output to the drive over the speed range). If, as is more normal, no more than a constant torque output is required by the drive, then the d.c. may be fed to a d.c./a.c. motor-generator set feeding the slip energy back into the supply system. Completely static conversion systems are also available, in which the motor-generator set is replaced by a static inverter.

Although the term 'slip recovery' is generally confined to schemes using slipring motors with some form of separate external recovery system, the a.c. commutator motors, when running below synchronous speed, also feed the slip power back into the supply system. A.C. commutator motors, however, have the additional advantage that the motors can also be run at super-synchronous speeds, a facility not available with the modified Kramer slip recovery system in which the power flow through the rectifying system is inherently unidirectional. The particular field of application of slip recovery systems is for drives which, because of their high speeds and outputs, are beyond the capability of straight commutator motors.

Variable frequency drives

Because public supply systems operate at a fixed frequency, cage-type induction motors normally operate as fixed speed motors. However, as the synchronous speed of an induction motor is proportional to the supply frequency, the speed

A 50-kVA cyclo-converter with a frequency range of 1–15 Hz.

of an induction motor can be varied if it is possible to control the supply frequency. In practice, this generally means providing special frequency changers for supplying the motors, and a frequency changer for an individual motor is of course expensive. More commonly converters are used to supply a group of motors whose speed is to be varied in unison. Where synchronism between the motors is required then, instead of induction motors, synchronous reluctance motors are used.

The supply voltage to a motor fed at variable frequency should vary as the frequency (*ie* should have a constant volts/Hz ratio) so as to maintain a substantially constant motor flux throughout the speed range. Driving an a.c. generator at a variable speed is a simple method of obtaining a variable frequency supply with a constant volts/Hz (Fig. 10). An additional advantage of the motor-generator set is that this ratio can be easily and cheaply adjusted by the control of the generator excitation. This is a useful facility for drives requiring a wide speed range where it is an advantage to increase the ratio at low speeds to compensate for voltage drops in the system (which become increasingly significant as the speed is reduced).

Three full power electromechanical energy conversions are involved in the use of a motor-generator set; from the constant-frequency electrical supply to mechanical power at the motor-generator set shaft; back to electrical power at the generator output terminals; and then the final conversion to mechanical output at the shaft of the induction motor.

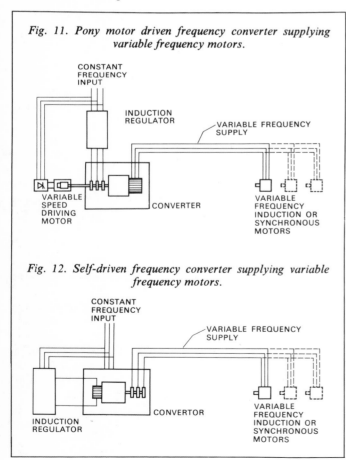

Fig. 11. Pony motor driven frequency converter supplying variable frequency motors.

Fig. 12. Self-driven frequency converter supplying variable frequency motors.

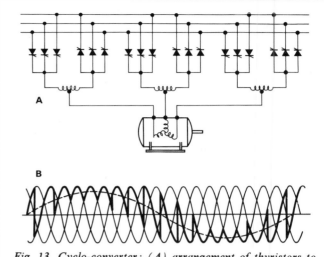

Fig. 13. Cyclo-converter: (A) arrangement of thyristors to feed variable frequency motor; (B) indication of how low frequency output wave is synthesised from the supply frequency input.

As such full power conversions must be both expensive and inefficient, it is often advantageous to use units in which the frequency conversion is accomplished electrically within the converter. Output frequency is still altered by changes in converter speed, but this can be accomplished using only a small driving motor whose size is related to the converter losses (rather than its output). The converter itself consists of an unwound stator within which rotates a rotor whose windings are connected to both sliprings and a commutator. The input to the converter is connected to the sliprings, and the output frequency appears at the commutator (at a constant voltage, but at a frequency dependent on the speed of rotation). As the converter output voltage is unaffected by speed, an induction regulator is required at the input to the converter to drop the output volts as the frequency is reduced, so as to maintain a constant volts/Hz. (Fig. 11).

Self-driven converters are also available which are basically stator-fed commutator motors in which the output is taken not as mechanical power at the shaft, but as slip frequency power from a set of sliprings connected to the rotor windings. The induction regulator in such an equipment controls the converter speed and therefore frequency, the output voltage inherently varying with the frequency to maintain a constant volts/Hz (Fig. 12).

The ready availability of thyristors now makes possible new methods of converting from constant to variable frequency by purely static means. Where the required frequency range is all well below the supply frequency, one can use the cyclo-converter system in which the required output frequency is synthesized by appropriate sequential switching of the motor supply terminals to successive supply phases. Two sets of thyristors are required, supplying respectively the positive and negative half-cycles of the alternating wave to the motor. The cyclo-converter is essentially a method of generating low frequencies, say from about half supply frequency downwards. The arrangement of thyristors permits power flow in either direction, *ie* regenerative

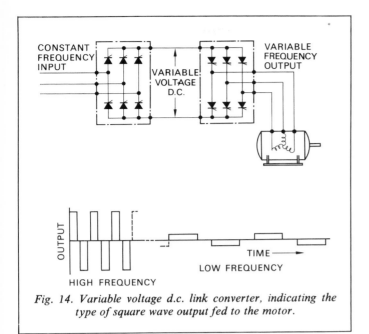

Fig. 14. Variable voltage d.c. link converter, indicating the type of square wave output fed to the motor.

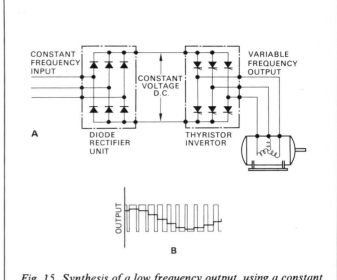

Fig. 15. Synthesis of a low frequency output, using a constant voltage d.c. link converter.

operation of the motor is possible permitting power flow back to the supply from the kinetic energy of a high inertia drive during rapid deceleration (Fig. 13).

Other static converters use an intermediate d.c. link between the constant-frequency and variable-frequency systems. Thyristor or diode bridges are used to rectify the constant-frequency supply to provide the d.c. to which the motor terminals are successively connected, thus providing a square wave input to the motor. Using a variable voltage link the motor voltage is determined by the d.c. link voltage, and the frequency by the frequency of switching of the square wave output from the converter (Fig. 14). Where a constant link voltage is used (*ie* using diodes rather than thyristors for obtaining the d.c.) high-speed switching at, say, 500 Hz is employed, and lower frequency waves are simulated by controlling the relative duration of the positive and negative pulses (Fig. 15).

Link converters can be made for much wider frequency ranges than any other type of converter. It is particularly valuable that their frequency range can include the supply frequency, and because frequencies much higher than supply frequency are easily obtained, higher motor speeds than otherwise available can be obtained. Most simple link converters require additional thyristors to give regenerative operation.

BIBLIOGRAPHY

Most textbooks dealing with the design of a.c. machines are worth consulting for additional information. A few are listed below, including some rather old but nevertheless useful volumes.

1. M G Say, The performance and design of alternating current machines.
2. J Hindmarsh, Electrical machines and their applications.
3. Liwschitz-Garik, Wipple, Alternating current machines.
4. R R Lawrence, Principles of alternating current machinery.
5. Miles Walker, The control of the speed and power factor of induction motors.

The following papers give further information on specific aspects:

Rawcliffe, Burbridge and Fong, Induction motor speed changing by pole amplitude modulation. *Proc. IEE* 105A August 1958. (Further papers in vol. 107A, 108A and 110).

R A F Craven, Motor control gear for A.C. ships. *Electrical Times* 24, 11, 1960.

G Wauchope, Recent development in squirrel cage induction motor starters. *Proc. IEE* Part II April 1944.

A H Pickering, The starting of electric motors. *Journ. Inst. Mining Electrical Engineers* December 1940.

J Griffin, Methods of overcoming poor inherent starting performance. *Electrical Review* 1, 10, 1965.

Alger, Ward, Wright, Split winding starting of 3-phase motors. *Trans. AIEE* 1951, No. 70 Part III.

Delfarge and Hartmann, Die untersynchrone Drehstromkaskade. *Conti Electro Bericht* April 1964.

D A Paice, Speed control of large induction motors by thyristor converters. *Trans. AIEE.* Vol. IGA 5 September 1969.

J C H Bone, Variable frequency drives. *Electrical Times* 28, 7, 1968.

E Friedlander, Speed control of large slipring induction motors. *IEE Conference Publication* No. 10 1965.

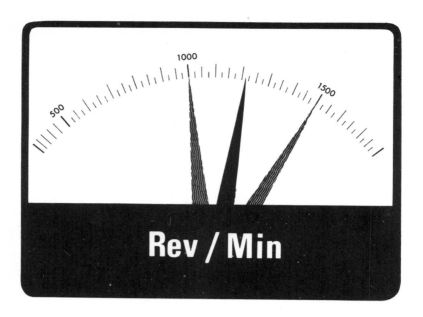

NOT TOO FAST NOT TOO SLOW BUT JUST RIGHT

For many industrial, power generation, and public services applications, it is vital that some driven auxiliaries operate at *exactly* the right speed for optimum performance. With thirty years' experience of the world's finest variable-speed a.c. motor, with specialist experience of variable-frequency multi-motor installations, and with ample field experience of both standard and special-purpose thyristor-fed d.c. motor drives, L.S.E. engineers are uniquely able to meet each and every variable-speed requirement. And examples of L.S.E. application engineering—both on conventional drives and with sophisticated automatically-speed-controlled equipments— are to be found in power stations, factories, process lines, cement works, pumping stations and collieries all over the world.

 **LAURENCE, SCOTT & ELECTROMOTORS LTD**
NORWICH NOR 85A   *Telephone: Norwich 28333*

Chapter 17
Motor Protection Devices

D Ramsden *CEng MIEE*
Bull Motors Ltd

Systems for the protection of electric motors have been available since motors themselves began to be used in industrial applications. However, until fairly recently, none of those systems was capable of affording fully comprehensive protection. This chapter outlines the types of protection system which are now commonly available and examines the advantages and disadvantages of each. Basically, the systems have been based on the protection of induction motors since these are such a widely used category in a vast range of industrial applications.

The need for protection
It is the individual requirements of an installation which determine the need for protection and the degree which is to be afforded. On large and medium sized machines protection devices are almost invariably used, but on small motors there are sometimes valid arguments against the provision of such devices. As an example one may consider certain multispindle textile applications where several thousand very small motors are used in a single plant. The failure rate of these motors is very low and the failure of an individual motor usually causes no serious production delay. Therefore the cost of installing individual protection for each motor would considerably exceed the value of the protection it afforded. In many cases the normal short circuit protection is all that is provided and this is perfectly valid on economic grounds. Such an installation is, however, a particular example rather than the general rule and in most cases protective devices are justified both technically and economically for one of the following reasons: a motor failure usually necessitates costly repairs; a failure can seriously disrupt production; protection of a particular machine is reflected in the indirect protection of other parts of the system; standby machines are not normally provided. In addition to the foregoing points it should be noted that the provision of protective devices is demanded by the regulations of certain Authorities.

Conditions which cause damage
The conditions which can cause damage to a motor fall basically into five categories, and although these are

different in nature, their effect on the motors is the same in all cases—they result in circumstances which give rise to overheating in the motor. Causes of overheating are: overload; misuse; misapplication; power supply; mechanical failure. Certain of these categories are difficult to distinguish between since conditions which the applications engineer might describe as misuse could, by the user, be described as misapplication. Nevertheless an attempt will be made to discriminate between the causes of overheating in induction motors.

Overload. This can be a sustained overloading of the driven equipment or can be fluctuating or cyclic overloading with stalling regarded as the ultimate degree of overload.

The designer of the driven equipment is not always in a position to anticipate accurately the final loading, particularly where duty cycle operation is involved. When certain types of machinery are new, friction losses can be abnormally high until the machinery is 'run-in'. Starting up under cold condition can be a problem, particularly on oil pumps when low temperature results in high oil viscosities. All these factors can result in overloading and consequent overheating of the motor.

The relevant British Standard Specification to which motor manufacturers work is on a CMR basis (*ie* continuous maximum rating) which does not require a motor to have any capacity or margin for continuous overload operation.

Misuse. There are four main divisions of misuse on induction motors: excessively frequent starting, particularly when high inertial loads are being driven; excessive inching or jogging; excessively frequent plug reversals. In each case the term excessively means beyond the degree for which the motor was designed or rated. The fourth category of misuse is bad maintenance.

Misapplication. This can take many forms. Incorrect power or rating of motor relative to load requirements, *ie* hp, torque, inertia, etc. Incorrect enclosure such as the use of ventilated motor in dirty or fluffy atmosphere, where vent openings become choked. Incorrect siting or mounting such that the motor cooling air is recirculated or drawn over, say, a hot pump. Motor sited or mounted so that it is subjected to radiated, convected or conducted heat from other equipment. Failure to notify makers of abnormal conditions such as high ambient temperatures or high inertias, etc. Failure to cater for loads of a reciprocating nature in terms of fitting adequate flywheels and matching motor slip characteristics.

Power supply. Single phasing (on 3-phase supply) due to blown fuse or defective contact or connection. Over voltage, increasing iron losses in motor. Under voltage, increasing copper losses in motor. Frequency variations. Increase in frequency can cause overloading on centrifugal fan and pump drives. Decrease in frequency can increase iron loss or copper loss and reduce ventilation. Inadequate supply capacity or inadequate cabling section can mean serious reduction in terminal voltage during starting with consequent loss of motor torque; this may result in excessively long starting times, failure to accelerate to full speed or even failure to start at all on drives with high starting torque requirements.

Mechanical component failure. Incorrect bearing lubrication either on motor, drive or driven equipment. Faulty or broken belts or couplings can result in excessive loading or eventual stalling of the motor.

INFLUENCE ON LINE CURRENT

It has been shown that there are numerous ways in which a motor winding can be subjected to overheating. Not all of these are reflected by corresponding increases in the line current drawn by the motor.

The causes of overheating which are reflected in line-current increases are:
Overload, etc., sustained or cyclic. Stall. Inching and plugging. Prolonged starting. Certain supply variations. Single phasing.

Causes of overheating not reflected in line current: High ambient. Radiated, convected or conducted heat from other equipment. Incorrect ventilation conditions (*ie* recirculation of air). Restriction of ventilation (choking of vent openings). Thermal lagging. Certain supply variations.

CURRENT-SENSING PROTECTIVE DEVICES

Protective devices which simply 'sense' currents or current changes cannot cope with overheating conditions which do not reflect in terms of changes in current. They can protect where changes in current occur, but the degree of protection afforded depends upon matching the current responsive characteristics of the device to the heating characteristics of the motor. To some degree these charac-

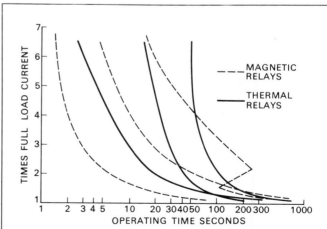

Fig. 1. Characteristic curves of a selection of thermal and magnetic overload relays (from different suppliers to the UK market).

teristics can be matched in terms of the stator winding of the motor when subjected to steady overload or stall conditions. They cannot however adequately cope with over-currents of a cyclic nature or where the over-currents are re-applied as soon as the protective device resets, since the protective device, whether thermal or magnetic in nature, cannot normally match the thermal inertia characteristics of the motor winding (*ie* the device re-sets much quicker than the motor cools down) nor can they take account of conditions in the rotor.

Thermostatic current-sensing devices

There is one exception to the last statement: the thermostatic device mounted in the motor wherein the bimetallic element senses the radiated heat from the windings and additionally is influenced by a small heater, the heat from which depends on the line current which passes through it. On some small single-phase types current flows through the bimetallic element which thus acts as its own current heater.

These devices are limited in use normally to fractional horsepower single-phase motors on which they are very effective. Three-phase versions are in very limited use, mainly in the United States on motors up to about 5 hp (3.7 kW). The three-phase version has never achieved popularity in the UK, partly because of the difficulty in physically accommodating such units and the variety of current ratings involved.

Current overloads

Apart from the exception discussed above, the current sensing protection devices are mounted in the control gear, since the contacts which form part of the device are essentially small and are arranged in the no-volt or contactor coil circuit of the control gear. They do not themselves directly break the current to the motor. These current-sensing protective devices are normally referred to as 'overloads' or 'overload trips' or 'overload relays' and are available in three distinctly different forms:
Thermal (bimetallic).
Thermal (solder-pot).
Magnetic (dash-pot).

Thermal and magnetic overloads can be very simple devices or can be quite sophisticated, dependent upon the characteristics and performance required from them (these are discussed later). They cannot however be regarded as devices which sense or monitor the temperature in the motor windings at any given time.

MAGNETIC OVERLOAD RELAYS

The magnetic overload relay is one of the simplest and most robust of the motor protective relays available. It is basically a solenoid having a coil capable of carrying the motor line current. A dashpot is attached to the plunger to provide the necessary time lag.

Design factors

The major problem encountered when designing a magnetic overload, and one which is not appreciated by many application engineers and users, is that the motor acceleration time which can be tolerated is considerably less than the tripping time as read from the overload characteristic curve. If at six times current the overload trips in 12 seconds, it will not be satisfactory for use with a machine which takes six times full-load current for 12 seconds to accelerate to full speed.

The reason is that, whilst the motor is accelerating, the relay plunger commences to rise in the coil and as it does so, the magnetic pull increases. At the end of the acceleration period when the current falls to the full load level the relay plunger, depending upon how far it has moved into the region of greater field strength within the solenoid, may continue to rise and eventually trip even though no over-current exists.

This is one reason why it is often said that magnetic overloads have insufficient time lag; in fact some magnetic overloads may have time lags which are too long to provide adequate protection at the higher currents. This is brought about because the relay has been designed to meet the maximum accelerating time specified in BS587:1957. In these cases it is usual to provide an adjustable relief valve to reduce the tripping time when the accelerating time is short. In practice the permitted safe starting time to avoid tripping during or after acceleration may well be only 20% of the total tripping time at the starting current read from the overload curve.

The nearer the setting is to normal full load, the worse this condition becomes, and if the relay were adjusted to operate actually at full load, the permissible accelerating time would be nil.

Performance

Fig. 2 shows the stator heating of the test machine when repeated attempts are made to start with locked rotor using a relay having a tripping time of 5 seconds and also with a relay having a tripping time of 20 seconds. In both cases the inherent resetting time of the relay was approximately 7–10 seconds.

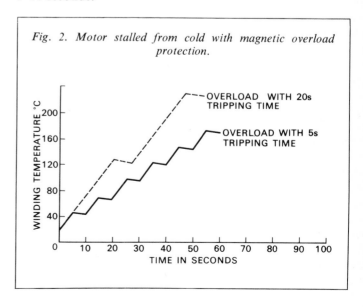

Fig. 2. Motor stalled from cold with magnetic overload protection.

In the case of the relay having a tripping time of 5 seconds, protection is probably adequate as the operator would, without doubt, realise after a few attempts to start, that something was wrong. Unfortunately, a relay having a lag of 5 seconds at start current would allow an acceleration time under healthy conditions of only approximately 1 second, and a relay having a long time lag would have to be used in most cases.

The other curve in Fig. 2 illustrates the stator heating with a relay time lag of 20 seconds at the stall current, which would permit an acceleration time of approximately 4–6 seconds. In this case protection is inadequate and the only way to make an improvement would be to increase the relay resetting time to the order of 10 minutes to allow the machine to cool before another attempt to start could be made. This is very difficult to achieve; and the user mostly would not tolerate such a delay as it would apply after every trip no matter what the degree of overload had been.

Fig. 3 illustrates the stator heating of the test machine under conditions of 25% overload using the relay having a time lag of 5 seconds at stall current, the time at 125% overload being approximately 500 seconds and at 150% overload approximately 180 seconds. The resetting time was 7–10 seconds. Due to the relatively short resetting time the curves follow closely the natural heating curves of the motor when repeated starts are made and the overload condition remains. Protection would be adequate under normal conditions of intelligent use; however, the relay cannot prevent abuse.

In the case of cyclic loading it will be seen from the characteristics in Fig. 1 that the magnetic overloads would not permit the loading illustrated in Fig. 12. A small degree of cyclic loading can be achieved but to determine the amount, full details of the relay characteristic must be known in order to prevent the plunger creeping and eventually tripping.

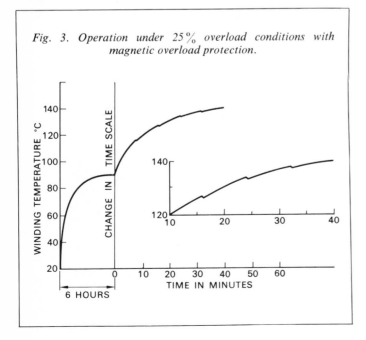

Fig. 3. Operation under 25% overload conditions with magnetic overload protection.

Magnetic overloads, being current-operated devices, cannot provide protection against causes of overheating which are not reflected in the line current, *eg* blocked ventilation passages, etc. This type of overload relay also cannot discriminate fully between a hot and a cold motor. The time lag of a hot overload is less than for a cold one, since the heat from the solenoid coil warms up the dashpot and the viscosity of the damping fluid falls. The thermal time constant of the relay, however, does not match that of the motor due to the differences in mass. In addition, since it has only a short memory (*ie* it resets quickly), it cannot provide adequate protection against such causes of overheating as repeated inching, etc.

THERMAL OVERLOAD RELAYS
Bimetallic
Thermal overloads are the most commonly used relays for the protection of induction motors in the UK. Basically they comprise three bimetal strips, each of which is heated by a small heater element connected in series with the motor supply.

Increase in the temperature of the bimetallic strips causes them to deflect due to the different coefficients of expansion of the two metals comprising the strip. Deflection of the strips operates a mechanism linked to the trip contacts and so arranged that the contacts are tripped when any one or all the phases heat up to predetermined temperature. In addition some types incorporate a differential arrangement which causes tripping should the currents become unbalanced due, for example, to a line fuse blowing or the supply voltage becoming severely unbalanced.

Being thermal in principle these relays have a natural characteristic which follows more closely the thermal characteristic of a motor winding. However, being physically small it is not possible to match completely the thermal inertia with that of the motor, particularly in the cooling cycle.

Fig. 4 shows the heating of the test motor with locked rotor, using a typical relay capable of enduring a start lasting approximately 5 seconds from cold. Whilst the protection is not ideal, it is better than that given by the equivalent magnetic relay, due to a reduced tripping time and an increased resetting time. The protection given does permit a degree of abuse, but is probably adequate.

Fig. 5 shows stator winding temperature, again of the test machine, at 25% overloaded; performance is adequate to protect the machine. Regarding cyclic loads, it can be seen, as in the case of magnetic overloads, that thermal overloads would not permit the duty cycle illustrated in Fig. 12. However, a duty cycle having a wider variation of loading can be achieved with the thermal relay than with the magnetic type. The applications engineer must have full knowledge of the characteristics of the relay being used to enable him to determine the degree of cyclic loading that can be achieved.

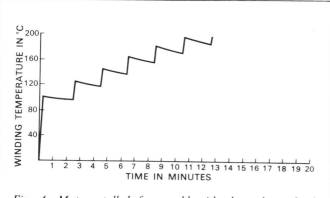

Fig. 4. Motor stalled from cold with thermal overload protection.

It has become common practice to incorporate ambient temperature compensation to thermal relays; *ie* the current/time characteristic is substantially the same within the normal limits of ambient temperature encountered. Assuming that the motor is in the same vicinity as the relay, this is quite wrong, as a motor is capable of giving more output without overheating at the lower ambients or conversely less output at the higher ambients.

In practice the problem is not simple, since in many cases the overload relay will be situated in a position where it has no knowledge of the temperature of the motor cooling air, so it can be wrong whether it is compensated or not. In practice it avoids unnecessary tripping and no cases are known where the failure of a machine has been attributed to compensation. Like the magnetic overload, protection is provided only against causes of motor overheating which are reflected in the current.

Thermal overloads are not so robust as the magnetic types; however, usually in the modern types all the delicate parts are enclosed within a moulded housing and are therefore protected against physical damage. When directly-connected thermal relays are used, particular care has to be taken in selecting the starter fuses to ensure that the fuses protect the heater elements under short circuit conditions.

Solder-pot type

The solder-pot type of overload utilises the change in condition of low-melting alloy of the solder type to operate a set of contacts. The hard solder retains the contacts by a catch, which is released when the solder melts. The solder is heated by the passage of current either directly through the solder itself or indirectly by heaters.

This type of relay is rarely used in the UK, but has been used with success in the USA for a number of years.

TEMPERATURE-SENSING PROTECTIVE DEVICES

These are devices which are in no way connected or related to the current in any part of the motor circuit. They are simply devices which sense or respond to the temperature of that part of the motor in which they are situated. Although primarily used in connection with the temperature of the motor windings, these devices can be, and are, used for sensing or monitoring temperatures of other parts of the motor, such as bearings, where the importance of the application or duty involved warrants this additional protection.

When located in motor windings, these devices cannot detect overloads as such, but only sense the integrated effect of these overloads or changes in current insofar as they affect the temperature of the windings. The devices are usually pre-set, so that somewhere in the control circuit, they open (or close) a pair of contacts when the device itself reaches the pre-set temperature.

One of the main criteria with such devices is that the temperature of the responsive element of the device shall follow the temperature of the windings within which it is located or embedded as closely as possible; *ie* with the minimum thermal lag or thermal inertia. It is therefore important that the responsive element be in as close contact as possible with the windings yet adequately insulated (electrically) from them. Equally the responsive element must have as small a mass as possible so that the quantity of heat energy (thermal inertia) corresponding to unit temperature change is a minimum.

Forms of temperature sensors

There are several ways in which electrical sensing devices, physically small enough to be located in the end windings of motors, can be used to effect the tripping of a no-volt coil circuit in motor control gear. The simplest is a thermostat, *ie* a device with contacts physically opened or closed at a pre-set temperature by the change in temperature of a bimetallic element.

Thermocouples consisting of a junction of two suitable dissimilar metals such as copper/constantan will produce a small voltage signal proportional to the temperature of the junction. Such thermocouples have substantially linear

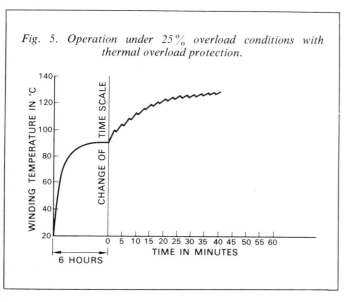

Fig. 5. Operation under 25% overload conditions with thermal overload protection.

147

voltage/temperature relations and are most frequently used for measuring or monitoring temperatures, but can be used in conjunction with amplifier/relay circuits to trip the supply to a motor when a predetermined temperature is reached.

A further system is to use a resistance element consisting of a fine wire laid between insulating layers of material, designed to have low thermal inertia. The resistance element is embedded in the motor windings and connected to a resistance sensing unit in the control gear. Whilst their response is good, and if applied correctly their degree of protection afforded is excellent, their use is usually limited to special application motors such as hermetically sealed refrigerator motors from low to quite high powers.

There are, however, semi-conductor materials which have a pronounced non-linear resistance/temperature characteristic which can be used with amplifiers or relays to produce very sharply defined tripping temperatures. Such non-linear semi-conductor devices are known as thermistors and are available with either negative or positive temperature coefficients.

There are therefore three temperature-responsive protective devices available: thermostats; thermocouples; thermistors. Before examining their relative advantages and disadvantages, it is necessary to consider the thermal factors of the motor itself.

MOTOR INSULATION AND MATERIALS

The most vulnerable part of a motor, when subjected to over-temperature, is normally the insulation in the stator (or rotor) windings. The specifications of the British Standards Institution, the International Electrotechnical Commission, Lloyds Register of Shipping, and other authorities assign permitted temperature rises to windings using various classes of insulant. (Table 1.)

THERMISTORS

Thermistors is the name given to a family of semiconductor (ceramic type) devices which exhibit significant changes in resistance with change in temperature; *ie* thermally sensitive resistors. Dependent upon the chemical formulation and processing treatment applied to the semiconductor material during manufacture, the shape of the characteristic curves of thermistors can be changed, and range from those which exhibit a reduction in resistance with increase in temperature (negative temperature coefficient—NTC) to those whose resistance increases with increase in temperature (positive temperature coefficient or PTC).

Earlier thermistors used in motor protection systems were of the NTC types, but for this application they have been superseded by PTC types which exhibit a more desirable characteristic curve form and other advantages. For instance, if the thermistor circuit be damaged or broken, the system will fail to safety, *ie* trip.

Characteristics

Based on barium titanate ($Ba\,Ti\,O_3$) the PTC thermistor now used for motor protection is formulated to produce negligible increase in resistance with increasing temperature until a point known as the Curie point is reached, whereafter with increasing temperature the resistance rises rapidly.

In the manufacture of thermistors the temperature at which the Curie point occurs can be selected by choice of the composition of the semi-conductor materials whilst the slope of the resultant curves is largely dependent upon the processing treatment (sintering, etc) applied to them. The user therefore has a wide choice of significant operating temperatures whilst still retaining the most desirable shape of curve.

Table 1. Analysis of temperature rises for windings of low tension rotating machines irrespective of size but below 1000 V (temperature in °C).

AMBIENT	BS2613 CLASS					AMBIENT	IEC34–1 CLASS					AMBIENT	**Lloyds Register of Shipping CLASS		
	A 40	E 40	B 40	F 40	H 40		A 40	E 40	B 40	F 40	H 40		A 45	E 45	B 45
BS2613 temp. rise by thermo.	55	65	75	95	115	IEC 34—1 temp. rise	50	65	70	85	105	Lloyds temp. rise by thermo.	40	55	60
BS2613 temp. rise by res.	60	75	80	100	125	IEC 34—1 temp. rise by res.	60	75	80	100	125	Lloyds temp. rise by res.	50	65	70
BS2613 total temp. by res	100	115	120	140	165	IEC 34—1 total rise by res.	100	115	120	140	165	Total temp. by res.	95	110	115
To get hot spot add*	10	10	10	15	15	To get hot spot add*	10	10	10	15	15	To get hot spot add*	10	10	10
Probable hot spot temp.	110	125	130	155	180	Probable hot spot temp.	110	125	130	155	180	Probable hot spot temp.	105	120	125
Hot spot temp. by BS2757	105	120	130	155	180	IEC 85 hot spot temp.	105	120	130	155	180	Recommended hot spot	105	120	130
VARIANCE	—5	—5	0	0	0	VARIANCE	—5	—5	0	0	0	VARIANCE	0	0	+5

*This is an allowance estimated by experience and is the difference between average and maximum figures.

**Lloyds Register of Shipping Rules for the Construction and Classification of Steel Ships 1966 quotes figures for Classes A, E, B only, Classes F and H being subject to special consideration. The figures shown are for unrestricted service.

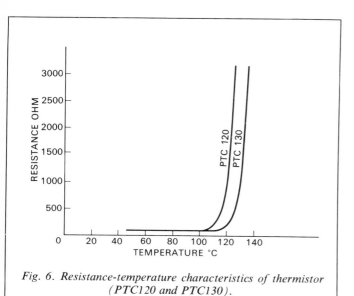

Fig. 6. Resistance-temperature characteristics of thermistor (PTC120 and PTC130).

Most motor manufacturers have standardised on PTC120 and PTC130 (Fig. 6) which are appropriate to Class 'E' and Class 'B' insulation systems though alternatives are available to suit other insulation systems or temperature requirements.

Thermistor control

The thermistors themselves carry very small current (a few milli-amps in the cold condition) and this is reduced to one tenth or less in the hot condition, due to the very high resistance involved. Very small thermistors cannot be used directly in the no-volt coil circuits nor simply in the circuits of a small interposed relay. It is necessary to provide additional circuitry which will respond to the sensing characteristics of the thermistor and which can provide a switching function.

Basically this circuitry can be entirely solid state using resistors, capacitors, transistors, zener diodes, etc., with a thyristor forming the final switching component. Such a circuit in fully encapsulated module form is illustrated in Fig. 7. Dependent upon supply voltages and VA requirements on the control gear no-volt circuit, additional small input transformer and/or output stage relay may be necessary.

The thermistor control unit is normally housed in the control gear which may be remote from the motor, and in such conditions the thermistor leads may run in the same conduit as the motor power cables. The control unit circuit design therefore must not be sensitive to any pick-up in the leads. The units shown in Fig. 7 were extensively tested during the development stage, using thirty-foot cable runs with all possible cable arrangements to confirm freedom from pick-up under transient switching conditions in the main cables.

Such thermistor control units, or amplifiers as they are sometimes called, may be adapted to or incorporated in various forms of automatic or specialised control gear arrangements. An example of the application of a thermistor control module embodied in a solid state reversing direct-on-line switching unit or starter as incorporated in sophisticated power station automatic valve actuator control is shown in Fig. 8. In this case the thermistor protection module has been specially designed so that, although primarily intended to operate with a sensing circuit consisting of three thermistors in series (one in each phase of the motor windings), it will function equally well with only two or even only one thermistor in the circuit with no more than 2°C difference in the tripping temperatures.

It should be appreciated that the system is such that whether four, three, two or one thermistor elements are in

Fig. 7. Thermistor and control modules: (B) wide range control module with transformer/relay; (A) thermistor; (C) low-voltage solid-state plug-in type control module.

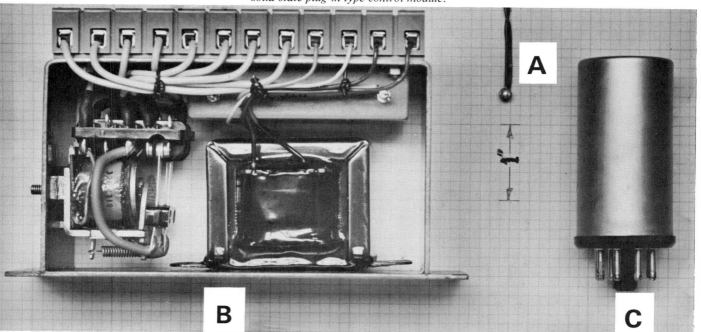

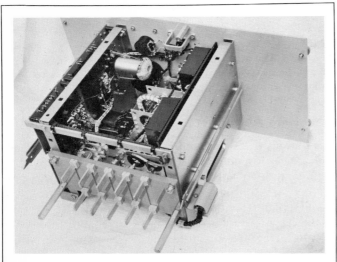

Fig. 8. Application of thermistor control module (LSE Ascron equipment: cylindrical unit, top left).

circuit, correct tripping features will apply even if only one or two of the thermistors is subjected to over-heating; *ie* as on single phasing. The purpose of this feature of the design is that whilst the thermistor itself is a very reliable component, the leads from the thermistor are very light (usually 7/0076 PTFE covered flexible) and when motors are stripped down for maintenance, damage to a lead may arise.

Thermistors handling larger currents are available, capable of operating directly a small interposing relay, but such thermistors are very expensive and more bulky to accommodate and the relay used is comparatively delicate due to the sensitivity required.

The advantages of thermistor protection are:
(a) The small physical size of the thermistor renders it eminently suitable for embedding inside the motor end winding prior to impregnation of the winding. (See Fig. 9(d).)
(b) Its small size results in minimum thermal inertia, minimum temperature gradients and lowest possible thermal time constant.
(c) As will be seen from the curves in the later section on tests, etc., the thermistor offers better protection than any other single system.

The disadvantages of thermistor protection are:
(d) The cost of the additional thermistor control unit (plus transformer and/or relay where required).
(e) The necessity to accommodate the thermistor control unit in control gear, but since this is very small, it will not normally present a problem.
(f) Two additional leads between motor and starter.

Where thermal protection devices are fitted, if they offer adequate protection as in the case of thermistors, then the normal current overloads in the control gear can be omitted and this can offset the disadvantages referred to under (d) and (e) above.

THERMOCOUPLES

Thermocouples, which consist of a junction between two dissimilar metals or alloys such as copper/constantan, produce a thermoelectric emf which increases substantially linearly with increase in temperature. The emf generated is in terms of only a few millivolts (approximately 4.25 mV at 100°C) the power being extremely small. Whilst this may be an adequate source for a sensitive galvanometer to measure temperature, it is not adequate for a protection system and some form of amplifier/relay is necessary.

There is also usually the necessity for a further (cold) junction to be accommodated somewhere in the circuit. As a result, this is not regarded as being a convenient way to protect machines, but is very suitable for measuring or monitoring temperatures in any part of a machine. Thermocouples can be very small and easily accommodated within the motor windings.

THERMOSTATS

Thermostats capable of being used with motor windings are available in several different forms and can initially be categorised as follows, in terms of types and method of fitting:

(1) Thermostats with non-sealed enclosures which are designed to be strapped or taped on to the external surfaces of stator end windings.

This type has several disadvantages: (a) The bimetallic element is separated from the winding by insulation and a comparatively large air distance involving a large temperature gradient. (b) It senses the temperature of the external surface of the end winding which is not the hottest point (c) On motors with ventilated enclosures (*eg* screen protected drip-proof, etc) the body of the thermostat is exposed to some degree to the internal air stream which will retard its operation. (d) Two additional leads between motor and starter are required.

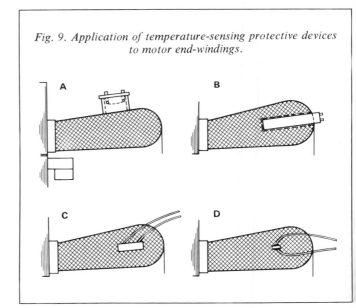

Fig. 9. Application of temperature-sensing protective devices to motor end-windings.

It has the following advantages: (e) Thermostats which are defective or damaged can be changed without interfering with the winding itself. (f) Can be fitted to existing motors.

(2) Thermostats of the non-sealed types which can be embedded in the end windings of the motor subsequent to winding impregnation. It is usual to preform a cavity in the end windings prior to impregnation and then insert the thermostat after impregnation. It is essential that the thermostat be a close fit in the cavity.

This type has the following disadvantages: (a) Such thermostats are bulky compared with miniature sealed types and consequently are somewhat difficult to accommodate on very small motors. (b) Thermal gradients are less than in (1) above, but greater than (3) below. (c) Cannot be fitted, by partially embedding, to existing motors. (d) Two additional leads between motor and starter are required.

The advantages are: (e) Thermostats are replaceable. (f) Are substantially unaffected by airstream. (g) Sense an average temperature, neither minimum nor maximum—more advantageous than (1) above, but inferior to (3) below.

(3) Thermostats of the miniature sealed types are usually of European manufacture, but are used extensively in the UK.

The disadvantages are: (a) The small size may limit contact voltage and current switching capacity. (b) They cannot be fitted in this manner (*ie* embedded) to existing motors. (c) Two additional leads between motor and starter.

The advantages are: (d) Small size permits easy accommodation. (e) Being sealed, they can be fitted in windings prior to impregnation, thus eliminating air spaces around the thermostat case and minimising temperature gradients. (f) Being small in terms of both case and bimetallic element, they have a lower thermal inertia than (1) or (2) above. (g) They can be fitted to an existing motor by taping onto end windings, but give inferior protection when used in this manner.

Thermostats all have the basic advantage that they are simple and, dependent on contact rating, can directly open or close the control gear no-volt circuit. Where the voltage or current parameters of the control gear no-volt circuit exceed the contact capacity of the thermostat, a simple relay is interposed in the circuit. No amplifier or other sophisticated equipment is necessary.

Though thermostats can be made with very close tolerances on preset operating temperature, the usual commercial tolerances for thermostats used as safety cut-outs in motors vary between $\pm 2°C$ to $\pm 6°C$. Fairly wide trip/reset temperature differentials tend to be used as this results in a much more stable and reliable thermostat snap action.

TEST PROGRAMME

In order to provide factual test data to substantiate the validity or otherwise of thermal protection, a test programme was instituted on a Class 'E' insulated 15 hp, 1500 rev/min motor of the ventilated (screen-protected drip-proof) type frame size C160M—normal squirrel-cage, delta connected. For the test, the motor was arranged to belt-drive a d.c. generator which had an inertia approximately three times that of the motor itself. Starting was direct-on-line.

An attempt was made simulate all the working conditions which could result in overheating of the motor, except where the motor is subjected to extraneous radiated, convected or conducted heat. Apart from the fact that these latter conditions are virtually impossible to simulate on a test bed, the range of other tests taken are sufficient to enable one to anticipate the results of these latter conditions.

The test programme included: (a) Various degrees of continuous overload. (b) Duty cycles involving various degrees of overload. (c) Stalling when motor was already at normal working temperature. (d) Stalling from cold condition. (e) Single phasing conditions. (f) Blocked ventilation conditions.

For these tests there were fitted: Three thermocouples in stator slots, one in each phase. Three thermocouples in stator end windings, one in each phase. Three thermostats of miniature form, identified on Fig. 10 etc as 'thermostat A', one in each phase and located as Fig. 9(c). Three thermostats identified as 'thermostat B', one in each phase, located as in Fig 9(b). Three thermistors type PTC120 located as in Fig. 9(d). All thermostats were 120°C setting.

It was found that on all tests, the temperatures recorded on the thermocouples showed highest on those located in the end windings and the temperatures shown on the curves are the highest values recorded.

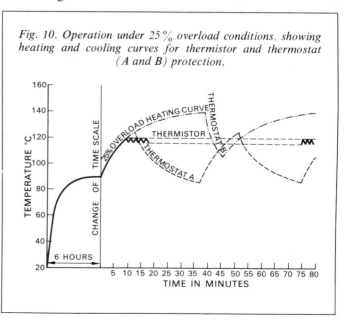

Fig. 10. Operation under 25% overload conditions, showing heating and cooling curves for thermistor and thermostat (A and B) protection.

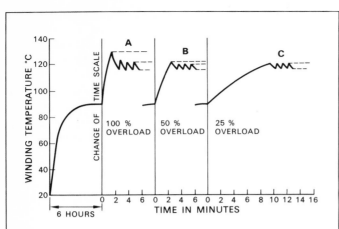

Fig. 11. Influence of degree of overloading on peak temperatures with thermistor (PTC120) protection: (A) peak temperature initially 130°C, stabilises at 122°C; (B) peak temperature initially 122°C, stabilises at 120°C; (C) maximum temperature stabilises at 120°C, no initial peak.

The conditions resulting in overheating of the motor can occur either when initially starting up, *ie* with the motor cold, or when the motor is already at normal steady working temperature or of course at any value between these two conditions. For the following tests the cold condition was taken at 20°C, and for the condition representing normal working temperature a figure of 90°C was taken, which represents 70°C rise (by thermocouple) above 20°C ambient.

Within the limits of BSS the motor could have had a permitted temperature rise of 75°C above an ambient of 40°C, *ie* an actual steady temperature of 115°C. The former figure of 90°C was used as this is more typical of average working conditions. If the latter figure of 115°C had been taken, the only difference in the results would have been to shorten initial tripping times.

BS2613 does not recognise thermocouples for winding temperature measurement on low tension machines, but it may be noted that tests taken on this machine at normal steady temperature showed that for 90°C actual temperature the corresponding figure calculated from resistance was two degrees lower at 88°C (68°C rise above 20°C ambient). For convenience all subsequent figures shown are measured by thermocouple.

Continuous applied overload

Fig. 10 shows operation under conditions of 25% overload applied after the motor had already reached normal working temperature. The time taken to attain a temperature of 120°C in the end windings was 9.5 minutes, at which stage the thermistor circuit tripped. Thermostat A operated at 124°C after 12.5 min but thermostat B took 38 min to trip at a winding temperature of 140°C.

After tripping in each case, the equipment was allowed to re-set naturally, restart and heat up to tripping point, etc, repeatedly, giving the curves shown. The shorter time constant and small differential (4°C) of the thermistor system resulted in the motor being back in operation only about one minute after it tripped and if the overload had been removed, as one would expect in normal service, the motor could have continued in operation with a lost time of only one minute approx. With thermostats A and B the lost times would correspondingly have been 24 min and 7 min respectively, due to their greater thermal inertia, longer thermal time constants, and inherently larger differentials.

Even though the thermistor protection system resets quickly under these conditions, the machine is still at a relatively high temperature. If restarted, immediately on re-set against a very high inertia load with consequent long accelerating time, there would be a likelihood of tripping during the accelerating period, even if overload, as such, no longer existed. Under such conditions an appropriate cooling down period may have to be allowed.

The rate of rise of temperature depends upon the current which is related to the degree of overload. Dependent upon the rate of temperature rise, there is a tendency to overshoot on the initial tripping. Fig. 11 illustrates the change in rate of temperature rise and degree of initial overshoot for conditions of 100%, 50% and 25% overload with thermistor protection. Although not shown on these curves, the degree of overshoot is greater with thermostats than with thermistors.

Duty cycle involving overloads

The temperature rise of a motor winding depends upon the various electrical losses within the machine, some of which are constant losses and are independent of load variations (*eg* iron loss). Others, such as stator and rotor copper losses, are variable and are related to the load, but not linearly.

Obviously then under cyclic load conditions, *ie* where load values are periodically changing as on machine tools, the

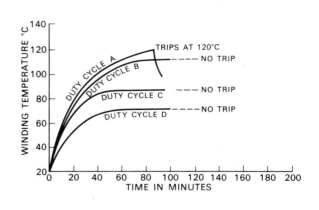

Fig. 12. Operation on duty cycle conditions, showing full use of capabilities of motor-thermistor protection; curves show maximum temperature attained during cycles; protection by thermistor type PTC120. Duty cycles: (A) 50% load 5 min, 150% load 4 min, repetitive; (B) 50% 5 min, 150% 3 min, repet; (C) 50% 5 min, 150% 2 min, repet; (D) 50% load 5 min, 150% load 1 min, repetitive.

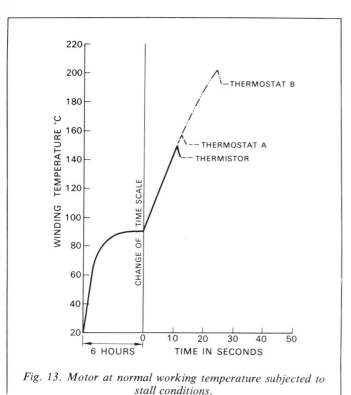

Fig. 13. Motor at normal working temperature subjected to stall conditions.

total heating applied to the motor is an integration of constant and variable losses over a period of time. In the case of a motor which can withstand a certain value of losses without overheating, on cyclic or varying load conditions with some periods of underloading, some periods of overloading can be withstood, providing the overall integrated losses do not exceed the value which the machine can withstand without overheating.

These conditions are shown in Fig. 12 for a motor, the load of which fluctuates between 50% underload and 50% overload. Four curves are shown illustrating different cycles of loading, from which it can be seen that quite appreciable lengths of time on 50% overload can be applied without overheating and tripping. Such conditions would be quite impossible using current overloads as they would invariably trip under these conditions.

This is a very good example of how thermal protection enables the full capabilities of the motor to be utilised. Though these are actual tests, and can be taken as typical of small/medium size motors, this should not be taken as an indication that this form of duty cycle overloading can be applied indiscriminately to all machines. Where such conditions are envisaged, the user should discuss the matter fully with the manufacturer to ensure that factors other than purely thermal ones are given adequate consideration.

Stalling motor at working temperature

Under stalling conditions, whether the motor is initially cold or already at working temperature, the rate of temperature rise is extremely rapid (in the order of 6°C per second on the machine tested). If the protective thermally-sensitive device is satisfactorily to sense the temperature rise of the windings under stall conditions, it must be capable of doing this very quickly as every second delay represents some 6°C additional temperature rise or overshoot.

Three parameters of the protective devices are therefore of paramount importance. (a) The device must have a small thermal inertia. (b) The device must have the minimum degree of thermal insulation from the windings compliant with adequate electrical insulation therefrom. (c) The device must be located in the hottest part of the windings. The thermistor is the most satisfactory on all three counts; thermostat performance depends on the type, design and physical size of the device.

Fig. 13 illustrates the degree of temperature overshoot due to thermal inertia and thermal gradient due to insulation, etc. Obviously the larger size, heavier thermal insulation, greater integral air space, and location of this type 2 thermostat (curve B) detracts from its efficient operation under stall conditions. The degree of temperature overshoot on thermistors and on thermostat A is quite low and presents no problems with respect to effect on insulation.

This initial temperature overshoot occurs on the first cycle only and if the system is such that the motor attempts to restart when the protective device resets, the peak temperature on subsequent cycles will be appreciably lower as shown in Fig. 14.

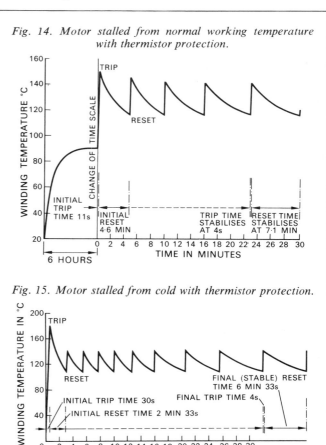

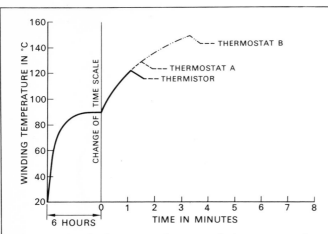

Fig. 16. Single-phasing conditions applied to motor when already at normal working temperature; temperatures recorded in phase of winding carrying maximum current.

Motor stalled from cold

Tests with thermistors only were taken with the machine on continuous stall and the results are as shown in Fig. 15. The initial peak is not repeated and as the temperature gradient throughout the motor settles down the length of time on decreases from 30 seconds to 4 and the length of time off increases from 2 min 33 s to 6 min 33 s.

Motor on single phasing conditions

If a 3-phase motor, due to any supply circuit fault, is subjected to single phasing, then if already stationary it will be incapable of starting and the conditions similar to the above paragraph will apply.

This test, therefore, was made on a motor already running at normal full working temperature with one phase then opened to simulate a line fuse blowing or a control gear or termination fault developing. As will be seen from Fig. 16, all three devices offer adequate protection with practically no temperature overshoot in the case of thermistor protection.

ULTIMATE IN PROTECTION

It will be seen from the test results referred to that thermal protection devices, in particular thermistors, offer very good motor protection, but looking at their weakest point (stall conditions) there is a degree of overshoot in winding temperatures beyond the pre-set protected temperature, particularly on cold stall. Thermostat protection here is somewhat vulnerable, in some cases, and perhaps inferior to current overloads which could trip in somewhat shorter time if suitably designed and correctly set.

If this is regarded in the light of ultimate protection, then in the case of thermostats there may be an argument in favour of retaining current overloads as an additional safeguard. There does not appear to be justification for this with thermistors on normal machines.

CONCLUSIONS

One can readily conclude that for most applications thermal protection, and in particular thermistors, offers considerable advantages over current protection. At present both systems have applications from purely economic aspects, since on relatively small machines the current protection systems are undoubtedly much cheaper, yet on the other hand the smaller the machine, the more effective is thermal protection.

The main object of this chapter is to present sufficient information on the systems of protection available today, so that users of motors can draw their own conclusions, taking into full account the aspects of economics, reliability, and operational requirements as applied to their own particular plant. It is hoped that this object has been achieved.

The author has based this chapter on an article by Ramsden and Dring, which appeared in the LSE Engineering Bulletin; he gratefully appreciates the permission of LSE Ltd to use the material and data.

Chapter 18

Installation and Maintenance

T J Williamson
Mather & Platt Ltd

The profitability of modern industry is directly related to high productivity, which in turn depends upon the reliability of machines; good reliability results from regular and high-standard maintenance. The annual shutdown was a traditional maintenance period, but progressive organisations have replaced this once-yearly event by preventive and planned maintenance programmes co-ordinated with production. The complexity of certain automated industrial equipment means that shutdowns for lengthy repairs are intolerable. For this reason higher degrees of protection are specified, recommended spares are stocked and replenished, and complete replacement units are considered against possible prolonged disruption of production.

It is obvious that any organisation with heavy capital investment in electrical plant should adopt efficient maintenance procedures. This chapter deals with such important items as storage, installation, commissioning, and general maintenance of electric drives. A suggested preventive maintenance schedule should ensure efficient operation of plant and minimize breakdowns in service.

Storage
Electrical machines in storage warrant the same degree of maintenance as machines which have already been commissioned. General machine deterioration during this period may be rapid if certain basic precautions are not taken.

Machines with ventilated openings must be stored in a clean dry building, whilst it is strongly recommended that totally enclosed machines intended to work outdoors are afforded similar protection, as this greatly reduces internal condensation. If enclosed machines have to be stored outside they must be protected from sun and rain by a good quality tarpaulin. On arrival at the storage depot machinery should by thoroughly examined for possible damage or component shortage. Paintwork damaged in transit should be touched up, whilst all the external bright surfaces of the machine must be treated with an anti-corrosive. Inspection covers and terminal box lids should be properly secured, and cable entry apertures sealed. Openings in machines leading to insulated parts should, where applicable, be rendered vermin-proof; perforated metal strips are best used for this purpose, as they permit free breathing.

If the stores are subject to any form of vibration it may be necessary to fit shaft clamps to stop static indentation of bearing races (Fig. 1). If shaft clamps are not used, then periodic rotation by hand frees bearings and changes the surface of contact. Vibration can also effect the surface of carbon brushes, and consequently these should be lifted from their points of contact.

Moisture, through internal condensation, can be prevented from affecting the insulation by heating the inside of the machine to a temperature above dew point. If the machine is fitted with anti-condensation heaters they should be connected and switched on. A suitably guarded 100-watt lamp placed conveniently in the machine offers similar protection for most machines not fitted with heating elements.

INSTALLATION AND ALIGNMENT

Atmosphere
Electrical machines should never be installed where atmospheric conditions are less favourable than those specified for the machine at the order stage. For instance, open ventilated motors should not be placed in a wet or dusty environment; similarly, standard totally enclosed machines usually are not suitable for flameproof areas. Great care must be taken when siting machines to ensure that air inlets and outlets are not blocked or impeded by surrounding structures or machines. The temperature of the cooling air should not exceed 40°C unless the motor has been specifically designed or derated to cater for such conditions.

Foundations
Motor foundations should be solid and level, with a natural frequency in excess of the frequency of vibration of the motor or drive at full speed. In the case of elevated machinery such as crane drives, it is important to ensure that the mounting platforms are rigid in both the longitudinal and transverse planes.

Axial and radial forces
Axial or radial thrust loads should never be imposed upon the driving shaft of an electric motor unless the bearing arrangement has been specifically designed to withstand such forces. Motors which are fitted with ball and roller bearings usually have no provision for endfloat and will take only a limited axial load. Consequently it is usual to fit flexible couplings, when it is intended to drive in line. The coupling must be chosen and adjusted to compensate for any axial misalignment of the shaft ends which may occur. If it is intended to use a belt or chain drive the user should get confirmation from the motor manufacturers that the bearings will withstand the radial forces exerted.

Fig. 1. Typical shaft clamp arrangement (courtesy Parsons Peebles Ltd.).

Couplings, pinions, pulleys
The rotors of electrical motors are dynamically balanced to ensure vibration free running. Couplings, pinions, and pulleys should be similarly balanced before mounting on the shafts of motors. Care must be taken not to damage bearings when fitting transmission parts to the shaft. Driving shafts usually have a tapped hole to take a stud and plate by means of which a coupling can be drawn on. Heating couplings to a temperature between 60°C and 80°C, and the application of a thin lubrication oil to the shaft surface, greatly assists this operation. On no account should transmission parts be hammered onto shafts as any heavy impact is likely to cause bearing damage.

Alignment
Direct drive. It is a common misconception that the use of flexible couplings alleviates the need for accurate alignment. The truth is that great care must always be taken during alignment as any substantial inaccuracy can produce excessive loads which quickly ruin bearings and flexible components of couplings. BS3170:1959 (Appendix) gives methods of checking the angular and parallel alignment of couplings. However, before commencing alignment rotate each shaft independently to check that the bearings run freely and that the shafts are not bent.

Where the working temperature of a unit has the effect of lifting the centreline of one machine in relation to the other, allowance in the height of the appropriate machine must be made at the time of alignment if this work is done with the unit cold.

Indirect drive. Indirect driving via belts or chains should only be applied if the shaft and bearings have been designed to withstand the resultant radial load. Pulleys must always be in parallel alignment to ensure that the belt will run true. The belt or belts should be just tight enough to prevent slipping; excessive tension must be avoided. With chain or gear drives the utmost care must be taken in the aligning of sprockets and gear wheels so that chains run true and gears are correctly meshed.

COMMISSIONING

Before running for the first time, the following checks are strongly recommended.

(a) Check that the supply voltage is correct and is in accordance with the motor nameplate.
(b) Rotate the shaft by hand to ensure that rotating parts are not fouling.
(c) Thoroughly inspect the terminal box making sure that gaskets, diaphragms and desiccators, where fitted, are in good condition; if not, replace them.
(d) Measure insulation resistance to earth. Check that the figure obtained is in accordance with the manufacturer's recommendations.
(e) Make certain that holding down bolts and location pins are secure.
(f) Check control gear for freedom of operation and circuit continuity. This is especially important on d.c. regulators.
(g) Inspect brushgear of wound rotor machines, making sure that brushes are correctly bedded and tensioned.
(h) Check that the bearings are lubricated.

Assuming the above checks meet with satisfaction of the commissioning engineer, the machine is ready to run. For the first run it is desirable to separate the motor from the drive.

Light run

If a driven unit is not to be damaged by rotation in an adverse direction, the electric motor must be subjected to an initial light run. Couplings and pinions must be parted, belts and chain drives removed, in order that the motor can run completely unloaded. The motor can then be switched on and the rotation checked. A light run for at least one hour is required to check bearings and vibration levels. If ambient conditions permit, cover plates on totally enclosed machines should be removed so that the internal heat can dispel moisture. (Note: Series wound d.c. machines must never be run unloaded. For such machines a controlled load start is required.)

Starting on load

Make quite sure that the driven unit is ready for operation. Reconnect the transmission from the motor to the driven unit. Adjust overloads to 110%, 115%, as required by the load. Switch on the machine, time the run up and check vibration during acceleration. Set timing devices accordingly.

Where practical, the load should be increased in equal increments up to full load conditions. Check vibration at each load. Vibration checks are required every hour or so during the initial heating cycle to ensure that alignment is not affected by the expansion of the shaft ends. Bearing temperatures should be logged at regular intervals until steady state condition is reached. Finally check that the current at full load does not exceed the nameplate figure.

Continuous stopping and starting should be avoided on squirrel cage induction motors. For short acceleration periods two starts in quick succession is a safe limit, followed by an interval to allow the machine to cool. If commissioning or subsequent operation requires a greater number of consecutive starts, suppliers should be consulted.

LUBRICATION OF BEARINGS

Ball and roller bearings are lubricated by grease or oil; the choice of lubricant depends upon the bearing speed. Grease lubrication is used for low and medium bearing speeds, whilst oil lubrication is used at higher speeds and on all machines which have sleeve bearings. Seals must always be fitted on oil lubricated bearings to stop the lubricant from seeping along the shaft. Grease has the advantage that it forms a seal suitable for most environmental conditions. The disadvantage of grease is that at higher bearing speeds it has to be added more frequently and the intervals between greasing may become too short to be practical.

Greases

There are three principal greases used on rolling bearings, based on lime, soda or lithium. Lime-based greases are good for working temperatures between $-20°C$ and $+55°C$, soda-based greases between $-20°C$ and $+70°C$, and lithium-based greases between $-35°C$ and $+100°C$. Because of the high operating temperatures now accepted in electrical machines, it is only natural that the lithium-based greases are most popular.

Modern greases have additives which reduce the rate of grease deterioration and inhibit rust, but the best of greases has its useful life impaired when contaminated by extraneous matter or when excessive temperatures are reached. Shaft seals are used to stop the entry of foreign matter likely to contaminate the grease. High grease temperatures may result from any of the following: (a) High ambient temperatures surrounding the machine. (b) Frequent starting, causing the grease to slump in the bearings, giving rise to excessive churning losses. (c) Too much grease in the bearing.

Because overgreasing can play such an important part in bearing performance, grease relief devices are becoming increasingly popular. In simple form the outer end cap of each bearing has an opening at its lowest point to allow grease to escape, reducing pressure built up and consequent overheating due to excessive greasing. Grease relief devices allow excess grease to be expelled by rotation when the machine is started. Because such devices cannot be overfilled the problem of excessive churning losses due to frequent starting can be eliminated in this way.

Oils

Oils are more uniform in their characteristics than greases and usually present less problems. They are generally selected on the basis of a suitable viscosity for a particular speed or temperature requirement. Oil lubrication is used on sleeve bearings and may be necessary on certain rolling element bearing applications. Sleeve bearings are most commonly found on large machines or machines requiring near-silent running.

Excessive oil level in a bearing reservoir can cause considerable churning, heating, formation of oil mist, and oxidation, thus reducing the working life of the oil. Fortunately it is easy to maintain exact oil quantities, as the oil reservoirs have transparent windows or dipsticks marked with optimum levels.

Maintenance

The greasing of bearings should be included in any planned maintenance schedule, but if such a scheme is not operated, systematic greasing should be done, and records kept. Recommended greasing periods are usually given on rating plates attached to the machines. Where machines are standing idle for long periods, the grease should be inspected regularly to check for hardening or separation. Different grades or makes of grease should not be mixed; if a change is required, the bearing, housing, grease pipe and grease gun should be thoroughly cleansed of old grease and repacked with new.

On oil-lubricated bearings it is important to maintain the specified level by 'topping up' at regular intervals. Oil should be changed at approximately 12-monthly intervals unless checks for sludging prove otherwise. If machines are operating in atmospheres which cause contamination of the lubricant more regular checks may be required.

SLIPRINGS, COMMUTATORS, BRUSHES

Sliprings and commutators are used on machines where some form of electrical connection to the rotating part is required. The electrical circuit is formed by carbon brushes sliding on the surface of the commutator or slipring. To avoid arcing and burning, which can result from the interruption of the circuit at this point, maximum brush area must be maintained in continuous contact with the rotating surface. Even if the brush contact is perfect there are numerous other factors governing satisfactory performance.

The ambient atmosphere can have a very critical effect on brush performance; certain gases such as chlorine, or sulphur, oil, and ammonia vapours hinder good commu-

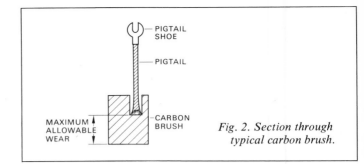

Fig. 2. Section through typical carbon brush.

tation. Dust causes scoring and grooving of the rotating surfaces and promotes rapid brush wear. Special consideration should be given to machines operating in such environments, especially if the source of pollution cannot be removed or reduced to tolerable limits.

Brush pressure has a marked influence on brush wear and current collection. Low pressure may produce heavy wear through failure to maintain intimate contact, resulting in higher electrical losses through increased contact resistance. Sparking can ensue with serious erosion of both brush and collector surfaces. High brush pressure can cause excessive abrasion and possible crushing of the brush surface. The increase in frictional losses may lead to roughening of the rings or commutators. Because electrical wear is usually greater than frictional wear, it is generally accepted that high pressure is less serious than low pressure.

Vibration can cause minute radial movement of the brushes in their holders and could lead to destructive arcing. Such vibration can result from external sources, bad balancing, faulty bearings or misalignment of the drive. Vibration must always be reduced to tolerable limits.

Brush maintenance

Brush performance can be adversely affected by any of the above conditions or can result from a combination of them. Certain problems which arise are easy to diagnose, others are extremely difficult. For this reason the controllable factors should become a regular feature of maintenance.

Pigtail shoes should always be kept clean and tight under fixing screws. Never allow brushes to wear to an extent that the pigtails become exposed and score the collector surface (Fig. 2). Different grades or makes of brush should not be mixed; if changes are considered necessary consult the

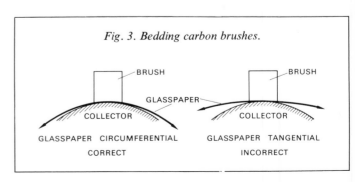

Fig. 3. Bedding carbon brushes.

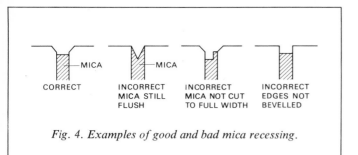

Fig. 4. Examples of good and bad mica recessing.

motor manufacturer. New brushes must always be correctly bedded, by inserting a piece of fine glass paper between the brush and contact surface, with the abrasive side to the brush. The tensioning spring is engaged on the brush, the glass paper is pulled taut and then moved back and forth until the brush has been correctly shaped (Fig. 3).

Holders

Brush holders should be set as near as possible to the collector surface; the distance from the bottom of the holder to the surface of the collector should not exceed 3 mm. Holders should be kept free from dust and dirt, allowing brushes to slide freely in the radial plane.

The clearance between the brush and holder should not be excessive, as this condition can cause brush chatter. The working parts of holders should always be free and in good condition. Holder supports should be rigidly fixed in their correct position and at the desired angle. Springs should be correctly tensioned so as to exert the recommended pressure on the brush.

Collectors

In normal service sliprings and commutators should take up a smooth surface of brown copper colour. This is often referred to as the 'skin' and presents the optimum condition for efficient current collection; consequently it should never be unnecessarily destroyed. Should the collector surface show signs of slight scoring, burning or unsatisfactory skin conditions, then light 'stoning' will remove these and present a good surface suitable for a fresh start.

Stoning should not be employed on commutators or rings in attempts to correct any of the following disorders: (a) Eccentricity. (b) Flats on collector surface. (c) Heavy scoring or pitting. (d) Heavy burning. (e) High mica separators on commutators. If any of these conditions arise it is necessary to remove the rotating members for machining. Sliprings present few problems and can be easily skimmed in a lathe. Commutators however, have to be undercut prior to skimming. Undercutting is a process by which mica separators are recessed below commutator surface level (Fig. 4).

During the critical initial running period frequent checks are required to ensure that brush performance is satisfactory. If performance is efficient checks at normal maintenance intervals will suffice.

TERMINAL BOXES

Connections to terminal boxes should not be made until the required direction of rotation has been decided and should then be in accordance with the diagram supplied by the manufacturer. It is important that incoming cables are correctly prepared and fitted. Cable terminations should be lugged so that excessive amounts of bare metal are not visible (Fig. 5). Undue strain should never be imposed on terminal posts, consequently if the machine is elevated ensure that cabling is adequately braced and supported. If the cables are to be encapsulated in a sealing chamber then the compound to be used must be compatible with the cable material and any other components within the chamber.

Should the direction of rotation be found to be incorrect after compounding, no attempt should be made to interchange the supply lines, as this subjects the insulation to strain at the crutch of the cables. In such instances the chamber has to be decompounded and all the joints remade.

Dirt and moisture provide suitable routes for tracking between terminals if allowed to accumulate on the base, also wet air within the enclosure can cause arcing. To minimise the ingress of such contaminants, all seals and gaskets must be maintained in good condition.

Failures in terminal boxes through poor installation and indifferent maintenance can prove very serious, especially if the machine is connected to a large distribution system which can sustain a high fault level. The increased size of distribution systems in power stations, oil refineries, steel works, and other large installations means that fault levels of 150 MVA or more at 3.3 kV are not uncommon.

If a failure should occur within the terminal enclosure of a machine connected to such a large system, then the tremendous fault energy released could cause pressure build up of sufficient magnitude to explode the box. To overcome such a hazard, modern machines entering areas with high fault level systems are fitted with BEAMA/CEGB approved boxes (Fig. 6). These boxes incorporate the following features: (a) All-steel construction designed to withstand a specific internal pressure. (b) A pressure relief diaphragm which ruptures under fault conditions. (c) A desiccator to absorb moisture within the enclosure

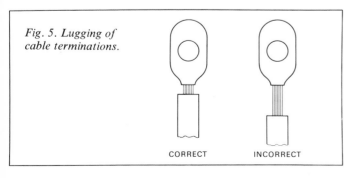

Fig. 5. Lugging of cable terminations.

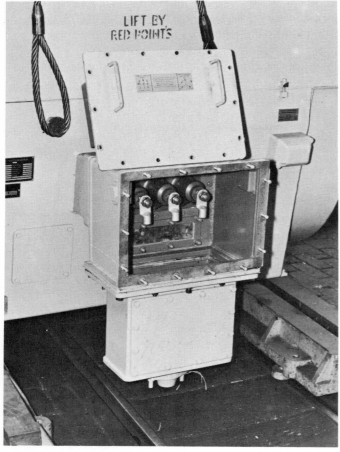

Fig. 6. A 3.3 kV BEAMA terminal box (courtesy Parsons Peebles Ltd.).

and also indicate whether the box is in fact dry. (d) Weatherproof gaskets to stop the ingress of dirt and moisture. (e) Cable seals. (f) Neoprene caps to cover terminals and cable lugs. (g) Cable spacing blocks. (h) Compounded sealing chamber. Satisfactory operation under fault conditions depends upon all the above items being in good condition and correctly positioned, hence such boxes may require more frequent maintenance.

WINDINGS AND INSULATION

Frequent operation at temperatures above the permitted limit drastically reduces the life of an insulation system. Excessive temperatures can be caused by sustained overloads, high ambient temperatures, frequent starting, single-phasing, jammed rotors, and blocked ventilation. If any of these conditions is likely to arise on the drive, then some form of thermal overload protection should be incorporated. Thermistor-actuated protection schemes are by far the most popular on low-tension motors. They respond quickly to increasing temperature and can offer split level protection. One set of thermistors can be used to trigger an alarm circuit when a certain temperature is reached, whilst the other set will trip the supply if the alarm is unanswered and the temperature is allowed to rise further. For a modest outlay these devices offer excellent insurance against a stator winding 'burn-out' disrupting production.

Moisture, dirt, oil carbon and metallic dust are other possible causes of insulation failure. A machine should therefore be kept as clean and dry as possible. Any dirt or dust visible on the windings should be dislodged by a soft brush and then removed from the surface by suction. High pressure air lines should *not* be used as they usually force dust further into crevices. Moisture can be absorbed by a machine to an extent dependent upon the environment in which it is stored. Moisture can penetrate the interspaces of the coreplate laminations, and between actual conductors and terminations, thus reducing the efficiency of the insulants.

Insulation resistance values can be dangerously misleading especially when taken cold, for when moisture is resting on the surface of the insulation, tests can show apparently safe values. If the machine is started in this state, tracking and eventual breakdown can occur as the machine warms and distributes the moisture. Periodic insulation tests are advisable and readings should be recorded, particularly on machines due for lengthy shutdown or storage. Such readings can form a guide to the causes of low insulation resistance and indicate when a motor requires attention. If the insulation resistance should fall during the idle periods until it is less than the values recommended in Table 1, the machine must be dried out.

Drying methods

There are two basic methods of drying out electrical machines: (a) Internal heating of the windings by means of a low current produced from a low voltage supply, is a very fast and effective method of drying a winding. This may be the only method that can be adopted, and if such is the case the motor manufacturer should be approached for suggestions. With this method there is always the danger of electrolytic action permanently damaging the insulation; it is particularly dangerous on high voltage machines. (b) External heating by a stream of air, usually from a fan heater, raises and maintains the temperature of the machine between 80°C and 90°C. Airflow is essential if pockets of moisture are to be dispersed. Porous sheeting can be used to conserve heat provided that it does not interfere with the air flow. Thermometers should be placed in the windings at relative points, so that recordings of temperature and insulation resistance at regular intervals can be plotted. A typical dry-out characteristic is shown in Fig. 7.

It is usual for the resistance to drop after drying begins and then to level out as the temperature stabilizes. Temperature

Table 1. Recommended minimum insulation resistance values.

bhp up to	Insulation resistance (megohms) at voltage range:		
	0–1000 V	1001–3000 V	3301–6600 V
100	1.0	5.0	—
500	1.0	5.0	5.0
1000	0.75	4.0	5.0
5000	0.5	2.0	2.0

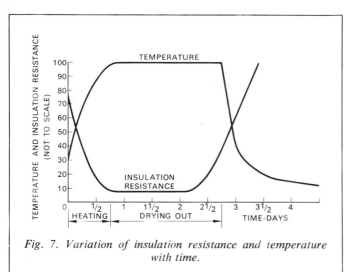

Fig. 7. Variation of insulation resistance and temperature with time.

should be closely controlled from this point, maintaining a constant level for maximum drying efficiency. It is from this point that reliable resistance readings can be taken. As the machine begins to dry, the readings rise and eventually level off at a steady value. After three hours at this value it is safe to assume that the windings are dry.

ROUTINE PREVENTIVE MAINTENANCE

Daily. (1) Check oil levels in pedestal reservoirs of sleeve bearings. Top up if necessary. (2) If winding thermocouples are fitted and relayed to instruments. record winding temperature (note: readings of bearing temperature indicators where fitted). (3) If machines incorporate a modern high tension terminal box examine the desiccator to ensure that the box is dry.

Weekly. (1) Examine air inlet and outlet grilles; if they are becoming blocked by foreign matter, stop the motor and clear them. (2) Record approximate weekly running hours.

Monthly. (1) Check condition of filters if fitted; clean or replace if blocked. (2) Check surface condition of sliprings or commutators, clean out dust or dirt that may have accumulated.

Six-monthly. (1) Check brushes for freedom of movement in holders, adjust spring tensions. (2) Check brush lengths; replace if worn beyond the specified point. (3) Check insulation resistance of windings to earth and between phases where applicable. (Note: If the machine is operated continuously, or intermittently at frequent intervals, the insulation check can be made at yearly intervals.)

Yearly. (1) Check alignment of both couplings. (2) Check the security of all holding down and fixing bolts. (3) Check for cleanliness of the main terminals, clean if necessary; renew terminal box lid gaskets if their condition is doubtful. (4) Drain and flush oil lubricated bearings; replenish with new oil. (5) If anti-condensation heaters are fitted check for continuity.

Two-yearly. In the absence of other instructions on the nameplate, flush out the grease lubricated bearings and repack with fresh grease.

General

Always observe manufacturer's recommended lubricating intervals. If the machine has at any time to be dismantled use such periods to inspect the condition of internal ventilation passages and general condition of the windings.

BIBLIOGRAPHY

R D Wolford. Environmental testing of insulation. *Insulation* Oct 1970.

J S Johnson. A maintenance inspection program for large machines. *AIEE* 1951 vol. 70.

Carbon brushes and electrical machines. Published: *Morganite Carbon Limited.*

Sliprings—materials and surface finishes. *Carbon and its uses,* issue no. 13 1971.

Carbon brushes for industrial machines. *Le Carbone (Great Britain) Ltd.*

F R Hutchings. A survey of the causes of failure of rolling bearings. Technical report vol. VI 1965; *British Engine Boiler Co Ltd.*

The application of lubricants. 1965 *Shell International Petroleum Co Ltd.*

J H Harris. The lubrication of rolling bearings. *Shell-Mex and BP Ltd.* 1967.

Ball and roller bearings. *SKEFKO General Catalogue GB.550.*

Installation and maintenance instruction for BEAMA/CEG B 3.3 kV terminal box. Publication no. 203, *BEAMA* 1964.

Development of BEAMA/CEGB motor terminal box. *The BEAMA Journal* June 1962.

BS3170:1959 Characteristics of flexible couplings for power transmission.

BS2613:1970 The electrical performance of rotating electrical machines.

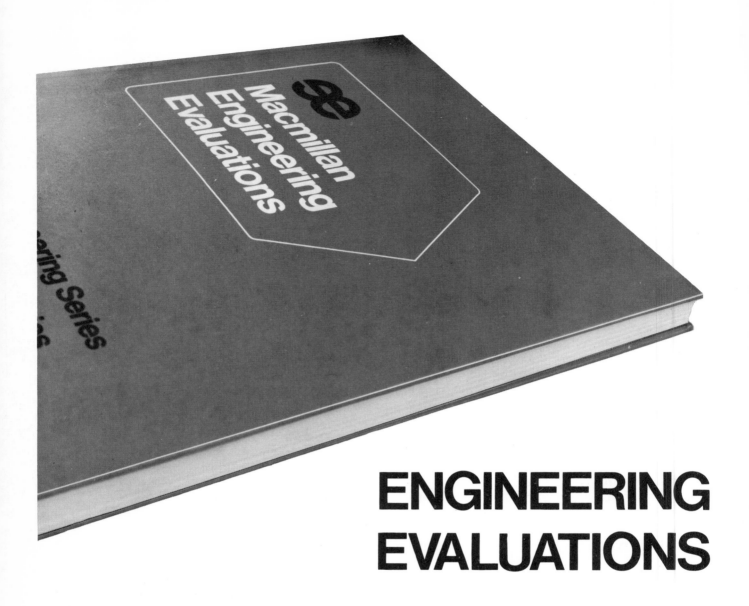

ENGINEERING EVALUATIONS

— a new series of technical books designed for the practising engineer.

These books, more practical than textbooks and more technical than manuals, are designed during discussions with industry and are then written by leading engineers working to carefully formulated editorial plans. The production and print operations are foreshortened and the end result is a *working* book providing the reader with a *current* and authoritative evaluation of its subject in the areas of theory development and application. Each book contains approximately twenty chapters, all commissioned exclusively for that publication.

In the first phase of Engineering Evaluations there are three series and within each series, four books. The series and books are as follows:

MECHANICAL ENGINEERING SERIES
Industrial Fuels
Mechanical Prime Movers
Electromechanical Prime Movers
Power Transmission

ELECTRONIC ENGINEERING SERIES
Electronics Design Materials
Component Reliability
Electronics Testing
Rectifier Circuits

PRODUCTION ENGINEERING SERIES
Castings
Metal Forming
Plastics Forming
Product Treatment and Finishing

Publishing begins March 1971 and at approximately monthly intervals thereafter, the books being designed to provide a progressive build-up of information in each series.

A4 size, heavily illustrated and casebound, these books sell at £3.50 each. An order form is included in this book.

Chapter 19
Enclosures and Motor Sizes

T J Williamson
Mather & Platt Ltd

This chapter is presented as a guide to engineers who require electrical prime movers for their mechanical drives. It deals with the various existing specifications, sizing requirements, and the ventilation and enclosure of electric motors. It is the ultimate responsibility of the final user to specify the operational requirements and prevalent environmental conditions in as much detail as possible in order that the motors eventually supplied will perform satisfactorily. For this reason the basic conditions which can influence the type and size of machine are outlined.

SPECIFICATIONS

In order to maintain a degree of uniformity throughout industry, electrical machinery is designed and manufactured to certain specifications. Such specifications can be related to machine dimensions, enclosure, performance or safety features. For instance, machinery required for use overseas is normally designed to meet the standards appropriate to the importing country. Where American influence is strong, there is a demand for machines made to NEMA inch dimensions. The majority of motor manufacturers supply ranges of machines to the following standards.

Dimensional standards
BS3979/1966. This is a current British Standard based on the recommended metric dimensions of the international standard publication IEC 72. This standard lists the height to shaft centre and fixing dimensions but does not list overall dimensions. Complete interchangeability is not guaranteed, because individual manufacturers have the right to develop their own ideas within the limits of the specification; however, main fixing dimensions are usually sufficient to allow replacement by motors of different manufacture. Outputs have been assigned to machines up to frame 315M. Above this frame preferred fixing dimensions are given and it is then left to individual companies to assign outputs.

The frame number itself relates to the height of rise of the motor shaft centre (mm) and is preceded by a letter designating the enclosure. The three letters specified are D, C, and E representing totally enclosed, ventilated, and flameproof enclosures respectively. A letter immediately following the number designates actual fixing dimensions.

BS2960. This is an older inch standard for motors which has now been superseded by BS3979. Nevertheless machines are still manufactured to this specification at the request of customers who committed themselves to earlier standardisation. Dimensionally these machines are similar to the American NEMA standard, but with different assigned outputs.

NEMA standard MGI. This is a very rigid specification and one that must be strictly adhered to if a machine is to be exported to the USA for resale. The specification assigns outputs to particular frames, gives overall dimensions in inches, starting performance and test procedure. Although such machines are completely interchangeable, the rigidity of the specification allows little room for design manoeuverability. Output allocations depend upon the insulation system which can be either class A or B. American standards do not recognise the Class E insulation system accepted in most European countries. The Canadian specification follows the US NEMA very closely.

Performance and testing standards

In order that machines may be acceptable to the majority of users, specifications have been drawn up to control motor performance within certain tolerances, to specify test procedure and state rules for the declaration of results. The current British Standard covering the electrical performance of rotating electrical machines is BS2613:1970. This specification describes eight standard 'duty type' ratings in the following classes.
(a) MCR—maximum continuous rating.
(b) STR—short time rating.
(c) ECR—equivalent continuous rating.
(d) DTR—duty type rating.

Starting characteristics are also classified permitting prospective motor users to specify a letter from A to G each of which relates to a particular starting characteristic expected from a cage induction motor. Permitted voltage variations, test procedure and performance tolerances are also given.

Environmental standards

The type of motor enclosure required is usually determined by the atmospheric conditions prevalent in the installation area. On large contracts several different manufacturers may be supplying equipment, consequently there must be no misunderstandings regarding enclosure. To overcome such difficulties BS2817:1957 gives lists of standard definitions and requirements for the enclosure of electrical apparatus.

Flameproof machines are suitable for use in certain hazardous atmospheres and are manufactured in accordance with BS229:1957. This specification classifies gases into groups which represent the degree of hazard. Flameproof motors require certification by an appropriate testing authority.

Customer specifications

Certain industries draw up their own specifications for electric motors which incorporate features to suit their individual requirements. Such specifications are common in the oil, chemical, steel, and marine industries. To quote a typical example, marine motors intended for use aboard ships are in accordance with the specification drawn up by Lloyds Register of Shipping.

MOTOR SIZE AND DESIGN

The assignment of a frame size to suit a particular application is the responsibility of the design engineer. The actual design process is to obtain dimensions and electrical winding details of a machine to satisfy a customer's requirements regarding output, performance, temperature rise, and general conditions of service. Electric motors can only be sized correctly if the information received is complete and correct. The various factors which can affect the physical size and electrical design of a motor are outlined below.

The supply

Supply details must be fully specified, including anticipated voltage variation, frequency fluctuation, fault level of the system, and where possible the supply impedance. The actual supply voltage will influence the basic design of the

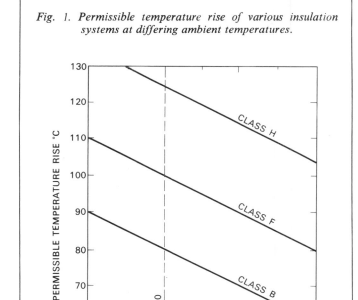

Fig. 1. Permissible temperature rise of various insulation systems at differing ambient temperatures.

electrical winding and also the insulation system. Considering a small machine, the output obtainable at say 415 V is substantially greater than at 3.3 kV. This is because the copper/insulation ratio is reduced for the high-tension machine which in turn leads to higher copper losses. On large machines the reverse can apply for high output, low voltage conditions impose difficult design problems due to the excessive number of slots required in the stator and rotor laminations.

Generally speaking high voltages should never be specified for machines below 150 kW (200 hp) whilst low voltage should not be specified for machines above 450 kW (600 hp). Variations in supply voltage, or a combination of voltage and frequency beyond the ± 5% allowed by BS2613:1970 could affect the frame size or class of insulation offered. High system impedances can seriously affect the starting performance of an induction motor and may necessitate a special design for certain drives.

Output

The engineer responsible for the design of the driven unit usually specifies the power output required from the electric motor. The physical size of motor required is then governed by the power output in relation to the specified speed. A 37 kW (50 hp) motor running at 3000 rev/min is considerably smaller than a 37 kW motor running at 500 rev/min. Although this can be partially due to a reduction in ventilation the main cause is the increase in output torque at the low speed. Power, torque and speed are related by:

$$\text{Power} = (\text{speed} \times \text{torque})/541.6 \qquad (1)$$

where power is expressed in kW, speed in rev/min, torque in kgf m (1 kW = approx. 0.746 hp).

Substituting the relevant figures in equation (1), the output torque for a 37 kW (50 hp) motor running at 500 rev/min is six times larger than for a similar power output at 3000 rev/min.

Because motor cost is directly related to overall size, slow speed machines are more expensive than high speed motors of the same power output. Gearboxes interposed between the motor and the driven unit permit the use of small high speed machines to drive a low speed unit and may prove economically more attractive than a single large low speed motor. When considering the extra financial outlay required for the gearbox, motor running costs should be taken into account; eg low speed machines usually have a lower efficiency and power factor than their high speed counterparts, resulting in higher electricity tariffs.

Speed

Single speed, multispeed, and variable speed drives are offered by motor manufacturers to cover a wide range of applications. The speed of an induction motor is related to the supply frequency and the number of poles for which it was wound. The basic relationship is:

$$N_s = (\text{frequency} \times 120)/\text{number of poles} \qquad (2)$$

where N_s is the synchronous speed of the motor, and frequency is expressed in Hz.

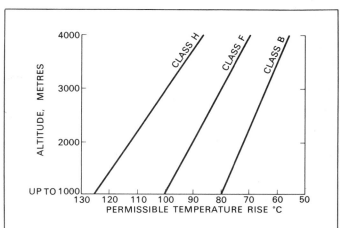

Fig. 2. Permissible temperature rise of various insulation systems at different altitudes, for air cooled machines in 40°C ambient.

The actual running speed on full load is usually about 1% to 5% lower than the synchronous speed, the difference being referred to as the slip. Dual speeds can be obtained from squirrel cage induction motors by adopting either of the following techniques.

(a) Insertion of two separate windings in the stator core. With this method it is obvious that the stator slot cannot be fully utilized for either speed, consequently a reduction of up to 50% in power output from the frame can result.
(b) Tapping the winding to produce pole changing. With this method both speeds utilize the full stator winding and necessitate a reduction in output of only 20–25%. Although the control gear required for this method may be more costly, the reduction in frame size and improved performance can make this technique financially attractive.

Slipring machines provide relatively cheap variable speed motors, speed variation being obtained by varying the value of external resistance in the rotor circuit. The extra resistance in the circuit can make this system inefficient at lower speeds and limited to drives such as fans and pumps where the torque is proportional to the speed squared. A.C. commutator motors and d.c. machines are more commonly used on variable speed applications.

Environment

Environmental conditions prevalent on the intended site should be fully specified in order that a suitable enclosure, insulation system, and frame size can be offered. Ambient temperatures above 40°C reduce the temperature rise permissible on a particular insulation system. This can be overcome by derating the machine or by incorporating a higher class of insulation. (Fig. 1.)

At site altitudes above 1000 metres cooling air is less effective, consequently machines relying on air as a coolant are designed for a reduced temperature rise. The reduction is calculated at the rate of 1% of the permissible temperature rise of an insulation system per 100 metres of altitude in excess of 1000 metres. (Fig. 2.)

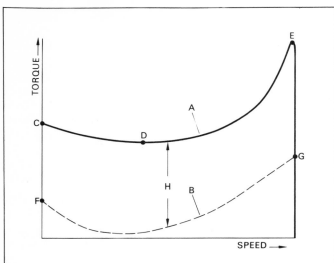

Fig. 3. Typical SCR induction motor starting characteristics against an assumed load: (A) motor torque-speed curve; (B) driven load torque-speed curve; (C) locked rotor or standstill motor torque; (D) minimum or run-up torque; (E) pull-out or stalling torque; (F) load 'stiction' torque; (G) full-load torque; (H) acceleration torque at a particular instant.

The enclosure is usually specified by the motor user who has first-hand knowledge of site conditions, although manufacturers are always ready to assist and make recommendations when required.

Starting characteristics

Ideally a torque-speed characteristic together with the moment of inertia of the driven unit should be supplied by the user to the motor manufacturer. With this information the manufacturer can calculate acceleration times, the heating effect during starting, and the permissible number of consecutive starts allowed on a particular application. High starting torques (200% full-load torque, FLT) and low starting current (450% full load current, FLC) can usually be achieved without derating a machine, but a combination of both conditions can necessitate an increase in frame size.

Induction motors can be started in various ways; the most widely adopted are:

(a) Direct on-line or full voltage starting of squirrel cage motors is usually adopted if the supply system is large enough to support the full short circuit current at the instant of start. This is the cheapest method of starting, and gives maximum starting and acceleration torque with minimum accelerating time.

(b) Star/delta starting is adopted if the supply system is limited. The starting current and torque in the star connection is approximately 33% of the values obtained in the delta connection. Acceleration of the load is more sluggish resulting in an acceleration time which is more than three times the direct on-line value. This method of starting can pose design problems if the driven unit cannot be unloaded or if the inertia to be accelerated is very large.

(c) Auto-transformer or reduced voltage starting is more flexible than a straight star/delta starter. For instance if a transformer tapping of 70% is used the starting current and torque are reduced to 49% of the direct-on-line values. Consequently, by selecting a suitable tapping any desired reduction in starting current can be achieved.

(d) Stator-rotor starting is adopted on slipring machines to produce superior starting performance to that obtained from a squirrel cage machine. By insertion of a suitable resistance in the rotor circuit starting torques of 200% FLT with a starting current of 200% FLC can be achieved. If the resistance is adjustable this performance can be maintained throughout the acceleration period.

Certain drives require a high locked rotor torque to overcome heavy 'stiction' at breakaway, however high starting torques should not be considered as a universal solution of problems (Fig. 3). The most important design consideration is to produce stator and rotor windings with sufficient thermal capacity to withstand the high copper losses incurred during the starting period. High inertia drives are particularly onerous because of the prolonged run-up

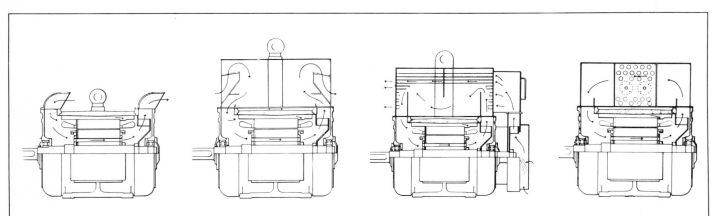

Fig. 4. Various enclosures obtainable from a standard ventilated frame (from left): ventilated drip-proof enclosure; ventilated weather-protected or NEMA II enclosure; totally enclosed CACA enclosure; totally enclosed CACW enclosure.

Fig. 5. Totally enclosed motor undergoing dust-tight tests in accordance with BS3807.

period and may necessitate special rotor design. If starting conditions appear unusually arduous the drive requirements must be specified fully to enable the motor manufacturer to produce a suitably designed motor.

Mechanical considerations

Shafts and bearings are designed and selected to withstand certain torques and loadings. If the motor is to be subjected to any axial or radial thrust load, full details of the mode of transmission will be required. For instance, belt and pulley power transmission will increase the shaft stresses by the addition of a bending moment and will also impose a radial thrust load on the bearings. Such loadings may necessitate special bearing arrangements and non-standard shafts or even an increase in frame size.

MOTOR VENTILATION AND ENCLOSURE

Heat can be dissipated from a hot body by radiation, convection and conduction; consequently surface area, heat flow, and a coolant to remove the heat generated by the losses in an electric motor are of prime importance to the design engineer. The physical size of the motor in relation to the total electrical losses to be dissipated will strongly influence the design of the ventilation system. At various positions within the motor, heat will be produced by the following losses. (a) Friction and windage losses. (b) Iron losses within the cores. (c) Stator and rotor copper losses. (d) Brush losses on machines employing sliprings or commutators. (e) Stray losses. The temperature rise of the motor is then dependent upon the amount of heat produced by these losses and the conditions of cooling, when the rates of heat production and dissipation are equal then the steady state temperature is reached.

Convection and radiation alone provide a natural form of cooling which is satisfactory for small fractional-power (< 750 W) motors or short time duty ratings. On fractional-power machines the surface area is very large in proportion to the heating losses which have to be dissipated, however the surface area required to cool a 5000 kW motor by radiation and convection alone would be several thousand times larger than that required for a 0.25 kW motor. Consequently, as motor sizes increase radiation and convection are augmented by moving the hot air currents into the path of a suitable coolant which conducts the heat away from the machine.

The coolant used depends upon several factors, such as the number of motors to be housed in one particular building, enclosure requirements, and coolant availability. The largest rotating machines are turbo-generators which rely on a very complex cooling system to disperse the tremendous losses incurred in output generation. These machines may rely on a combination of water-cooled stator windings and gases such as hydrogen to cool the stator and rotor core surfaces. Fortunately industrial motors rarely present such severe ventilation problems and are cooled by air or water.

Water has certain advantages over air cooling, such as noise reduction, permissible increase in output, removal of the bulk of the heat from the motor to a source external to the building in which it is housed, and its suitability for use in severe environments. The disadvantages of water cooling are the increased cost of the motor, the need for a water pumping set and for a substantial water supply, and the possibility of the coolant freezing, *eg* in the winter months.

Fans used for motor ventilation are usually designed to move some 5.6–9.3 m^3/min (60–100 ft^3/min) of air per kW of electrical loss through the machine. On low speed applications it may prove impossible to achieve sufficient air flow with fans mounted on motor shafts and in such cases separate forced ventilation units are adopted.

Enclosures

Basically there are two forms of enclosure, ventilated and totally enclosed. The term ventilated enclosure is applied to any machine whose inner surfaces are in contact with the ambient air. The term totally enclosed refers to a machine constructed so that the free exchange of air between the inner and outer surfaces is prevented. Although totally enclosed machines have no openings for ventilation from the external air, they can have internal air circuits which allow movement of the warm air onto a radiating surface. Both totally enclosed and ventilated enclosures have many variations to suit differing environmental conditions.

Fig. 6. Frame E355L (BS3979) 185-kW (250-hp) 2-pole a.c. induction motor; totally enclosed flameproof construction.

In certain cases the enclosure designation is followed by words relating to further protection requirements. Typical examples are dust-proof, dust-tight, waterproof, and water-tight. The basic difference is that with the suffix 'proof', the specified material must not interfere with satisfactory operation, while the suffix 'tight' refers to an enclosure constructed so that the specified material cannot enter.

Ventilated enclosures

Protected type. With this enclosure, protection is given against accidental or inadvertent contact with the rotating or live parts. The openings need not prevent deliberate access to internal parts.

Screen protected. With the screen protected enclosure the possibility of contact with internal parts is minimised by the use of wire mesh or perforated metal screens. The maximum openings permitted on these screens are such that finger contact with rotating or live parts should not be possible. The windings of machines can be impregnated to withstand attack by certain chemicals, consequently screen protected motors can be modified to cope with various atmospheric conditions, though they are not usually suitable for damp or dusty environments.

Drip proof is by far the most popular enclosure for small and medium sized machines. They are constructed so that liquid or solid particles that fall from an angle within 15° of the vertical cannot enter the machine in any amount sufficient to interfere with satisfactory operation. Most manufacturers incorporate drip-proof features in their ranges of ventilated motors, so that they standardise on one construction which is also suitable for the other enclosures outlined above. A very common enclosure for motors driving pumps in water pumping stations.

Splash-proof and hose-proof. For situations where the motor is likely to be drenched in water a splash-proof or hose-proof enclosure is necessary. Splash-proof motors have to be made so that liquids or solids impinging at any angle between the vertical and 10° below the horizontal cannot enter in sufficient amounts to prevent satisfactory operation.

A hose-proof enclosure must guarantee satisfactory operation after hosing down from any accessible direction. Such enclosures may involve the use of quite elaborate air ducting if a ventilated motor is required. Most totally enclosed machines meet the hose-proofing requirements, but can prove far more costly than the ventilated equivalent. Hose-proof motors are widely used in the food industries, where hygiene regulations stipulate regular hosing down of working surfaces and floors.

Pipe or duct ventilated. A ventilated machine can be used in unfavourable environments by providing cooling air from a source external to the room in which the motor is housed. This is achieved by providing ducting to the motor inlets and outlets; the air can then be drawn through by the motor fan alone or can be boosted by a fan at the extremity of the ducting. By forced ventilation the ducting can be pressurised to stop the ingress of contaminants at ducting joints. The rubber industry, with its heavy concentration of carbon black, is a typical environment for such an enclosure.

Weather protected or NEMA type II. Ventilated machines can be used outdoors by adoption of ducting or canopies suitably designed to stop the direct ingress of driving rain or snow. Fig. 4 shows a typical NEMA II ventilation arrangement. Because such machines are continuously open to the atmosphere anti-condensation heaters should be considered standard equipment.

Total enclosures

By totally enclosing a motor, large amounts of dust can be kept away from the internal parts of the machine, and dust-proofing becomes fairly straightforward; however

Fig. 7. Totally enclosed CACA motor; 2MW (2700hp) 4160 V, 3-phase, 60Hz, 3575 rev/min; driving boiler feed pumps in a Canadian power station.

Fig. 8. Section through totally-enclosed fan-cooled induction motor.

small particles of dust (dependent upon fineness) can still enter the machines. Rigorous attention to construction is still required to render the motor dust-tight. The construction may necessitate special seals and gaskets, and decreased machining tolerances. In addition, since most surface cooled machines rely on external fins for cooling, these will have to be regularly cleaned. (Fig. 6.)

The following is a list of enclosures based on totally enclosed principles, commonly specified by motor users.

Plain totally enclosed motors. The disturbance of air by a fan in heavily dust-laden atmospheres can cause irritation to personnel employed in the area. Consequently if pipe ventilated or similar enclosures cannot be considered, the only solution is to adopt a plain totally enclosed machine. The frame can be a normal ventilated motor with the openings completely blocked off or more likely a frame with fins which increase the surface area. Such machines rely on radiation and convection, and have to be heavily de-rated if continuous operation is required. As short time rated machines they are quite economical propositions, and are commonly found on crane hoist drives.

Totally enclosed fan cooled motors. An external fan mounted on the motor shaft blows air along the radiating surface between external cooling fins or channels to remove heat. Such machines can be easily weatherproofed to allow operation outdoors. Outputs of motor units rise much more rapidly than their corresponding surface areas consequently there is an output limit above which the provision of enough cooling fins on the carcass becomes expensive and difficult. For this reason closed air circuit machines, outlined below, are used on the higher outputs.

Totally enclosed closed air circuits. Closed air circuits can employ air-to-air heat exchangers (CACA) or air to water heat exchangers (CACW); Fig. 4. The construction is basically the same as a ventilated machine, the significant difference being the exchanger which covers both the inlet and outlet orifices. The internal air is circulated and passed to the heat exchanger where it is cooled before re-entering the motor. The use of water as a coolant through the exchanger allows the machine to be rated up to its full ventilated output; the use of air as a coolant will probably necessitate a 15–20% derating.

Several electrical machines within a building can substantially increase the ambient temperature and cause discomfort to personnel; water cooled machines remove the heat to an external source and so reduce this effect. They are generally quieter, more efficient and suitable for most environmental conditions, and these factors could influence the extra financial outlay required for the piping and pumping equipment necessary.

Totally enclosed flameproof motors. The basis of a flameproof enclosure design is not necessarily to stop the ingress of the flammable gas into the machine, but rather to prevent the explosion caused by the electrical ignition of such gases from spreading to the gas-laden atmosphere outside the machine. For this reason the frame is more rigidly constructed than for a standard totally enclosed motor and internal clearances are reduced to a minimum. The validity of such machines to cope with these circumstances depends on certification by an appropriate testing authority. There are various locations within the oil, chemical, and mining industries where such enclosures are imperative.

BIBLIOGRAPHY

BS2613:70. Specification for the electrical performance of rotating electrical machines.
BS3979:66. Specification for dimensions of electric motors (metric sizes).
BS2960:64. Dimensions of 3-phase electric motors.
BS229:57. Flameproof enclosures of electrical apparatus.
BS269:27. Methods of declaring efficiency of electrical machinery.
BS2817:57. Types of enclosure of electric apparatus.
BS3807:64. Method for the type testing of enclosures for electrical apparatus for use in onerous conditions.
NEMA publication M62-1969. Standards for motors and generators.
M G Say. Performance and design of alternator current machines.
A E Clayton and N H Hancock. Performance and design of d.c. machines.
Selecting three-phase induction motors. *Electrical Times* supplement 13 November 1969.

Guide to Equipment Makers

FRACTIONAL POWER MOTORS

a.c.	A
d.c.	B
geared	C
variable speed	D
flameproof	E

AC–Delco Division of General Motors Ltd	A B
ACEC	A B C D E
AEG (Great Britain) Ltd	A B C D E
Acadex Engineering Co Ltd	A
Acral Ltd	A C
Alcho Electric Co Ltd	A B
ASEA (Great Britain) Ltd	A B C D E
BKB Electric Motors Ltd	B
BVC Electronic Developments Ltd	A B D
Bastat Electrical Equipment Ltd	A B E
Batwin Electric Motors Ltd	A B C E
Bauer (Great Britain) Ltd	A C D
Beanwy Electric Ltd	A B E
Bosch Ltd	B
Boxall (T A) Ltd	A B
British Brown-Boveri Ltd	A B C D E
Brook Motors Ltd	A B C D E
Bull Motors Ltd	A B
Bulpitt & Sons Ltd	A B
CAV Ltd	B
CGE Internationale UK Ltd	A B
Carter Electrical Co Ltd	A B C
Citenco Ltd	A B C D E
Comtex Ltd	A C D E
Conyers (W E) Ltd	B C D E
Crompton Parkinson Ltd	A
Croydon Engineering Co Ltd	A B C D E
Decca Reid & Sigrist Ltd	A B D
Dynamotive Engineering Co Ltd	B C
EMI Electronics Ltd	A B C
Electrical Power Engineering Co (Birmingham) Ltd	A B C D E
Electro Dynamic Construction Co Ltd	A B D E
English Electric–AEI Machines Ltd (Newcastle under Lyme)	A B C D E
Fractional HP Motors Ltd	A B C
Green (Horace) & Co Ltd	A B C E
Harland Simon Ltd	A D
Hawker Siddeley Dynamics Ltd	A B C D E
Highfield Gear & Engineering Co Ltd	A C
Hoover Ltd	A B C
Ingleby Motors Ltd	A
International General Electric Co of New York Ltd	A B C D E
Klaxon Ltd	A B C E
Laurence, Scott & Electromotors Ltd	A B C D E
Lewis Electric Motors (Maidenhead) Ltd	A C D E
MG Electric (Colchester) Ltd	A B
Mackie (W) & Co Ltd	A B D
Medical Engineering Co Ltd	A B C D E
Millett (H G) & Co Ltd	A C E
Mitsubishi Shoji Kaisha Ltd	A C E
Modern Techniques Ltd	A
Morley Electrical Engineering Co Ltd	A C E
Mortley (G E) Sprague & Co Ltd	A B
Motor Gear Engineering Co Ltd	A B C
Mycalex (Motors) Ltd	A B C
Nelson Transformer & Rotary Equipment Ltd	A B
Newman Industries Ltd	A C D E
Newton Derby Ltd	A B C
Normand Electrical Co Ltd	A B C D E
Oliver Pell Control Ltd	A B C
Parvalux Electric Motors Ltd	A B C D
Plessey Co Ltd	B
Sadi (Great Britain) Ltd	A C D
Salon (Nelson) Ltd	A B
Sew-Tric (Holdings) Ltd	A B
Simon Equipment Ltd	A B D
Siba Electric Ltd	B
Small Electric Motors Ltd	A B D
Smiths Industries Ltd	A B C
Smith Bros & Webb Ltd	A C
STC (Electromechanical Division) Ltd	A
Stone-Platt Crawley Ltd	A B
TKS (Aircraft De-icing) Ltd	B
Teddington Aircraft Controls Ltd	B
Tefco Vari-speed Motor Co	A B D
Thomson (S) & Sons (Liverpool) Ltd	A B C D E
Tuscan Engineering Co Ltd	A B C D E
Unimatic Engineers Ltd	A B C D
Vactric Control Equipment Ltd	A B C
Vickalectric Engineering Co Ltd	B

A.C. COMMUTATOR MOTORS

ACEC
ASEA (Great Britain) Ltd
British Brown–Boveri Ltd
Bauer (Great Britain) Ltd

Comtex Ltd
Electro Dynamic Construction Co Ltd
English Electric–AEI Machines Ltd (Rugby)
English Electric–AEI Machines Ltd (Warley)
Greenwood & Batley Ltd
Laurence, Scott & Electromotors Ltd
Lewis Electric Motors (Maidenhead) Ltd
Neco Electrical Ltd
Nelson Engineering Co Ltd
Newman Industries Ltd
Reyrolle (A) and Co Ltd
Siemens (United Kingdom) Ltd
Thomson (S) & Sons (Liverpool) Ltd

A.C. MOTORS (to 50 hp; 37 kW)

ACEC
AEG
ASEA (Great Britain) Ltd
BKB Electric Motors Ltd
Batwin Electric Motors Ltd
Bauer (Great Britain) Ltd
British Brown–Boveri Ltd
British Jeffrey Diamond Ltd
Brook Motors Ltd
Brush Electrical Engineering Co Ltd
Bull Motors Ltd
CGE Internationale UK Ltd
Conyers (W E) Ltd
Crompton Parkinson Ltd
Demag Material Handling Ltd
Doxford & Sunderland Shipbuilding &
    Engineering Co Ltd
Dynadrive Ltd
Electric Construction (Wolverhampton) Ltd
Electro Dynamic Construction Co Ltd
English Electric–AEI Machines Ltd (Bradford)
    (medium industrial machines)
English Electric–AEI Traction Ltd (Manchester)
    (railway traction motors)
English Electric–AEI Machines Ltd (Warley)
    (small industrial motors)
Green (Horace) & Co Ltd
Greenwood & Batley Ltd
Harland Engineering Co Ltd
Hawker Siddeley Dynamics Ltd
Houchin Ltd
Howells Stanway Ltd
Ingleby Motors Ltd
International General Electric Co of New York Ltd
Jennings Electric (Belfast) Ltd
Laurence, Scott & Electromotors Ltd
Macfarlane Engineering Co Ltd
Mather & Platt Ltd
Mawdsley's Ltd
Mitsubishi Shoji Kaisha Ltd
Morley Electrical Engineering Co Ltd
Motor Gear Engineering Co Ltd
Nelson Engineering Co Ltd
Newman Industries Ltd
Newton Derby Ltd
Norco Electrics Ltd
Normand Electrical Co Ltd

Reyrolle (A) and Co Ltd
Sadi (Great Britain) Ltd
Scott (Hugh J) & Co (Belfast) Ltd
Siemens (United Kingdom) Ltd
Smith Bros & Webb Ltd
Stone–Platt Crawley Ltd
Tefco Vari-Speed Motor Co
Thomson (S) & Sons (Liverpool) Ltd
Thorn Automation Ltd
Tuscan Engineering Co Ltd
Wright Electric Motors (Halifax) Ltd

A.C. MOTORS (over 50 hp; 37 kW)

ACEC
AEG
ASEA (Great Britain) Ltd
British Brown–Boveri Ltd
British Jeffrey Diamond Ltd
Brook Motors Ltd
Brush Electrical Engineering Co Ltd
Bull Motors Ltd
CGE Internationale UK Ltd
Crompton Parkinson Ltd
Doxford & Sunderland Shipbuilding &
    Engineering Co Ltd
Electric Construction (Wolverhampton) Ltd
English Electric–AEI Machines Ltd (Bradford)
English Electric–AEI Traction Ltd (Manchester)
English Electric–AEI Machines Ltd (Warley)
English Electric–AEI Machines Ltd (Rugby)
Green (Horace) & Co Ltd
Greenwood & Batley Ltd
Harland & Wolff Ltd
Harland Engineering Co Ltd
Howells Stanway Ltd
International General Electric Co of New York Ltd
Jennings Electric (Belfast) Ltd
Laurence, Scott & Electromotors Ltd
Mather & Platt Ltd
Mawdsley's Ltd
Mitsubishi Shoji Kaisha Ltd
Morley Electrical Engineering Co Ltd
Nelson Transformer & Rotary Equipment Ltd
Newman Industries Ltd
Parsons Peebles Ltd (Edinburgh)
Parsons Peebles Ltd (Witton)
Richardsons Westgarth (Hartlepool) Ltd
Siemens (United Kingdom) Ltd
Thorn Automation Ltd

SYNCHRONOUS MOTORS (Industrial)

ACEC
ASEA (Great Britain) Ltd
British Brown–Boveri Ltd
Brush Electrical Engineering Co Ltd
Bull Motors Ltd

CGE Internationale UK Ltd
Croydon Engineering Co Ltd
Doxford & Sunderland Shipbuilding &
   Engineering Co Ltd
Electric Construction (Wolverhampton) Ltd
English Electric–AEI Machines Ltd (Rugby)
English Electric–AEI Turbine Generators Ltd
   (Stafford)
Greenwood & Batley Ltd
Harland & Wolff Ltd
Harland Engineering Co Ltd
International General Electric Co of New York Ltd
Klaxon Ltd
Laurence, Scott & Electromotors Ltd
Lewis Electric Motors (Maidenhead) Ltd
Macfarlane Engineering Co Ltd
Mackie (W) & Co Ltd
Mawdsley's Ltd
Millett (H G) & Co Ltd
Newman Industries Ltd
Normand Electrical Co Ltd
Parsons Peebles Ltd (Edinburgh)
Parsons Peebles Ltd (Witton)
Parvalux Electric Motors Ltd
Redman Heenan Froude Ltd
Richardsons Westgarth (Hartlepool) Ltd
Siemens (United Kingdom) Ltd
Small Electric Motors Ltd

LINEAR MOTORS

Brook Motors Ltd
Bull Motors Ltd
English Electric–AEI Machines Ltd
Millett (H G) & Co Ltd
Morris (Herbert) Ltd

D. C. MOTORS (to 50 hp; 37 kW)

ACEC
ASEA (Great Britain) Ltd
Bastat Electrical Equipment Ltd
Batwin Electric Motors Ltd
Beard Electrical Ltd
British Brown–Boveri Ltd
British Jeffrey Diamond Ltd
Brush Electrical Engineering Co Ltd
Bull Motors Ltd
CAV Ltd
CGE Internationale UK Ltd
Conyers (W E) Ltd
Croydon Engineering Co Ltd
Doxford & Sunderland Shipbuilding &
   Engineering Co Ltd
Dynadrive Ltd
Electrical Power Engineering (Birmingham) Ltd
Electric Construction (Wolverhampton) Ltd
Electro Dynamic Construction Co Ltd
English Electric–AEI Traction Ltd (Manchester)
English Electric–AEI Machines Ltd (Warley)
Green (Horace) & Co Ltd

Greenwood & Batley Ltd
Harland Engineering Co Ltd
International General Electric Co of New York Ltd
Laurence, Scott & Electromotors Ltd
Macfarlane Engineering Co Ltd
Mackie (W) & Co Ltd
Mawdsley's Ltd
Nelson Engineering Co Ltd
Nelson Transformer & Rotary Equipment Ltd
Newton Derby Ltd
Normand Electrical Co Ltd
Parsons Peebles Ltd (Witton)
Reyrolle (A) and Co Ltd
Scott (Hugh J) & Co (Belfast) Ltd
Siemens (United Kingdom) Ltd
Small Electric Motors Ltd
Stone–Platt Crawley Ltd
Tefco Vari-speed Motor Co
Thomson (S) & Sons (Liverpool) Ltd
Tuscan Engineering Co Ltd
Westinghouse Brake & Signal Co Ltd
Wright Electric Motors (Halifax) Ltd

D. C. MOTORS (over 50 hp; 37 kW)

ACEC
ASEA (Great Britain) Ltd
British Brown–Boveri Ltd
British Jeffrey Diamond Ltd
Brush Electrical Engineering Co Ltd
Bull Motors Ltd
CGE Internationale UK Ltd
Crompton Parkinson Ltd
Doxford & Sunderland Shipbuilding &
   Engineering Co Ltd
Electrical Power Engineering Co (Birmingham) Ltd
Electric Construction (Wolverhampton) Ltd
English Electric–AEI Machines Ltd (Rugby)
English Electric–AEI Traction Ltd (Manchester)
Greenwood & Batley Ltd
Harland & Wolff Ltd
Harland Engineering Co Ltd
International General Electric Co of New York Ltd
Laurence, Scott & Electromotors Ltd
Mawdsley's Ltd
Nelson Transformer & Rotary Equipment Ltd
Parsons Peebles Ltd
Siemens (United Kingdom) Ltd
Thomson (S) & Sons (Liverpool) Ltd

FLAMEPROOF EQUIPMENT

Motors	A
Controls	B

ACEC	A B
AEG (Great Britain) Ltd	A B
Allen Bradley UK Ltd	B
Allen West (Brighton) Ltd	B
ASEA (Great Britain) Ltd	A B

BVC Electronic Developments Ltd	A
Bastat Electrical Equipment Ltd	A
Batwin Electric Motors Ltd	A
Beanwy Electric Ltd	A
British Brown–Boveri Ltd	A B
British Jeffrey Diamond Ltd	A
Brook Motors Ltd	A
Brownings Electric Co Ltd	A
Brush Electrical Engineering Co Ltd	A B
CGE Internationale UK Ltd	A
Cableform Ltd	B
Citenco Ltd	A
Comtex Ltd	A
Conyers (W E) Ltd	A
Crompton Parkinson Ltd	A
Davis (John) & Son (Derby) Ltd	B
Donovan Electrical Co Ltd (The)	B
Electrical Remote Control Co Ltd	B
Electrical Power Engineering Co (Birmingham) Ltd	A
Electro Dynamic Construction Co Ltd	A
English Electric–AEI Machines Ltd (Bradford)	A
Erskine, Heap & Co Ltd	B
GEC–Elliott Industrial Controls Ltd	B
Green (Horace) & Co Ltd	A
Greenwood & Batley Ltd	A
Heyes & Co Ltd	B
Howells Stanway Ltd	A
International General Electric Co of New York	A B
Klaxon Ltd	A
Laurence, Scott & Electromotors Ltd	A
M & C Switchgear Ltd	B
Mather & Platt Ltd	A
Millet (H G) & Co Ltd	A
Mitsubishi Shoji Kaisha Ltd	A B
Morley Electrical Engineering Co Ltd	A B
Neco Electrical Co Ltd	A
Nelson (W H) Ltd	B
Newman Industries Ltd	A
Normand Electrical Co Ltd	A
Parsons Peebles Ltd (Edinburgh)	A
Reyrolle Belmos Ltd	B
Siemens (United Kingdom) Ltd	A B
Simmons Electrical Services (Sutton) Ltd	B
Simplex–GE Ltd	B
Solenoids & Regulators Ltd	B
Square D Ltd	B
Switchgear & Cowans Ltd	B
Telemecanique Electrique (Great Britain) Ltd	B
Thomson (S) & Sons (Liverpool) Ltd	A
Tuscan Engineering Co Ltd	A
Vlasto Clark & Watson	B
Wallacetown Engineering Co Ltd (The)	B
Wright Electric Motors (Halifax) Ltd	A

MINIATURE MOTORS

AB Electronic Components Ltd
AC–Delco Division of General Motors Ltd
ACEC
AEG (Great Britain) Ltd
AK Fans Ltd
Adwest Engineering Ltd
Appliance Components Ltd
Auriema Ltd
BVC Electronic Developments Ltd
Bauer (Great Britain) Ltd
British Sonceboz Co Ltd (The)
Brown (S G) Ltd
CDC (Great Britain) Ltd
Carter Electrical Co Ltd
Chamberlain & Hookham Ltd
Clarke, Chapman & Co Ltd (Electronics Division)
Conyers (W E) Ltd
Crouzet Ltd
Drayton Controls Ltd
EMI Electronics Ltd
Elmic Ltd
Ether Ltd
Evershed & Vignoles Ltd
Ferranti Ltd
Francis Clocks Ltd
GEC–Elliott Precision Controls Ltd
Imhof–Bedco Ltd
Impex Electrical Ltd
International General Electric Co of New York Ltd
Jones (Walter) & Co (Engineers) Ltd
Klaxon Ltd
Landis & Gyr Ltd
Latham (H J) Ltd
Lewis Electric Motors (Maidenhead) Ltd
Limit Sales (Holdings) Ltd
Lloyd (J J) Instruments Ltd
Marubeni-iida Co Ltd
Micromotor Ltd
Midgley Harmer Ltd
Miles (F G) Engineering Ltd
Millett (H G) & Co Ltd
Moore Reed & Co Ltd
Mortley (G E) Sprague & Co Ltd
Muirhead Ltd
Mycalex (Motors) Ltd
Nelson Engineering Co Ltd
Oliver Pell Control Ltd
Park Bros Ltd
Parvalux Electric Motors Ltd
Plessey Co Ltd
Precision Electrical Equipments Ltd
Printed Motors Ltd
Rank Precision Industries Ltd
Rank Pullin Controls Ltd
Salon (Nelson) Ltd
Sangamo Weston Ltd
Siemens (United Kingdom) Ltd
Small Electric Motors Ltd
Smiths Industries Ltd (Brighton)
Smiths Industries Ltd (Wembley)
Specto Ltd
Sterling Instruments Ltd
Sperry Gyroscope Division Sperry Rand Ltd
Tapley Meters Ltd
Teddington Aircraft Controls Ltd
Torrin Corporation Ltd
Trident Engineering Ltd
Trist (Ronald) Controls Ltd
Unigears Ltd
Unimatic Engineers Ltd
Union Carbide UK Ltd
Vactric Control Equipment Ltd
Wynstruments Ltd (Electric Motor Division)

SYNCHRONOUS MOTORS
(Instrument)

ACEC
AEG (Great Britain) Ltd
BVC Electronic Developments Ltd
British Sonceboz Co Ltd (The)
Brown (S G) Ltd
Carter Electrical Co Ltd
Chamberlain & Hookham Ltd
Crater Controls Ltd
Crouzet Ltd
EMI Electronics Ltd
Educational Measurements Ltd
Electrical Remote Control Co Ltd
Ether Ltd
Evershed & Vignoles Ltd
Ferranti Ltd
Francis (Clocks) Ltd
GEC–Elliott Precision Controls Ltd
Garrard Engineering Ltd
Imhof–Bedco Ltd
Impex Electrical Ltd
International General Electric Co of New York Ltd
Landis & Gyr Ltd
Lewis Electric Motors (Maidenhead) Ltd
Lloyd (J J) Instruments Ltd
Marubeni-iida Co Ltd
Mitsubishi Shoji Kaisha Ltd
Mortley (G E) Sprague & Co Ltd
Plessey Co Ltd
Pye of Cambridge Ltd
Sangamo Weston Ltd
Smiths Industries Ltd (Brighton)
Sterling Instruments Ltd
Sperry Gyroscope Division Sperry Rand Ltd
Unimatic Engineers Ltd
Union Carbide UK Ltd
Vactric Control Equipment Ltd

SERVOMOTORS

ACEC
Auriema Ltd
BVC Electronic Developments Ltd
Bauer (Great Britain) Ltd
Brown (S G) Ltd
Carter Electrical Co Ltd
Croydon Engineering Co Ltd
Danfoss (London) Ltd
Daystrom Ltd
EMI Electronics Ltd
English Electric–AEI Machines Ltd (Bradford)
Ether Ltd
Evershed & Vignoles Ltd
GEC–Elliott Precision Controls Ltd
International General Electric Co of New York Ltd
Jones (Walter) & Co (Engineers) Ltd
Klaxon Ltd
Mackie (W) & Co Ltd
Marubeni-iida Co Ltd
Morecambe Electrical Equipment Ltd

Moore Reed & Co Ltd
Muirhead Ltd
Nelson Engineering Co Ltd
Newton Derby Ltd
Park Bros Ltd
Parvalux Electric Motors Ltd
Printed Motors Ltd
Pye of Cambridge Ltd
Rank Pullin Controls Ltd
Sadi (Great Britain) Ltd
Servomex Controls Ltd
Small Electric Motors Ltd
Smiths Industries Ltd (Wembley)
Sperry Gyroscope Division Sperry Rand Ltd
Trident Engineering Ltd
Unimatic Engineers Ltd
Union Carbide UK Ltd
Vactric Control Equipment Ltd
Westinghouse Brake & Signal Co Ltd
Woden Transformer Co Ltd

STEPPER MOTORS

ACEC
Auriema Ltd
B & K Instruments Ltd
Brown (S G) Ltd
Crouzet Ltd
Danfoss Ltd
English Electric–AEI Machines Ltd (Bradford)
Evershed & Vignoles Ltd
GEC–Elliott Precision Controls Ltd
Imhof–Bedco Ltd
Impex Electrical Ltd
International General Electric Co of New York Ltd
Helga Electronics Ltd
JAC Electronics Ltd
Kempston Electrical Ltd
Marubeni-iida Ltd
Moore Reed & Co Ltd
Muirhead Ltd
Scott (James) (Electronic Agencies) Ltd
Smiths Industries Ltd (Wembley)
Sperry Gyroscope Division Sperry Rand Ltd
Unimatic Engineers Ltd
Vactric Control Equipment Ltd

SPECIAL MOTORS

Custom production	A
Silent	B
Submersible	C
Torque	D
Various	E

AEG (Great Britain) Ltd	B D E
Allam (E P) Ltd	E
Auriema Ltd	D
BKB Electric Motors Ltd	A
BVC Electronic Developments Ltd	A

Brook Motors Ltd	B D
Brush Electrical Engineering Co Ltd	A
Bull Motors Ltd	A
Carter Electrical Co Ltd	E
Citenco Ltd	B
Conyers (W E) Ltd	B
Crompton Parkinson Ltd	A
Croydon Engineering Co Ltd	C D
Doxford & Sunderland Shipbuilding & Engineering Co Ltd	A
Electrical Power Engineering Co (Birmingham) Ltd	A
Electric Construction (Wolverhampton) Ltd	A
Electro Dynamic Construction Co Ltd	A D
English Electric–AEI Machines Ltd	A D E
Ferranti Ltd	D E
Forco Electrical Services Ltd	A
GEC–Elliott Precision Controls Ltd	D E
Green (Horace) & Co Ltd	B
Honeywell Ltd	D
Houchin Ltd	A E
Howells Stanway Ltd	B
International General Electric Co of New York Ltd	B C D E
Jennings Electric (Belfast) Ltd	A
Jones (Walter) & Co (Engineers) Ltd	B
Laurence, Scott & Electromotors Ltd	C E
MG Electric (Colchester) Ltd	A
Macfarlane Engineering Co Ltd	A
Magco Moxey (Division of Winget Ltd)	E
Mawdsley's Ltd	A
Millett (H G) & Co Ltd	E
Mitsubishi Shoji Kaisha Ltd	C E
Moore Reed & Co Ltd	D
Nelson Engineering Co Ltd	C
Newton Derby Ltd	A
Normand Electrical Co Ltd	B E
Parsons Peebles Ltd (Witton)	A
Parvalux Electric Motors Ltd	B D
Siemens (United Kingdom) Ltd	A E
Sirco Controls Ltd	E
Small Electric Motors Ltd	B C D E
Sumo Pumps Ltd	C
Tuscan Engineering Co Ltd	A
Unimatic Engineers Ltd	D
Vactric Control Equipment Ltd	D
Westinghouse Brake & Signal Co Ltd	D
Woden Transformer Co Ltd	D

BRAKES

Electro-magnetic	A
Electro-mechanical	B
Integral with motor	C
Thrustor	D

ACEC	A B C D
Amphenol Ltd	A
ASEA (Great Britain) Ltd	A B C D
Aurora Gearing Co (Wilmot North) Ltd	A
Barnsley (John) & Sons Ltd	A B
British Brown–Boveri Ltd	A B C D
Brooke & Green Ltd	A B
Brookhirst Igranic Ltd	A B
Carruthers (J H) & Co Ltd	A
CGE Internationale UK Ltd	C
Clark Clutch Co Ltd	A B
Coventry Gauge & Tool Co Ltd	A
Crofts Engineers Ltd	A B C
Crouzet Ltd	C
Danfoss (London) Ltd	A
Daval (S) & Sons Ltd	A
Demag Material Handling Ltd	C
Dewhurst & Partner Ltd	B
ESL Engineers (Basildon) Ltd	A
English Electric–AEI Machines Ltd (Bradford)	B
English Electric–AEI Machines Ltd (Rugby)	B
English Electric–AEI Machines Ltd (Warley)	D
Escomat Ltd	A
Eurodrive Ltd	A B C
Evershed & Vignoles Ltd	A
Ferranti Ltd	A
Foulds (E A) Ltd	A
GEC–Elliott Precision Controls Ltd	A
Graseby Instruments Ltd	A
Holme (Edward) & Co Ltd	B
International General Electric Co of New York Ltd	A B C D
Klaxon Ltd	C
MTE Contactor Ltd	A
Mitsubishi Shoji Kaisha Ltd	C
Morecambe Electrical Equipment Ltd	A B
Parsons Peebles Ltd (Witton)	A B D
Pye of Cambridge	A
Redman Heenan Froude Ltd	A
Renold Ltd	A
Siemens (United Kingdom) Ltd	A B C
Smiths Industries Ltd (Wembley)	A
Square D Ltd	B
Twiflex Couplings Ltd	B
Unimatic Engineers Ltd	A
Westinghouse Brake & Signal Co Ltd	B

STARTERS, hand operated

Air break	A
Oil immersed (hv)	B

ACEC	A B
Acme Electrical Manufacturing Co (Tottenham) Ltd	A
Airedale Electrical & Manufacturing Co Ltd	A
Allen Bradley UK Ltd	A
Allen West (Brighton) Ltd	A B
Arrow Electric Switches Ltd	A
ASEA (Great Britain) Ltd	A B
Asquith Electrics (Colne) Ltd	A
Automatic Electric Controls Ltd	A
B & R Relays Ltd	A
Bailey Meters & Controls Ltd	A
Baldwin & Francis Holdings Ltd	B
Bray (E N) Ltd	A
British Brown–Boveri Ltd	A B
Brook Motors Ltd	A
Brookhirst Igranic Ltd	A
CAV Ltd	A

Crabtree (J A) Ltd	A
Craig & Derricott Ltd	A
Crater Controls Ltd	A
Crouzet Ltd	A
Cubigear Control Equipment Ltd	A
Danfoss (London) Ltd	A
Davies (Fred W) & Son Ltd	A B
Davis (John) & Son (Derby) Ltd	A
Dewhurst & Partner Ltd	A
Diamond H Controls Ltd	A
Doxford & Sunderland Shipbuilding & Engineering Co Ltd	A
EDC Electricals Ltd	A
EMB Co Ltd	A
Electrical Remote Control Co Ltd	A B
Electric Construction (Wolverhampton) Ltd	A B
Electro Dynamic Construction Ltd	A
Electro-Mechanical Manufacturing Co Ltd	A B
Ellison (George) Ltd	A B
Erskine Heap & Co Ltd	A B
GEC–Elliott Industrial Controls Ltd (Stoke-on-Trent)	A B
GEC–Elliott Industrial Controls Ltd (Rugby)	A B
Hampson Industries Ltd	A
Harland & Wolff Ltd	A
Heyes & Co Ltd	A
Herbert Controls Ltd	A
Holme (Edward) & Co Ltd	A
Huggett Electrical Ltd	A
International General Electric Co of New York Ltd	A B
Klockner–Moeller Ltd	A
Laurence, Scott & Electromotors Ltd	A B
Lounsdale Electric Ltd	A
M & C Switchgear Ltd	A B
MEM (Midland Electric Manufacturing) Co Ltd	A
Mitsubishi Shoji Kaisha Ltd	A
Morecambe Electrical Equipment Co Ltd	A
Morley Electrical Engineering Co Ltd	A B
Nelson (W H) Ltd	A
Precision Electrical Equipments Ltd	A
Reyrolle Belmos Ltd	A B
Santon Ltd	A
Shawford Control Gear Co Ltd	A
Siemens (United Kingdom) Ltd	A B
Simmons Electrical Services (Sutton) Ltd	A
Solenoids & Regulators Ltd	A
South Wales Switchgear Ltd	A
Smith Bros and Webb Ltd	A
Square D Ltd	A
Stone–Platt Crawley Ltd	A B
Switchgear & Cowans Ltd	A B
Thomson (S) & Sons (Liverpool) Ltd	A
Vlasto Clark & Watson Ltd	A
Wallacetown Engineering Co Ltd (The)	A
Walsall Conduits Ltd	A
Whipp & Bourne Ltd	A
Yorkshire Switchgear & Engineering Co Ltd	B

STARTERS, automatic contactor

ACEC
Acme Electrical Manufacturing Co (Tottenham) Ltd
Airedale Electrical and Manufacturing Co Ltd
Aish & Co Ltd
Allen Bradley UK Ltd
Allen West (Brighton) Ltd
Arrow Electric Switches Ltd
Arcontrol (Electro Panels) Ltd
ASEA (Great Britain) Ltd
Asquith Electrics (Colne) Ltd
B & R Relays Ltd
Boulting (W A) Ltd
Bray (E N) Ltd
British Brown–Boveri Ltd
Brook Motors Ltd
Brookhirst Igranic Ltd
Brush Electrical Engineering Co Ltd
Cableform Ltd
Cabtree (J A) & Co Ltd
Dane Electric Controls Ltd
Danfoss (London) Ltd
Davis (John) & Son (Derby) Ltd
Dewhurst & Partner Ltd
Don (William) Ltd
Donovan Electrical Co Ltd (The)
Dorman Smith Control Gear Ltd
Doxford & Sunderland Shipbuilding & Engineering Co Ltd
Drayton Controls Ltd
EDC Electricals Ltd
EMB Co Ltd
Electrical Remote Control Co Ltd
Electric Construction (Wolverhampton) Ltd
Electro Dynamic Construction Co Ltd
Electro-Mechanical Manufacturing Co Ltd
Ellison (George) Ltd
Elliston, Evans & Jackson Ltd
Erskine, Heap & Co Ltd
Field & Grant Ltd
GEC–Elliott Industrial Controls Ltd (Stoke on Trent)
GEC–Elliott Industrial Controls Ltd (Rugby)
Hackbridge Faraday Electrics Ltd
Hannah Switchgear Ltd
Harland & Wolff Ltd
Harland Engineering Co Ltd
Holme (Edward) & Co Ltd
Huggett Electrical Ltd
International General Electric Co of New York Ltd
Kempston Electrical Co Ltd
Klockner–Moeller Ltd
Laurence, Scott & Electromotors Ltd
M & C Switchgear Ltd
MEM (Midland Electric Manufacturing) Co Ltd
MTE-Contactor Ltd (Leigh-on-Sea)
MTE-Contactor Ltd (Systems) Ltd (Wolverhampton)
Mann (Albert) Engineering Co Ltd
Mawdsley's Ltd
Milne & Longbottom Ltd
Mitsubishi Shoji Kaisha Ltd
Morris (John) Electrical Engineering Co Ltd
Nelson (W H) Ltd
Reyrolle Belmos Ltd
Siemens (United Kingdom) Ltd
Shawford Control Gear Co Ltd
Simmons Electrical Services (Sutton) Ltd
Simplex–GE Ltd
Smith Bros and Webb Ltd
Solenoids & Regulators Ltd
South Wales Switchgear Ltd

Square D Ltd
Starkstrom Ltd
Switchgear & Cowans Ltd
Switchgear & Instrumentation Ltd
Thomson (S) & Sons Ltd
UK Solenoid Ltd
Vlasto Clark & Watson Ltd
Wallacetown Engineering Co Ltd (The)
Watford Electric Co Ltd
Whipp & Bourne Ltd
Zone Controls Ltd

MOTOR CONTROL PANELS & CENTRES

ACEC
Acme Electrical Manufacturing Co (Tottenham) Ltd
Acrastyle Ltd
Airedale Electrical & Manufacturing Co Ltd
Allen Bradley UK Ltd
Allen West (Brighton) Ltd
Arcontrol (Electro Panels) Ltd
Arrow Electric Switches Ltd
ASEA (Great Britain) Ltd
Asquith Electrics (Colne) Ltd
Automatic Electric Controls Ltd
Baldwin & Francis (Holdings) Ltd
Bennett (Allen) Ltd
Berco Controls Ltd
Bon Automation Ltd
Boulting (W A) Ltd
Bray (E N) Ltd
British Brown—Boveri Ltd
Brookhirst Igranic Ltd
Brush Electrical Engineering Ltd
Chandos Products (Scientific) Ltd
Clarke, Chapman & Co Ltd
Control Engineering Ltd
Controlec Ltd
Contropanels Ltd
Crabtree (J A) Ltd
Dane Electric Controls Ltd
Dewhurst & Partner Ltd
Don (William) Ltd
Donovan Electrical Ltd (The)
Doxford & Sunderland Shipbuilding & Engineering Co Ltd (The)
EDC Electricals Ltd
EMB Co Ltd
Electric Construction (Wolverhampton) Ltd
Electrical Apparatus Co Ltd (The)
Electrical Remote Control Co Ltd
Electrical Systems Ltd
Electro Dynamic Construction Co Ltd
Electro-Magnetic Control Co
Electro-Mechanical Manufacturing Co Ltd
Ellison (George) Ltd
Elliston, Evans & Jackson Ltd
Erskine, Heap & Co Ltd
Field & Grant Ltd
Fisher Controls Ltd
GEC—Elliott Industrial Controls Ltd
    (Rugby and Stoke-on-Trent)
Hannah Switchgear Ltd
Harland & Wolff Ltd
Herbert Controls & Instruments Ltd
Holme (Edward) & Co Ltd
Huggett Electrical Ltd
Industrial Control Systems (1969) Ltd
Industrial Switchgear Ltd
International General Electric Co of New York Ltd
Klockner—Moeller Ltd
Laurence, Scott & Electromotors Ltd
M & C Switchgear Ltd
MEM (Midland Electric Manufacturing) Co Ltd
MTE Contactor (Systems) Ltd
MacGeoch (William) & Co (Birmingham) Ltd
Marshall Richards Barcro Ltd
Mertech Electrical Equipment Ltd
Milne & Longbottom Ltd
Mitsubishi Shoji Kaisha Ltd
Millett (H G) & Co Ltd
Morecambe Electrical Equipment Ltd
Nelson (W H) Ltd
Newton Derby Ltd
Ottermill Engineering Ltd
Phillipson (A F) & Co Ltd
Power Control Panels
Precision Electrical Equipments Ltd
Rank Precision Industries Ltd
Rank Pullin Controls Ltd
Redler Industries Ltd
Reyrolle Belmos Ltd
Siemens (United Kingdom) Ltd
Shawford Control Gear Co Ltd
Simmons Electrical Services (Sutton) Ltd
Simplex—GE Ltd
Solenoids & Regulators Ltd
South Wales Switchgear Ltd
Spectar Engineering Ltd
Square D Ltd
Starkstrom Ltd
Switchgear & Instrumentation Ltd
Teddington Auto Controls Ltd
Telemechanique Electrique (Great Britain) Ltd
Thomson (S) & Sons (Liverpool) Ltd
Vlasto Clark & Watson Ltd
Wallacetown Engineering Co Ltd (The)
Watford Electric Co Ltd
Zone Controls Ltd

ELECTRONIC MOTOR CONTROL GEAR

ACEC
AEI Semiconductors Ltd
Abbey Electronics & Automation Ltd
Allen Bradley UK Ltd
Allen West (Brighton) Ltd
BEP Controls Ltd
Baldwin & Francis (Holdings) Ltd
Bennett (Allen) Ltd
Berco Controls Ltd
Bray (E N) Ltd
British Brown—Boveri Ltd
Brush Electrical Engineering Co Ltd
Burndept Electronics Ltd
Cableform Ltd

Clarke, Chapman & Co Ltd
Communications & Electronics Ltd
Control Technology Ltd
Conyers (W E) Ltd
Coppas International (UK) Ltd
Cressall Manufacturing Co Ltd
Davian (Instruments) Ltd
Dawe Instruments Ltd
Daystrom Ltd
Dennis (C) & Co Ltd
Dewrance Controls Ltd
Dix–Kerwood Ltd
Don (William) Ltd
Donovan Electrical Co Ltd (The)
EDC Electricals Ltd
EIH Electronics Ltd
East Coast Electronics Ltd
Electric Construction (Wolverhampton) Ltd
Electrical Remote Control Co Ltd
Electronic & Mechanical Engineering Co Ltd
Electronic Power Controls Ltd
Electro Voice Products Ltd
Eltromet Ltd
Eurotherm Ltd
Express Transformers & Controls Ltd
Failsafe Automation Ltd
Farge Electronics Ltd
Findlay Irvine Ltd
Fotherby Willis Electronics Ltd
Fry (Alan) Controls Ltd
GEC–Elliott Industrial Controls Ltd (Stoke-on-Trent)
Gilfillan (R) & Co Ltd
Graham & White Instruments Ltd
Harland Engineering Co Ltd
Harland Simon Ltd
Herbert Controls & Instruments Ltd
Hird Brown Ltd
Hirst Electric Industries Ltd
Houchin Ltd
ITT Components Group (Harlow)
International General Electric Co of New York Ltd
Irwin & Partners Ltd
JAC Electronics Ltd
Joyce, Loebl & Co Ltd
Kempston Electrical Co Ltd

Klockner–Moeller Ltd
Laurence, Scott & Electromotors Ltd
MEL Equipment Ltd
MTE-Contactor Ltd (Leigh-on-Sea)
Magnetic Devices Ltd
Mann (Albert) Engineering Co Ltd
Marconi Instruments Ltd
Mawdsley's Ltd
Medical Engineering Co Ltd
Millett (H G) & Co Ltd
Mitsubishi Shoji Kaisha Ltd
Monitor & DP Controls Ltd
Morecambe Electrical Equipment Ltd
Morris (John) Electrical Engineering Co Ltd
Neco Electrical Ltd
Nelson (W H) Ltd
Orbit Controls Ltd
Paktronic Engineering Co Ltd
Parmeko Ltd
Photain Controls Ltd
Plessey Co Ltd (The)
Precision Electrical Equipments Ltd
Rank Precision Industries Ltd
Rank Pullin Controls Ltd
Sapphire Research & Electronics Ltd
Saunders Electronics Ltd
Servomex Controls Ltd
Siemens (United Kingdom) Ltd
Solid State Controls Ltd
Square D Ltd
Symot Ltd
Telemechanique Electrique (Great Britain) Ltd
Texas Instruments Ltd
Thorn Automation Ltd
Thornfield Laboratories Timperley Engineering
 (WPM) Ltd
Thricis Electronics Ltd
Trist (Ronald) Controls Ltd
Vosper Electric
Wallacetown Engineering Co Ltd
Watford Electric Co Ltd
Westinghouse Brake & Signal Co Ltd
Witton Electronics Ltd
Zenith Electric Co Ltd
Zone Controls Ltd

ADDRESSES

AB Electronic Components Ltd
5/6 Argyle Street
London W1V 1AD
01-262 6671

AC–Delco Division of General Motors Ltd
Watling Street
Dunstable Beds
Dunstable 64264

ACEC
(Ateliers de Constructions
 Electriques de Charleroi)
Cromwell House
Fullwood Place
London WC1
01-242 2932

AEG (Great Britain) Ltd
Chichester Rents
Chancery Lane
London WC2A 1NH
01-242 9944

AEI Semiconductors Ltd
Carholme Road
Lincoln
Lincoln 29991

AK Fans Ltd
Upper Park Road
London NW3
01-586 0266

Abbey Electronics & Automation Ltd
Delamere Road
Cheshunt Herts
Waltham Cross 20161

Acadex Engineering Co Ltd
Paycocke Road
Basildon Essex
Basildon 20261

Acme Electrical Manufacturing Co
 (Tottenham) Ltd
Tarrif Road
Tottenham London N17
01-808 2702

Acral Ltd
Bush Fair
Harlow Essex
Harlow 25186

Acrastyle Ltd
North Lonsdale Road
Ulverston North Lancs
Ulverston 3232

Adwest Engineering Ltd
The Aerodrome
Reading Berks
Sonning 2351

Airedale Electric and
    Manufacturing Co Ltd
7 Clayton Wood Bank
Ring Road
West Park Leeds 16
Leeds 59711

Aish & Co Ltd
Vanguard Works
Poole Dorset
Poole 4141

Alcho Electric Co
Industrial Estate
Edenbridge Kent
Edenbridge 3434

Allam (E P) & Co Ltd
Arterial Road
Eastwood
Leigh-on-Sea Essex
Southend-on-Sea 525243

Allen Bradley UK Ltd
Denbigh Road
Bletchley Bucks
Bletchley 5884

Allen West (Brighton) Ltd
Moulsecoombe Way
Brighton BN2 4QE Sussex
Brighton 66666

Amphenol Ltd
Thanet Way
Whitstable Kent
Whitstable 4345

Appliance Components Ltd
Cordwallis Estate
Maidenhead Berks
Maidenhead 25151

Arcontrol (Electro Panels) Ltd
Borough Green Kent
Borough Green 3151

Arrow Electric Switches Ltd
Brent Road
Southall Middlesex
01-574 2442

ASEA (Great Britain) Ltd
41 Strand
London WC2
01-930 1671

Asquith Electrics (Colne) Ltd
Walton Street
Colne Lancs
Colne 4949

Auriema Ltd
23/31 King Street
London W3
01-992 5388

Aurora Gearing Co (Wilmot North) Ltd
Edmund Road
Sheffield S2 4EF
Sheffield 24385

Automatic Electric Controls Ltd
John Morris Building
Vulcan Road
Bilston Staffs
Bilston 43341

BEP Controls Ltd
Eclipse Mill
Feniscowles
Blackburn Lancs
Blackburn 22363

B & K Instruments Ltd
59 Union Street
London SE1
01-407 4567

BKB Electric Motors Ltd
St Georges Works
Camden Street
Birmingham 1
021-236 8292

B & R Relays Ltd
South Road
Temple Fields
Harlow Essex
Harlow 25231

BTU Engineering Corporation
42 Queens Road
Farnborough Hants
Farnborough 46264/5

BVC Electronic Developments Ltd
Leatherhead Surrey
Ashstead 6121

Bailey Meters & Controls Ltd
Purley Way
Croydon CR9 4HE Surrey
01-688 4191

Baldwin & Francis (Holdings) Ltd
Eyre Street
Sheffield S1 3GP
Sheffield 28241

Barnsley (John) & Sons Ltd
Netherton
Dudley Worcs
Cradley Heath 66886

Bastat Electrical Equipment Ltd
Church House
Felpham
Bognor Regis Sussex
Bognor 5229

Batwin Electric Motors Ltd
331 Sandycombe Road
Richmond Surrey
01-940 0157/8

Bauer (Great Britain) Ltd
St Andrews House
Portland Street
Manchester 1
061-236 0035

Beanwy Electric Ltd
Rushey Lane
Tyseley Birmingham 11
021-706 6363

Beard Electrical Ltd
Vulcan Road
Bilston Staffs
Bilston 42310

Bennett (Allen) Ltd
12 Orgreave Road
Sheffield S13 9NN
Woodhouse 3281

Berco Controls Ltd
Queensway
Enfield Middx
01-804 2411

Bignell (M C) Ltd
Chantry Works
Yiewsley Middx
West Drayton 3601

Bon Automation Ltd
Town Cross Avenue
Bognor Regis Sussex
Bognor 4089

Bosch Ltd
Rhodes Way
Radlett Road
Watford WD2 4LB Herts
Watford 44233

Boulting (W A) Ltd
Winwick Street
Warrington Lancs
Warrington 38771

Boxall (T A) & Co Ltd
20 Balcombe Road
Horley Surrey
Horley 2337

Bray (E N) Ltd
Britannia Road
Waltham Cross Herts
Waltham Cross 28822

British Brown–Boveri Ltd
Glen House
Stag Place
London SW1
01-828 9422

British Jeffrey Diamond Ltd
Stennard Works
Wakefield Yorks
Wakefield 75133

British Manufacture & Research Co Ltd
Springfield Road
Grantham Lincs
Grantham 2101

British Sonceboz Co Ltd (The)
Victoria Road
South Ruislip Middx
01-845 5626

Brook Motors Ltd
Empress Works
Huddersfield Yorks
Huddersfield 22150

Brookes & Green Ltd
Gomshall
Guildford Surrey
Shere 2056

Brookhirst Igranic Ltd
Elstow Road
Bedford Beds
Bedford 67433

Brown (S G) Ltd
Dukes Avenue
London W4
01-994 7494

Brownings Electric Co Ltd
Boleyn Castle
Upton Park London E13
01-552 1212

Brush Electrical Engineering Co Ltd
Falcon Works
Loughborough Leics
Loughborough 3131

Bull Motors Ltd
Foxhall Works
Ipswich Suffolk
Ipswich 78122

Bulpitt & Sons Ltd
St George's Works
132 Icknield Street
Birmingham 18
021-236 8292

Burndept Electronics (E R) Ltd
St Fidelis Road
Erith Kent
Erith 39121

CAV Ltd
Warple Way
Acton London W3
01-743 3111

CDC (Great Britain) Ltd
Terminal House
Grosvenor Gardens
London SW1
01-730 8271

Cableform Ltd
Green Lane
Romiley SK6 3JQ Cheshire
Woodley 2246

Carruthers (J H) & Co Ltd
College Milton
East Kilbride Glasgow
East Kilbride 20591

Carter Electrical Co Ltd
Eastern Avenue
Romford Essex
Romford 42525

CGE Internationale UK Ltd
CGE House
Alma Road
Windsor Berks
Windsor 69151

Chamberlain & Hookham Ltd
24 Watling Street
Motherwell Lanarkshire
Motherwell 64571

Chandos Products (Scientific) Ltd
High Street
New Mills
Stockport Cheshire
New Mills 2345

Citenco Ltd
Manor Way
Boreham Wood Herts
01-953 3666

Clark Clutch Co Ltd
Bourne Works
Whyteleafe Surrey
01-668 1763

Clarke, Chapman & Co Ltd
   (Electronics Division)
Kingsway
Team Valley Trading Estate
Gateshead NE11 0LD Co Durham
Low Fell 878064

Communications & Electronics Ltd
Stanley Green Road
Poole Dorset
Poole 3107

Comtex Ltd
Senate House
Tyssen Street
Dalston London E8 2ND
01-249 3451

Contraves Industrial Products Ltd
Times House
Ruislip Middx
Ruislip 39649

Control Engineering Ltd
Factory 9
Industrial Estate West
Witham Essex
Witham 3348

Control Technology Ltd
44 Meeching Road
Newhaven Sussex
Newhaven 3535

Controlec Ltd
JWB Works
Chalfont St Peter
Gerrards Cross Bucks
Gerrards Cross 85544

Contropanels Ltd
99 Banknock Street
Carntyne Industrial Estate
Glasgow E2
041-778 4301

Conyers (W E) Ltd
Greystone Works
Trawden
nr Colne Lancs
Colne 5174

Coppas International (UK) Ltd
Wandle House
Riverside Drive
Mitcham Surrey
01-640 0553

Coventry Gauge & Tool Co Ltd
PO Box 39
Fletchamstead Highway
Coventry Warwicks
Coventry 0203 75521

Crabtree (J A) & Co Ltd
Walsall Staffs
Walsall 21202

Craig & Derricott Ltd
Royal Works
Coleshill Street
Sutton Coldfield Worcs
021-355 1161

Crater Controls Ltd
Lower Guildford Road
Knaphill
Woking Surrey
Brookwood 2571

Cressall Manufacturing Co Ltd (The)
Cheston Road
Aston Birmingham
021-327 3571

Crofts (Engineers) Ltd
Thornbury
Bradford 3 Yorks
Bradford 665251

Crompton Parkinson Ltd
Crompton House
39–41 Bridge Street
Northampton NN1 1NY

South Eastern	01-668 4255
Eastern	Hoddesdon 67100
S Wales & S W	Newport 73311
Midlands	021-357 5471
N Western	061-428 3681
Yorkshire	Leeds 58761
Northern	Gateshead 73645
Scottish	E Kilbride 34831
N Ireland	Belfast 24255

Croydon Engineering Co Ltd
Commerce Way
Purley Way
Croydon Surrey
01-688 4125

Cubigear Control Equipment
Eagle Road
Guildford Surrey
Guildford 2328

Dane Electric Controls Ltd
Taylor Lane
Denton Manchester
061-336 2681

Danfoss (London) Ltd
6 Wadsworth Road
Perivale
Greenford Middx
01-998 2041

Daval (S) & Sons Ltd
1 Wadsworth Road
Perivale
Greenford Middx
01-998 1011

Davian (Instruments) Ltd
47 Rowelfield
Luton Beds
Luton 30270

Davies (Fred W) & Son Ltd
Electrical Maintenance Works
Huddersfield Yorks
Huddersfield 31276

Davis (John) & Son (Derby) Ltd
PO Box 38
Alfreton Road
Derby
Derby 41671

Dawe Instruments Ltd
Concord Road
Western Avenue
London W3
01-992 6751

Daystrom Ltd
Bristol Road
Gloucester GL2 6EE
Gloucester 29451

Decca Reid & Sigrist Ltd
Golf Course Lane
Hinkley Road
Leicester LE3 1UA
Leicester 872101

Demag Material Handling Ltd
Beaumont Road
Banbury Oxon
Banbury 50821

Dennis (C) & Co Ltd
Derby Works
Carey Place
Watford Herts
Watford 22321

Dewhurst & Partner Ltd
Melbourne Works
Inverness Road
Hounslow Middx
01-570 7791

Dewrance Controls Ltd
Pit Hey Place
Pimbo
Skelmersdale Lancs
Skelmersdale 24270

Diamond H Controls Ltd
Vulcan Road North
Norwich NOR 85N
Norwich 45291

Dix—Kerwood Ltd
Forge Lane
Mucklow Hill
Halesowen Worcs
021-550 3121

Don (William) Ltd
Kerry Hill
Horsforth
Leeds Yorks
Horsforth 4286

Donovan Electrical Co Ltd (The)
Control Gear Division
Northcote Road
Birmingham 33
021-783 3071

Dorman Smith Control Gear Ltd
Progress Works
Shelley Road
Preston Lancs
Preston 52755

Doxford & Sunderland Shipbuilding &
   Engineering Co Ltd
PO Box 1
Pallion
Sunderland Co Durham
Sunderland 5473

Drayton Controls Ltd
Horton Road
West Drayton Middx
West Drayton 4012

Dynadrive Ltd
South Circular Road
Bangor
Northern Ireland
Bangor 4301

Dynamotive Engineering Co Ltd
Dymec Works
Ystradgynlais
Swansea SA9 1DE South Wales
Glantawe 3278

EDC Electricals Ltd
Bristol Road
Bridgewater Somerset
Bridgewater 2882

EIH Electronics Ltd
Green Lane
Dronfield Sheffield S18 6LL
Dronfield 4441

EMB Co Ltd
Moor Street
West Bromwich Staffs
021-553 1171

EMI Electronics Ltd
Hayes Middx
Hayes 3888

ESL Engineers (Basildon) Ltd
6 Winstanley Way
No 2 Industrial Area
Basildon Essex
Basildon 3350

East Coast Electronics
36 Bessingby Road
Bridlington Yorks
Bridlington 5545

Educational Measurements Ltd
Warsash
Southampton SO3 6HP
Locksheath 84221

Electric Construction
   (Wolverhampton) Ltd
Bushbury Engineering Works
Wolverhampton Staffs
Wolverhampton 27831

Electric Motor Developments
   (Halstead) Ltd
Factory Lane
Halstead Essex
Halstead 2443

Electrical Power Engineering Co
   (Birmingham) Ltd
Kitts Green
Birmingham 33
021-783 2261

Electrical Remote Control Co Ltd
PO Box 10
The Fairway
Bush Fair
Harlow Essex
Harlow 24285

Electrical Systems Ltd
Birmingham Road
Saltisford
Warwick
Warwick 41581

Electro Dynamic Construction Co Ltd
St Mary Cray
Kent BR5 2ND
Orpington 27551

Electro-Magnetic Control Co
41 Rochdale Road
Leyton London E17
01-539 5421

Electro-Mechanical Manufacturing
   Co Ltd
Eastfield
Scarborough Yorks
Clayton Bay 441/3

Electronic & Mechanical Engineering
   Co Ltd
Forge Lane
Mucklow Hill
Halesowen Worcs
021-550 3121

Electronic Power Controls Ltd
Patrick Gregory Road
Linthouse Lane
Wedensfield
Wolverhampton Staffs
Wolverhampton 732616

Electro-Voice Products
Maple Cross Industrial Estate
Rickmansworth Herts
Rickmansworth 75381

Ellison (George) Ltd
Wellhead Lane
Perry Bar
Birmingham 22B
021-356 4562

Elliston, Evans & Jackson Ltd
Bath Road
Bridgwater Somerset
Bridgwater 2056

Elmic Ltd
Southend Arterial Road
Romford RM3 0XB Essex
Ingrebourne 46521/2

Eltromet Ltd
6 Hunting Gate
Wilbury Way
Hitchin Herts
Hitchin 3664

English Electric—AEI Machines Ltd
Phoenix Works
Bradford BD3 8JZ
Bradford 665221

English Electric—AEI Machines Ltd
Lower Milehouse Lane
Newcastle-under Lyme ST5 9BQ Staffs
Newcastle-under-Lyme 561421

English Electric—AEI Machines Ltd
Mill Road
Rugby Warwicks
Rugby 2121

English Electric—AEI Machines Ltd
Blackheath Works
Rowley Regis
Warley Worcs
021-559 1500

English Electric—AEI Traction Ltd
Trafford Park
Manchester M17 1PR
061-872 2431

English Electric—AEI Turbine
   Generators Ltd
(Generator Division)
Stafford
Stafford 3232

Erskine, Heap & Co Ltd
Lancashire Switchgear Works
Manchester M7 9SP
061-832 4561

Escomat Ltd
26a Cleaver Square
London SE11
01-735 8647

Ether Ltd
Caxton Way
Stevenage Herts
Stevenage 4422

Eurodrive Ltd
PO Box 1
Normanton Yorks
Normanton 3855

Eurotherm Ltd
Broadwater Trading Estate
Worthing Sussex
Worthing 31681

Evershed & Vignoles Ltd
Acton Lane
Chiswick London W4
01-994 3670

Express Transformers & Controls Ltd
44/46 Beddington Lane
Croydon Surrey
01-684 2443

Failsafe Automation Ltd
Clifton Works
Manor Road
Manchester M19 3EJ
061-224 6224

Farge Electronics Ltd
Water Street
Portwood
Stockport SK1 2BU Lancs
061-480 7155

Ferranti Ltd
Ferry Road
Edinburgh EH5 2XS
031-332 2411

Field & Grant Ltd
Kent Street
Birmingham 5
021-692 1341

Findlay Irvine Ltd
Bog Road
Penicuik Midlothian
Penicuik 2111

Fisher Controls Ltd
Brearley Works
Luddendenfoot
Halifax Yorks
Calder Valley 2711

Forco Electrical Services Ltd
Southernhay East
Exeter EX1 1PE
Exeter 72639

Fotherby Willis Electronics Ltd
Broadgate Lane Industrial Estate
Horsforth
Leeds LS18 4SF Yorks
Leeds 5636

Foulds (E A) Ltd
Albert Works
Clifton Street
Colne Lancs
Colne 126

Fractional HP Motors Ltd
Millmarsh Lane
Brimsdown
nr Enfield Middx
01-804 4775

Francis (Clocks) Ltd
Lower Guildford Road
Knaphill
Woking Surrey
Brookwood 2571

Fry (Alan) Controls Ltd
Booton
Norwich Norfolk NOR 76X
Cawston 221

Garrard Engineering Ltd
Newcastle Street
Swindon Wilts
Swindon 5381

GEC—Elliott Industrial Controls Ltd
Kidsgrove
Stoke-on-Trent Staffs
Kidsgrove 3511

GEC—Elliott Industrial Controls Ltd
Mill Road
Rugby Warwicks
Rugby 2121

GEC—Elliott Precision Controls Ltd
Century Works
Lewisham London SE13
01-692 1271

Gilfillan (R) & Co Ltd
Southdown View Road
Broadwater Trading Estate
Worthing
Worthing 8719

Globe Engineering Co Ltd
Huddersfield Road
Brighouse Yorks
Brighouse 2899

Graham & White Instruments
82 London Road
St Albans Herts
St Albans 59373

Grantham Electrical Engineering Co Ltd
Harlaxton Road Works
Grantham Lincs
Grantham 2691

Graseby Instruments Ltd
Hook Rise
South Tolworth
Surbiton Surrey
01-397 5311

Green (Horace) & Co Ltd
Station Works
Cononley
Keighley Yorks
Cross Hills 2201

Greenwood & Batley Ltd
Armley Road
Leeds Yorks
Leeds 20011

Hackbridge Faraday Electrics Ltd
Moulscoomb Way
Brighton Sussex
Brighton 685696

Hampson Industries Ltd
Bull Lane
West Bromwich Staffs
West Bromwich 2071

Hannah Switchgear Ltd
Roman Ridge Road
Sheffield S9 1FH
Sheffield 388471

Harland Engineering Co Ltd
BEP Works
Alloa Clackmannanshire
Alloa 2100

Harland Simon Ltd
Bond Avenue
Bletchley Bucks
Bletchley 5331

Harland & Wolff Ltd
Queens Island
Belfast BT3 9DU
Belfast 58456

Hawker Siddeley Dynamics Ltd
Manor Road
Hatfield Herts
Hatfield 62300

Helga Electronics Ltd
Ann Street
Southbank Teeside
Eston Grange 4691

Heyes & Co Ltd
Water-Heyes Electrical Works
PO Box 60
Wigan Lancs
Wigan 4366

Herbert Controls & Instruments Ltd
Shaftmoor Lane
Birmingham 28
021-777 2274

Highfield Gear & Engineering Co Ltd
Nile Street
Huddersfield Yorks
Huddersfield 24466

Hird Brown Ltd
Lever Street
Bolton Lancs
Bolton 27311

Hirst Electric Industries Ltd
Gatwick Road
Crawley Sussex
Crawley 25721

Holme (Edward) & Co Ltd
Moss Lane
Altrincham Cheshire
061-928 2694

Hoover Ltd
Western Avenue
Perivale
Greenford Middx
01-997 3311

Honeywell Ltd
Great West Road
Brentford Middx
01-568 9191

Houchin Ltd
Garford Works
Chart Road
Ashford Kent
Ashford 23211

Howells Stanway Ltd
Vale Place Works
Hanley Stoke-on-Trent ST1 5DG
0782 24611

Huggett Electrical Ltd
Park Street Mews
Bath Somerset
Bath 26271

ITT Components Group Europe
(Standard Telephones & Cables Ltd)
Edinburgh Way
Harlow Essex
Harlow 26811

Imhof—Bedco Ltd
Ashley Works
Cowley Mill Road
Uxbridge Middx
Uxbridge 37123

Impex Electrical Ltd
Market Road
Richmond Surrey
01-876 1047

Industrial Control Systems (1969) Ltd
76/90 Clarke Road
Northampton NN1 4OE
Northampton 32417

Industrial Switchgear Ltd
Enfield Industrial Estate
Redditch Worcs
Redditch 67336

Ingleby Motors Ltd
Sweet Street
Leeds 11 Yorks
Leeds 20275

International General Electric Co
of New York Ltd
296/302 High Holborn
London WC1
01 242 6868

Irwin & Partners Ltd
294 Purley Way
Croydon Surrey CR9 4QL
01 686 6441

JAC Electronics Ltd
Blackwater Station Estate
Camberley Surrey
Camberley 5309

Jennings Electric (Belfast) Ltd
Deramore Works
Durham Street
Belfast BT12 4GG
Belfast 44292

Jones (Walter) & Co (Engineers) Ltd
Newlands Park
Sydenham London SE26
01-778 6264/6

Joyce, Loebl & Co Ltd
Princesway
Team Valley
Gateshead NE11 0UJ Co Durham
Gateshead 877891

Kempston Electrical Co Ltd
Shirley Road
Rushden Northants
Rushden 4351

Klaxon Ltd
Warwick Road
Tyseley Birmingham
021-706 1654

Klockner—Moeller Ltd
Griffin Lane
Aylesbury Bucks
Aylesbury 4481

Landis & Gyr Ltd
Elgee Works
Victoria Road
North Acton London W3
01-992 5311

Latham (H J) Ltd
Towerfield Road
Shoeburyness
Southend Essex
Shoeburyness 3209

Laurence, Scott & Electromotors Ltd
Norwich NOR 85A
Norwich 28333
Telex 97323

Lewis Electric Motors (Maidenhead) Ltd
Moor Works
Blackamoor Lane
Maidenhead Berks
Maidenhead 21216

Limit Sales (Holdings) Ltd
64 Essex Road
Islington London N1
01-226 2481

Litton Precision Products Ltd
95 High Street
Slough Bucks
Slough 28267

Lloyd (J J) Instruments Ltd
Brook Avenue
Warsash
Southampton SO3 6HP
Locksheath 4221

Lounsdale Electric Ltd
Lounsdale Works
Meikeriggs Road
Paisley Renfrewshire
041-889 8911

M & C Switchgear Ltd
Kelvinside Works
Kirkintilloch Glasgow
041-776 2216

MEL Equipment Co Ltd
Manor Royal
Crawley Sussex
Crawley 28787

MEM (Midland Electric Manufacturing
  Co Ltd)
Reddings Lane
Tyseley Birmingham 11
021-706 3300

MG Electric (Colchester) Ltd
Hawkins Road
Colchester Essex
Colchester 6397

MTE-Contactor Ltd
20 Progress Road
Leigh-on-Sea Essex
Southend-on-Sea 524281

MTE-Contactor (Systems) Ltd
Moorfield Road
Wolverhampton Staffs
Wolverhampton 25911

Macfarlane Engineering Co Ltd
Netherlee Road
Cathcart Glasgow S4
041-637 2255

MacGeoch (William) & Co
  (Birmingham) Ltd
Warwick Works
46 Coventry Road
Bordesley Birmingham 10
021-772 3371

Mackie (W) & Co Ltd
Station Approach
St Mary Cray
Orpington Kent
Orpington 27551

Magco Moxey (Division of Winget Ltd)
Lake Works
Portchester
Fareham Hants
Cosham 70472

Magnetic Devices Ltd
Exning Road
Newmarket Suffolk
Newmarket 3451

Mann (Albert) Engineering Co Ltd
Basildon Industrial Estate
Basildon Essex
Basildon 20421

Marconi Instruments Ltd
(Sanders Division)
Gunnels Wood Road
Stevenage Herts
Stevenage 2311

Marshall Richards Barco Ltd
Woodside Road
Eastleigh SO5 4YA Hants
Eastleigh 2701

Marubeni-iida Co Ltd
164 Clapham Park Road
London SW4
01-720 1911

Mather & Platt Ltd
Park Works
Manchester M10 6BA
061-205 2321

Mawdsley's Ltd
Zone Works
Dursley Glos
Dursley 2921

Medical Engineering Co Ltd
Stanley Works
Belgrave Road
South Norwood London SE25
01-653 2245

Mertech Electrical Equipment Ltd
Lower Dagnall Street
St Albans Herts
St Albans 60355

Micromotor Ltd
227 Turnpike Link
Croydon CR0 5NW Surrey
01-688 5891

Midgley Harmer Ltd
Wellington Road
Kensall Green London NW10
01-969 2311

Miles (F G) Engineering Ltd
Old Shoreham Road
Shoreham-by-Sea Sussex
Shoreham-by-Sea 4511

Millett (H G) & Co Ltd
18 Mannings Heath Road
Parkstone
Poole Dorset
Parkstone 5163

Mitsubishi Shoji Kaisha Ltd
Bow Bells House
Bread Street
London EC4M 9BQ
01-248 3292

Milne & Longbottom Ltd
Elm Works
Mere Lane
Rochdale Lancs
Rochdale 49897

Modern Techniques Ltd
Wedmore Street
London N19
01-272 3114

Monitor & DP Controls Ltd
Kings Road
Wallsend Northumberland
Wallsend 625211

Moore Reed & Co Ltd
Walworth
Andover Hants
Andover 4155

Morecambe Electrical Equipment Co Ltd
Westgate Works
Morecambe Lancs
Heysham 52474

Morris (Herbert) Ltd
PO Box No 7
Loughborough Leics
Loughborough 3123

Morris (John) Electrical Engineering Ltd
Vulcan Road
Bilston Staffs
Bilston 43341

Morley Electrical Engineering Co Ltd
Stanningley
Pudsey Yorks
Pudsey 3034

Mortley (G E) Sprague & Co Ltd
Lyons Crescent
Tonbridge Kent
Tonbridge 2358

Motor Gear Engineering Co Ltd
Chadwell Heath
Romford RM6 4ES Essex
01-590 7788

Muirhead Ltd
Croydon Road
Elmers End
Beckenham Kent
01-650 4888

Mycalex (Motors) Ltd
Ashcroft Road
Cirencester GL7 1QY Glos
Cirencester 2551

Neco Electrical Co Ltd
Walton Road
Eastern Road
Cosham Hants
Portsmouth 71711

Nelson (W H) Ltd
Nelbest Works
Bellshill Lanarks
Bellshill 2633/4

Nelson Engineering Co Ltd
Netherfield Works
Netherfield Road
Nelson Lancs
Nelson 62545

Nelson Transformer &
  Rotary Equipment Ltd
Fir Tree Mills
Higham
Burnley Lancs
Padiham 72716

Newman Industries Ltd
Yate Bristol BS17 5HG
Chipping Sodbury 3311

Newton Derby Ltd
Alfreton Road
Derby
Derby 47676

Norco Engineering Ltd
Burrell Road
Haywards Heath Sussex
Haywards Heath 51771

Normand Electrical Co Ltd
Walton Road
Cosham
Portsmouth Hants
Cosham 71711

Oliver Pell Control Ltd
Cambridge Row
Burrage Road
Woolwich SE18
01-854 1422

Orbit Controls Ltd
Alstone Lane Industrial Estate
Cheltenham Glos GL51 8JQ
Cheltenham 26608

Ottermill Engineering Ltd
Heron House
Wembley Hill Road
Wembley Middx
01-903 2166

Paktronic Engineering Co Ltd
Cambridge Street
Grantham Lincs
Grantham 2730

Park Bros Ltd
Bankfield Works
Ordnance Street
Blackburn BB1 3AF Lancs
Blackburn 52561

Parmeko Ltd
Percy Road
Aylestone Park Leicester
Leicester 832287

Parsons Peebles Ltd
East Pilton
Edinburgh EH5 2XT
031-552 6261

Parsons Peebles Ltd
Electric Avenue
Witton Birmingham 6
021-327 1941

Parvalux Electric Motors Ltd
Wallsdown Road
Bournemouth BH1 8PU Hants
Bournemouth 52575/8

Phillipson (A F) & Co Ltd
Tariff Road
Tottenham London N17
01-808 2441

Photain Controls Ltd
Randalls Road
Leatherhead Surrey
Leatherhead 2776

Plessey Co Ltd (The)
Kembrey Street
Swindon Wilts
Swindon 6211

Power Control Panels (Cambridge) Ltd
36 Kingston Street
Cambridge
Cambridge 56410

Precision Electrical Equipments Ltd
PO Box 3
Rednal Birmingham
021-445 1887

Printed Motors Ltd
Upper Street
Fleet
Aldershot Hants
Fleet 3414

Pye of Cambridge
St Andrews Road
Cambridge CB4 1DL
Cambridge 58985

Quinton Crane Electronics Ltd
High Street
Melbourne DE7 1GL Derbys
Derby 2266

Rank Precision Industries Ltd
(Broadcast Division)
Watton Road
Ware Herts
Ware 3939

Rank Pullin Controls Ltd
Phoenix Works
Great West Road
Brentford Middx
01-560 1212

Redler Industries Ltd
Dudbridge Works
Stroud Glos
Stroud 3611

Redman Heenan Froude Ltd
Shrub Hill Road
Worcester
Worcester 23461

Renold Ltd
Renold House
Wythenshawe
Manchester M22 5WL
061-437 5221

Reyrolle Belmos Ltd
Bellshill Lanarks
Blantyre 3171

Reyrolle (A) Parsons Ltd
Hebburn Co Durham
Hebburn 832441

Richardsons Westgarth (Hartlepool) Ltd
Hartlepool Co Durham
Hartlepool 66678

Sadi (Great Britain) Ltd
East Street
Commerce Road
Brentford Middx
01-580 8056

Salon (Nelson) Ltd
Hollin Bank
Brierfield
Nelson Lancs
Nelson 65217

Sangamo Weston Ltd
Great Cambridge Road
Enfield Middx
01-366 1100

Santon Ltd
Somerset Works
Newport NPT 0XU Mon
Newport 71711

Sapphire Research & Electronics Ltd
Sapphire Works
Ferndale Glos
Ferndale 782

Saunders Electronics Ltd
Faraday Road
Hinckley Leics
Hinckley 4028

Scott (Hugh J) & Co (Belfast) Ltd
Volt Works
Belfast BT6 8GN
Belfast 57225

Scott (James) (Electronic Agencies) Ltd
90 West Campbell Street
Glasgow C2
041-221 3866

Servomex Controls Ltd
Crowborough Sussex
Crowborough 2151

Sew-Tric (Holdings) Ltd
Honeypot Lane
Stanmore HA7 1JZ Middx
01-952 5261

Shawford Control Gear Co Ltd
715 Tudor Estate
Abbey Road
Park Royal London NW10
01-965 7527

Siba Electric Ltd
Frimley Road
Camberley Surrey
Camberley 3252

Siemens (United Kingdom) Ltd
Great West House
Great West Road
Brentford Middx
01-568 9133

Simmons Electrical Services (Sutton) Ltd
Diceland Works
22 Dicelands Road
Banstead Surrey
Burgh Heath 54006

Simon Equipment Ltd
Bond Avenue
Bletchley Bucks
Bletchley 5331

Simplex—GE Ltd
PO Box 2
Blythe Bridge
Stoke-on-Trent ST11 9LL Staffs
Blythe Bridge 3551

Sirco Controls Ltd
Sweynes Industrial Estate
Ashington Road
Rochford Essex
Southend-on-Sea 545125

Small Electric Motors Ltd
Churchfields Road
Beckenham Kent
01-650 0066

Smith Bros & Webb Ltd
Stratford Road
Sparkbrook Birmingham 11
021-772 2714

Smiths Industries Ltd
(Appliance Control Division)
Crowhurst Road
Brighton BN1 8AN Sussex
Brighton 506363

Smiths Industries Ltd
(Aviation Division)
Kelvin House
Wembley Park Drive
Wembley HA9 0NH Middx
01-452 3333

Solenoids & Regulators Ltd
Moseley Road
Birmingham 12
Calthorpe 0532

Solid State Controls Ltd
Brunel Road
Acton London W3
01-743 7665

South Wales Switchgear Ltd
Blackwood Mon
Blackwood 3001

Spectar Engineering Ltd
Stourport Road
Kidderminster Worcs
Kidderminster 4711

Specto Ltd
Vale Road
Windsor Berks
Windsor 61474

Sperry Gyroscope Division
  Sperry Rand Ltd
Bracknell Berks
Bracknell 3222

Square D Ltd
Cheney Manor
Swindon Wilts
Swindon 6222

Starkstrom Ltd
Mosley Street Works
Blackburn Lancs
Blackburn 57232

STC (Electromechanical Div)
(Standard Telephones & Cables Ltd)
West Road
Harlow Essex
Harlow 26811

Sterling Instruments Ltd
Sterling Works
Crewkerne Somerset
Crewkerne 2222

Stone—Platt Crawley Ltd
Gatwick Road
Crawley Sussex
Crawley 27711

Sumo Pumps Ltd
PO Box 2
Crawley Green Road
Luton Beds
Luton 31144

Switchgear & Cowans Ltd
Elsinore Road
Old Trafford Manchester 16
061-872 2881

Switchgear & Instrumentation Ltd
Bowling Park Works
Bradford 4 Yorks
Bradford 34221

Symot Ltd
17 Market Place
Henley-on-Thames Oxon
Henley 4455

TKS (Aircraft De-icing) Ltd
162/164 Uxbridge Road
London W7
01-567 7274

Tapley Meters Ltd
Belvidere Works
Totton Southampton
Totton 3232

Teddington Aircraft Controls Ltd
Cefn Coed
nr Merthyr Tydfil Glam
Merthyr Tydfil 3261

Teddington Autocontrols Ltd
Windmill Road
Sunbury-on-Thames Middx
Sunbury 85500

Tefco Vari-speed Motor Co
Tefco Works
Greenacres Road
Oldham Lancs
061-624 8798

Telemechanique Electrique
  (Great Britain) Ltd
Henwood
Ashford Kent
Ashford 21311

Texas Instruments Ltd
Manton Lane
Bedford
Bedford 67466

Thomson (S) & Sons (Liverpool) Ltd
Gladstone Works
491 Hawthorne Road
Bootle Liverpool
051-922 2697

Thorn Automation Ltd
Brereton Road
Rugeley Staffs
Rugeley 3271

Thornfield Laboratories Timperley
  Engineering (WPM) Ltd
Dallimore Road
Wythenshawe Manchester M23 9NX
061-998 3414

Thricis Electronics Ltd
46 The Ridgeway
Watford WD1 3TN
Watford 42643

Torrin Corporation
Greenbridge Industrial Estate
Swindon Wilts
Swindon 21387

Trident Engineering Ltd
Shute End
Wokingham Berks
West Forrest 6444

Trist (Ronald) Controls Ltd
Bath Road
Slough Bucks
Slough 25401

Tuscan Engineering Ltd
Industrial Estate
Bridgend Glam
Bridgend 2973

Twiflex Couplings Ltd
The Green
Twickenham Middx
01-894 1161

UK Solenoid Ltd
Hungerford Berks
Hungerford 2427

Unigears Ltd
Henwood Estate
Ashford Kent
Ashford 21661

Unimatic Engineers Ltd
16 Coverdale Road
Cricklewood London NW2
01-459 2145

Union Carbide UK Ltd
(Electronics Division)
PO Box 2LR
8 Grafton Street
London W1 A2 LR
01-629 8100

Vactric Control Equipment Ltd
Garth Road
Morden Surrey
01-337 6644

Vickalectric Engineering Co Ltd
Bodmin Road
Wyken
Coventry Warwicks
Coventry 322293

Vlasto Clark & Watson
London Road Bridge
Stockton Heath
nr Warrington Lancs
Warrington 61617

Vosper Electric
(Industrial & Marine Controls
   Division of Vosper Ltd)
Castle Trading Estate
Portchester
Fareham Hants
Cosham 79481

Wallacetown Engineering Co Ltd (The)
Viewfield Road
Ayr
Ayr 6763

Walsall Conduits Ltd
Dial Lane
Hill Top
West Bromwich Staffs
021-557 1171

Watford Electric Co Ltd
Whippendell Road
Watford Herts
Watford 28201

Westinghouse Brake & Signal Co Ltd
82 York Way
London N1
01-837 6432

Witton Electronics Ltd
Tame Road
Witton Birmingham 6
021-327 3214

Whipp & Bourne Ltd
Castleton
Rochdale Lancs
Rochdale 32051

Woden Transformer Co Ltd
Oxford Street
Bilston Staffs
Bilston 42681

Wright Electric Motors Ltd
Pellon Lane
Halifax Yorks
Halifax 60201

Wynstruments Ltd
   (Electric Motor Division)
Staverton Airport
Gloucester GL2 9QW
Churchdown 3264

Yorkshire Switchgear &
   Engineering Co Ltd
Meanwood
Leeds 6 Yorks
Leeds 57121

Zenith Electric Co Ltd
Pinecrest Works
Cranfield Road
Wavendon Bucks
Woburn Sands 2531

Zone Controls Ltd
PO Box 37
Dibdale Road
Dudley Worcs
Dudley 56881

Glossary of Terms

ampere (A): base SI unit of electric current; defined as constant current which, if maintained in two straight parallel conductors (of infinite length and negligible cross-section) placed 1 m apart in vacuum, would produce between the conductors a force of 2×10^{-7} newton per metre of length.

Armature: (elec machine) stationary or rotating part in which emf is produced (in the case of a generator) or the torque is produced (in the case of a motor); includes winding through which the main current of the machine passes, and part of the magnetic circuit on which the winding is mounted.

Armature reaction: (elec) mmf produced by armature currents in the magnetic circuit of an electric machine.

atto (a): prefix used to denote sub-multiple of SI units; value 10^{-18}.

Auto-transformer starting: (elec) reduced voltage starting of an induction motor which is more flexible than a straight star-delta starter (q.v.); by selecting a suitable transformer tapping, any desired reduction in starting current can be achieved.

Back-emf: (elec) the induced voltage in a d.c. motor armature; its direction opposes the current flow in the motor conductors.

Cage: (elec) term used in electrical engineering, basically derived from construction of a conductor system in the form of a cage of bars connected to end-pieces or rings; cage motor, cage rotor, squirrel-cage motor, cage winding (the bars or conductors forming the winding pass through slots in the core).

candela (cd): unit of luminous intensity; one of the six base SI units; the luminous intensity, in the perpendicular direction, of a surface of $1/600\,000\text{m}^2$ of a black body at the temperature of freezing platinum under a pressure of $101\,325\text{N}/\text{m}^2$.

Capacitor motor: (elec) a split-phase motor with a capacitor normally in series with the auxiliary primary winding; term not used alone, in practice (but see capacitor start and run; capacitor start; two-value capacitor motor).

Capacitor start and run motor: (elec) capacitor motor in which the auxiliary primary winding and series-connected capacitor remain in circuit for both starting and running.

Capacitor start motor: (elec) capacitor (split-phase) motor in which the auxiliary primary winding connected in series with a capacitor is in circuit only during the starting period.

centi (c): prefix used to denote sub-multiple of SI units; value 10^{-2}; eg cSt (centistokes, measure of viscosity); preferably limited to use where other recommended prefixes are inconvenient.

Closed air circuit enclosure: (elec) a total enclosure type, but basically a ventilated housing with openings connected to a closed-cycle heat exchanger; heat exchange may be air-to-air (CACA) or air-to-water (CACW); see enclosure.

Cogging: (elec servomotor) see slot effect.

Compound motor: (elec) motor which contains both series and shunt connected field windings.

Control synchro (transmitter): (elec) high impedance version of the torque transmitter (q.v.).

Copper loss: (elec) loss occurring from current in windings (eg rotor or stator) of electrical machine; is proportional to product of resistance and the square of the current.

coulomb: derived SI unit; the quantity of electricity transported in 1 second by a current of 1 ampere; symbol C.

Cumulative compounding: (elec) fluxes produced by the two field windings of a compound motor assist each other.

Current: (elec) passage of electricity through a material or medium by movement of electrons; expressed as amperes (A).

Current density: (elec) the concentration or density of an electric current; A/m^2.

deca (da): prefix used to denote multiple of SI units; value 10^1; preferably limited to use where other recommended prefixes are inconvenient.

deci (d): prefix used to denote sub-multiple of SI units; value 10^{-1}; eg dm decimetre; preferably limited to use where other recommended prefixes are inconvenient.

Delta connection: (elec) windings in a 3-phase electrical machine are connected in series, the 3-phase supply being taken from or supplied to the junctions; vector diagram of current or voltage in the windings is in the form of a triangle (or Greek delta); see star connection.

Detent torque: (elec) the torque that an unexcited permanent magnet motor can resist before starting to rotate.

Differential compounding: (elec) fluxes produced by the two field windings of a compound motor oppose each other.

Direct starting: (elec) also known as on-line or full voltage starting; simplest way of starting a cage motor by applying full short circuit current at instant of start; gives maximum starting / accelerating torque with minimum accelerating time.

Drip-proof enclosure: (elec) machine housing made so that liquid/solid particles that fall from a small angle (eg 15°) from vertical cannot enter in sufficient amount to interfere with satisfactory operation; see ventilated enclosure.

DTR: (elec) duty type rating; term used eg in specifications for performance of rotating electrical machines.

Duct ventilated machine: (elec) a ventilated-enclosure (q.v.) machine can be used in adverse environments by providing clean cooling air from an external source to the inlet and outlet openings via ducting; various forms of fan or compressor forced-ventilation can be incorporated; see enclosure.

ECR: (elec) equivalent continuous rating; term used eg in specifications for performance of rotating electrical machines.

Eddy current: (elec) produced by voltages induced in a conductor when a magnetic field passing through the conductor varies; side effects caused by eddy currents include uneven current distribution and heating, which can affect efficiency of electric machines.

Electrical loading: (elec machine) usually expressed as product of number of armature conductors and current per conductor, divided by the air-gap circumference; ampere conductors per metre.

emf: (elec) electromotive force, which tends to produce a movement of electricity around an electric circuit.

Enclosure: (elec) the structure within which an electrical machine is housed; two basic types are ventilated (inner surfaces are in contact with ambient atmosphere) or totally-enclosed (free exchange of atmosphere between inner and outer surfaces is prevented); see total and ventilated enclosures for further subdivisions of types of enclosure.

Fan-cooled total enclosure: (elec) a totally-enclosed housing provided with an external fan (and finning, etc) to assist in cooling; may be weatherproofed for use in the open.

Faraday's law: (elec) when the magnetic flux linked with an electric circuit is changed, an emf is induced in the circuit; the size of the emf is proportional to the rate of change of flux; the direction of the emf is given by Lenz's law (q.v.).

femto (f): prefix used to denote sub-multiple of SI units; value 10^{-15}.

Flameproof enclosure: (elec) machine housing designed to prevent spread to the external atmosphere of any flammable gas that has been ignited within an electrical machine (rather than preventing ingress of flammable gas into the housing); normally a totally enclosed type built to rigid specifications; see enclosure.

FLC: (elec) commonly used abbreviation for full-load current.

FLT: (elec) commonly used abbreviation for full-load torque.

Flux: (magnetic) a concept of the flow of a magnetic field; flux units are expressed in webers (Wb); flux symbol is Φ.

Flux density: (magnetic) the concentration or density of a magnetic field; expressed as tesla (T) = weber/m^2; symbol is B.

Following error: (elec) for a stepping motor, the error in angular position between different steps in static (not dynamic) conditions; results in practice because motor steps may be slightly unequally spaced so the rotor may reach a (slightly) different position when approaching the same step from the opposite direction.

Fractional horsepower (FHP): term used particularly in electrical engineering with reference to FHP motor, defined as a motor having a continuous rating not exceeding 1 hp (0·75 kW) per 1 000 rev/min; although it is inconsistent with the adoption of SI units, the term is likely to continue in use in British, Commonwealth and North American industry.

Gas conductor: (elec) also called plasma (q.v.); electrical conductivity depends on concentration of electrons and their freedom of passage through the gas; production of enough electrons (ionisation) for practical uses may need very high temperatures (as in plasma arc) or 'seeding' (the introduction of small amounts of readily ionised elements); see magnetohydrodynamics.

Generator effect: (elec) an emf is induced in a conductor which is moved through a magnetic field; the emf induced by a constant magnetic field produced by the field windings of a generator is proportional to the rate at which the conductor cuts the flux (ie the generator effect or cutting rule).

giga (G): prefix used to denote multiple of SI units; value 10^9; eg GHz, gigahertz.

Hall effect: (elec) a magnetic field applied at right angles to a current flowing in a conductor (esp semiconductor) produces a voltage along a third axis at right angles to the other two; Hall voltage $E = RIB/t$ where R is Hall coefficient, B flux density, I current (a.c. or d.c.), t thickness met by the field; because voltage varies as $1/t$. Hall elements are made in the form of thin plates.

Hall element: (elec) a thin plate of semi-conductor material used to exploit the Hall effect (q.v.).

Hall effect synchro: (elec) a component similar to a 2-phase synchro, but using semiconductors and the resultant Hall effect (q.v.).

hecto (h): prefix used to denote multiple of SI units; value 10^2; eg hpz, hectopièze (esp in French technical use) = 100 times MTS unit of pressure = 1 bar; preferably limited to use where other recommended prefixes are inconvenient.

henry: derived SI unit; the inductance of a closed circuit in which an electromotive force of 1 volt is produced when the electric current varies uniformly at the rate of 1 ampere per second (also applies to emf in one circuit produced by a varying current in a second circuit, ie mutual inductance); symbol H.

Holding torque: (elec) for an excited stepping motor, the torque that develops before the motor is pulled out of step and starts to rotate.

Hose-proof enclosure: (elec) machine housing made so that the machine operates satisfactorily after hosing down from any accessible direction; may involve elaborate ducting in ventilated type; see enclosure.

Iron loss: (elec) loss of energy caused by alternating flux in the iron of the magnetic circuit; see eddy current.

joule: derived SI unit; the work done when a force of 1 newton is applied over a displacement of 1 metre in the direction of the force; symbol J.

kelvin (K): unit of thermodynamic temperature; 1/273·16 of the thermodynamic temperature of the triple point of water (the ice point is 273·15 K); units of kelvin and Celsius temperature interval are identical.

Kelvin bridge: (elec) derivation of the Wheatstone bridge (q.v.), and similarly used to compare resistors.

kilo (k): prefix used to denote multiple of SI units; value 10^3; eg kV, kilovolt.

Kirchoff's laws: (elec) used in electrical network calculations; (1) if a closed path is traversed in an electric network, the algebraic sum of voltage drops across individual circuit elements in the direction of traverse is zero; (2) the algebraic sum of all currents directed to a junction point in a network is zero; (laws are true for instantaneous values, and for vector addition of a.c. quantities).

Lenz's law: (elec) the direction of the emf induced in a circuit by a change in magnetic flux linked with the circuit tends to oppose the change in the inducing flux; *ie* any electromagnetic circuit resists an attempt to change its existing state; see Faraday's law.

Linear induction motor: (elec) basically a (primary) row of coils supplied with a.c. to provide a straight-line magnetic field, which can move a secondary winding cut by the field; in practice either the primary or secondary parts can move, and either may be short or long, depending on the length of travel or motion required; uses include linear actuators and vehicle (*eg* railway) traction.

Locked rotor torque: (elec) minimum measured torque which an electric motor develops at rest for all angular positions of its rotor with rated voltage applied at rated frequency.

Losses: (elec) within an electrical machine there are losses of efficiency in components or caused by other effects; *eg* friction and windage losses; iron loss in cores; stator and rotor copper loss; brush loss (on slipring or commutator machines); stray losses.

Magnetic field: (elec) for a straight conductor carrying a current I the magnetic field at a distance r is $2I/r$; the lines of force tend to move a N magnetic pole in the direction of a right-hand screw advancing in the direction of the current; the field around a straight conductor is in the form of concentric lines of force.

Magnetic field force: (elec) a current-carrying conductor at an angle to a magnetic field is acted on by a force; at right angles, the force is proportional to field strength and current (and is at right angles to the conductor and to the field).

Magnetic loading: (elec machine) usually expressed as product of flux/pole and number of poles, divided by the product of the air-gap circumference and the active armature core length; Wb/m^2.

Magnetising current: (induction motor) the non-power component of supply current which is directly responsible for setting up the air-gap field, and is only indirectly affected by the loading.

Magnetohydrodynamic (MHD): effect produced by the interaction between moving electrically-conducting fluids and magnetic fields; can be used (1) to generate electricity; (2) as a source of power derived from a system which produces a rapidly moving fluid stream.

Magnetostriction: change in dimensions of a solid placed in a magnetic field; relatively large in ferromagnetic or similar materials; can cause problems (*eg* strain; hum in transformer cores from alternating changes in length); or can be used in transducers, delay lines and filters.

Maximum following rate: (elec) for a stepping motor, see maximum pull-out rate.

Maximum pull-in rate: (elec) for a stepping motor, maximum switching rate at which the motor runs without losing a step (*eg* expressed in steps/s).

Maximum pull-out rate: (elec) for a stepping motor, the switching rate at which an unloaded motor starts to drop out of step as the stepping rate is gradually and smoothly increased.

Maximum synchronising rate: (stepping motor) see maximum pull-in rate.

MCR: (elec) maximum continuous rating; term used *eg* in specifications for performance of rotating electrical machines.

mega (M): prefix used to denote multiple of SI units; value 10^6; *eg* MW, megawatt.

MHD generator: a magnetohydrodynamic (q.v.) unit equivalent to an electric motor, but with a fluid as the moving conductor; motion may be linear and conductor speeds can be very much greater than in conventional rotary machine; can operate at high temperatures, *eg* in closed cycle units using nuclear or fossil fuels as energy source.

micro (μ): prefix used to denote submultiple of SI units; value 10^{-6}; *eg* μs, microsecond.

milli (m): prefix used to denote submultiple of SI units; value 10^{-3}; *eg* mm, millimetre.

mmf: (elec) magnetomotive force; the product of current and winding turns which causes a flux to be produced in a coil.

'M' motor: (elec) a type of stepping motor used for many years in naval instruments; limited in industrial applications to low powers and speeds; superseded by analogue and, later, semiconductor-based devices.

Multiple stator: (elec stepping motor) design which provides large number of steps (phases) in a given diameter by placing several stators (displaced by a suitable number of degrees) in line, all acting on the same rotor; resolution increases in proportion to number of stators used.

Multipolar synchro: (elec) a synchro (q.v.) which behaves as if it were geared up with perfect gearing (*ie* without backlash *etc* which occur in mechanical gearing); transmitter synchro is wound so that during 360° of movement the voltage pattern undergoes, say, N cycles, giving a gearing of N-times

nano (n): prefix used to denote submultiple of SI units; value 10^{-9}; *eg* nm, nanometre.

nit: commonly used term for unit of luminance, candela per square metre, cd/m^2.

newton: derived SI unit; the force which, when applied to a body having a mass of 1 kilogram, gives it an acceleration of one metre per second; symbol N.

ohm: derived SI unit; resistance between two points of a conductor when a constant potential difference of 1 volt, applied between these two points, produces in this conductor a current of 1 ampere, the conductor not being the source of any electromotive force; symbol Ω.

Ohm's law: (elec) fundamental law in electric circuitry; the current through any circuit element is proportional to the voltage across it; the ratio of voltage to current is the resistance, which may be constant (at a given temperature) for materials such as metallic conductors (*ie* for which the ratio of voltage to current is virtually constant).

Pancake synchro: (elec) see slab synchro.

Permanent slip capacitor motor: (elec) a capacitor start and run motor (q.v.).

pico (p): prefix used to denote submultiple of SI units; value 10^{-12}; *eg* pF, picofarad.

Pipe ventilated: (elec) see duct ventilated.

Plain total enclosure: (elec) simple form of totally enclosed housing, *eg* a normal ventilated motor with openings sealed; may be finned to assist cooling, and possibly derated for continuous operation.

Plasma: (elec) term commonly used for an ionised gas which acts as an electrical conductor; *eg* in plasma arc welding; operating fluid in magnetohydrodynamic systems (q.v.); see gas conductor.

Potentiometer: (elec) instrument for measuring potential differences (or for comparing resistors); basically a resistor which can be tapped at short intervals (ideally at any point) and which carries a d.c. current; the known potential difference in the instrument resistor is balanced against the potential difference to be measured.

Power factor: (elec) ratio of total power (W) through an electric circuit to total equivalent volt-amperes (VA); *ie* W/VA; equals cos ϕ, where ϕ is the phase angle (1) between voltage and current in a singlephase circuit or (2) between phase voltage and phase current in a balanced 3-phase circuit.

Protected enclosure: (elec) prevents accidental contact with rotating or live parts of machine, but does not prevent deliberate access to internal parts; see ventilated enclosure.

Pull-in torque: (elec) for a stepping motor, the torque at which the motor can step absolutely synchronously when started from rest.

Pull-out torque: (induction motor) the highest torque that an induction motor can develop while running at rated voltage and frequency.

Pull-out torque: (stepper motor) at a certain motor stepping rate, torque can be increased (to pull-out value) until the unit drops out of step and comes to rest (or hunts).

Pull-out torque: (synchronous motor) torque value at which the motor drops out of synchronism.

Pull-over: (elec) a rotor is said to pull-over if it comes into contact with (*ie* mechanically rubs) the stator when the motor is running, due to unbalanced forces in the unit.

Pull-up torque: (induction motor) the least torque developed by the motor between zero speed and the speed which corresponds with pull-out torque (q.v.) when the motor is supplied at the rated voltage and frequency.

Reactor start split-phase motor: (elec) a split-phase motor designed for starting with a reactor normally in series with the main primary winding. (Auxiliary primary circuit is opened and reactor short-circuited or otherwise made ineffective when the motor has reached an appropriate speed.)

Repulsion induction motor: (elec) a repulsion motor with an additional rotor cage winding.

Repulsion motor: (elec) single-phase induction motor with a primary winding, generally on the stator, connected to the power source, and a secondary winding, generally on the rotor, connected to a commutator, of which the brushes are short-circuited and can occupy different positions.

Repulsion start induction motor: (elec) a repulsion motor in which the commutator bars are short-circuited or otherwise connected at an appropriate speed to give the equivalent of a cage winding.

Resistance start split-phase motor: (elec) split-phase motor having a resistance connected in series with the auxiliary primary winding. (Auxiliary circuit is opened when the motor has reached an appropriate speed.)

Resolver: (elec) a form of synchro (q.v.) which uses sinusoidal relation between shaft angle and output volts, *eg* to solve trigonometrical problems in analogue computers and other applications, including data transmission, phase shifting, radar sweep resolution.

Response range: (elec) for a stepping motor, the range of speeds within which it can be started, stopped, reversed; normally specified for no-load conditions; may alter considerably under load.

Screen protected enclosure: (elec) a protected (q.v.) housing in which wire mesh or screen is used to minimise chance of contact with internal parts during use of a machine, *eg* by a finger; see ventilated enclosure.

Series motor: (elec) motor in which the field winding is connected in series with the armature.

Servomotor: (elec) typically 2-phase induction motors; commonly used in closed-loop servo systems; speed-torque curve is approximately linear.

Servomotor acceleration: (elec) theoretical acceleration at stall of an unloaded servomotor is the ratio of maximum stall torque to rotor inertia; in practice, torque falls off as speed rises, and acceleration diminishes.

Servomotor time constant: (elec) time taken for an unloaded servomotor to reach 63·2% of its final speed = ratio of inertia to (viscous friction x g).

Servomotor torque constant: (elec) ratio of stall torque to control voltage; *ie* torque and voltage are (approximately) proportional.

Servomotor velocity constant: (elec) ratio of no-load speed to control volts; assesses the ability of a servomotor to follow a uniformly-moving command signal; is useful in determining angular lag which occurs in a servosystem between command signal and output shaft.

Servomotor viscous friction: (elec) approximately ratio of stall torque to no-load speed; in practice, damping is usually less at lower control voltages, and at zero control volts may be half the ratio quoted above.

Shaded pole motor: (elec) single-phase induction motor that has one or more auxiliary short-circuited windings displaced in magnetic position from the main winding, all these windings being on the primary core, usually the stator.

Shunt motor: (elec) motor in which the field winding is connected across the supply, in parallel with the armature circuit.

Single stator: (elec stepping motor) type of stator with multi-phase windings distributed in slots; similar to induction motor stators.

Slab synchro: (elec) a synchro (q.v.) consisting of a wound stator and wound rotor, usually specially made to be fitted into a particular application, *eg* transmitting a gyro position; slab is the UK term, pancake the equivalent North American term.

Slewing range: (elec) for a stepping motor, the range of switching rates within which the motor can run in one direction, within a certain maximum rate of increase, and can follow the switching rate without losing a step; maximum and minimum points of slewing range are known respectively as pull-out and pull-in speeds.

Slot effect: (elec servomotor) reluctance of rotor to move away from one of a number of positions; overcome when a certain control voltage is exceeded.

Splash-proof enclosure: (elec) machine housing made so that liquids or solids impinging at any angle between vertical and below horizontal (*eg* 10°) cannot enter in sufficient amounts to interfere with operation; see ventilated enclosure, hose-proof enclosure.

Split-phase motor: single-phase induction motor with an auxiliary primary winding, displaced in magnetic position from, and connected in parallel with the main primary winding; there is a phase displacement between the currents in the two windings. (The auxiliary circuit is normally assumed to be opened when the motor has reached an appropriate speed.)

Star connection: (elec) method of connecting the phases of a 3-phase supply; *eg* the beginnings of 3 windings of an a.c. machine may be connected together (the junction being known as the star point); see delta connection.

Star-delta starter: (elec) device for starting an induction motor; (1) connects stator windings in star for starting; (2) reconnects windings in delta when the motor has reached necessary speed; see star connection, delta connection.

Star-delta starting: (elec) adopted to start an induction motor if the supply system is limited; acceleration time typically over three-times the direct on-line starting value; see direct starting.

Stator-rotor starting: (elec) used on sliping machines to get better starting performance than from a cage machine; a suitable resistance inserted in the rotor circuit can give starting torque of 200% FLT with starting current of 200% FLC (maintained throughout the acceleration period if the resistance is adjustable).

Stepper (stepping) motor: (elec) a d.c. excited machine in which the supply is directed to each winding in turn

191

to as to advance the rotor in a series of steps.

Stepping accuracy: (elec) for a stepping motor, generally expressed as % of stepping angle (in practice of the order of 5%); for stepping motors, the errors in individual steps are not cumulative.

Stepping angle: (elec) for a stepping motor, the angle through which the rotor shaft turns when the stator winding is switched sequentially by one step; expressed in degrees as the reciprocal of the number of steps per revolution, N; *ie* 360:N.

Steps per revolution: (elec) in a stepper motor, steps/rev depend on number of stator and rotor poles; can be defined as number of discrete steps made by a fixed point on the output shaft in one direction of rotation before returning to its original position.

Steromotor: (elec) motor with permanent-magnet rotor which rotates eccentrically inside a stator to provide inherent hypocyclic gearing; has low effective inertia and provides small stepping angles useful in control systems.

STR: (elec) short time rating; term used *eg* is specifications for performance of rotating electrical machines.

Synchro: (elec) essentially a transformer in which the coupling between windings may be varied by rotating one winding; in construction resembles a scaled-down electric motor; *ie* stator, rotor, with sliprings and brushes as electrical connections; a device used to transmit data electrically from one place to another with high degree of accuracy.

Synchro differential: (elec) consists of a 3-phase stator and 3-phase rotor, and performs same function as a mechanical differential in a shaft drive but in a synchro (q.v.) control system.

Synchronous motor: (elec) an alternating current motor in which the speed on load and the frequency of the system to which it is connected are in a constant ratio.

Synchronous speed: (induction motor) the speed of rotation of the fundamental component of the stator-produced field.

Tachometer generator: (elec) provides an a.c. voltage whose amplitude is proportional to the speed of rotation; the voltage is added to the output of a control transformer in an a.c. control system to provide damping and make the system stable; the tachometer generator is similar in appearance to a 2-phase servomotor, with design differences.

tera (T): prefix used to denote multiple of SI units; value 10^{12}; *eg* Tm, terametre.

tesla (T): unit of magnetic flux density; equals 1 weber per square metre of circuit area; $T = Wb/m^2$.

Thermoelectric generation: practical application of the emf produced when a junction between two dissimilar conductors is heated (*ie* the Seebeck effect); used to provide *eg* less than 1 kW of electric power as in control systems, solar generators, waste-heat generators; see thermoelectric refrigeration.

Thermoelectric refrigeration: practical use of the cooling effect produced when a current is passed (in a certain direction) through a junction between two dissimilar conductors (*ie* the Peltier cooling effect); for small-scale (instruments, *etc*) or relatively large scale (air-conditioning) cooling; see thermoelectric generation.

Torque chain: (elec) term describing the linking of a pair of synchros (q.v.), acting as torque transmitter and torque receiver; enables angular movements to be transmitted over a distance via lengths of cables (and sometimes through sliprings).

Torquer: (elec) short segment of stator iron with teeth on inside or outside diameter, and with conventional 2-phase servomotor winding on it; as for slab synchro (q.v.) is provided for special applications, *eg* in gyros.

Torque receiver: (elec) see synchro, torque chain; torque synchro.

Torque synchro: (elec) a synchro unit used to transmit torque; when the stators of two similar units are connected, an alternating field can be set up in the second stator resembling that in the first unit; rotating the shaft of the first (transmitter) unit moves both fields and causes a similar rotor in the second (receiver) unit to rotate correspondingly.

Torque transmitter: (elec) see synchro, torque synchro, torque chain.

Total enclosures: (elec) housings of electrical machines which prevent external atmosphere (air) from reaching inner surfaces; types include plain, fan-cooled, closed air circuit, flameproof.

Transducer: device to convert a parameter that is difficult to measure into a parameter that is more readily measurable; may be mechanical or electrical in principle, or combined in a system; may make use of *eg* piezo-electric effect, magnetostriction, photo-electric effect, etc.

Transformer effect: (elec) a varying magnetic field applied through fixed conductors produces in them an emf which is proportional to the rate of change of flux-linkages (the product of the number of conductors, or turns in a winding, and the flux linking them); known as the transformer effect or threading rule.

Tubular motor: (elec) a form of linear induction motor (q.v.) in which the secondary is bounded by the primary in more than one dimension; the long-secondary short-primary type is most commonly used.

Two-value capacitor motor: (elec) a capacitor motor that uses different values of capacitance for starting and running.

Universal motor: (elec) a motor which can be operated by either direct current or single-phase alternating current of normal supply frequencies.

Variable reluctance rotor: (elec stepping motor) made from magnetically 'soft' material and provided with teeth so that the rotor assumes definite positions when the stator windings are energised; provides no holding torque when the stator is not energised; gives less damping than a permanent magnet rotor.

Ventilated enclosures: (elec) housings of electrical machines which allow external atmosphere to reach inner surfaces but give various degrees of protection; *eg* protected, screen protected, drip-proof, splash / hose-proof, pipe or duct ventilated, weather protected.

Vernier stator construction: (elec stepping motor) a single or multiple stator is given a few more teeth than its rotor; the windings of the vernier motor are energised so that each step causes the rotor to advance one rotor tooth pitch.

volt: (elec) derived SI unit of difference of electrical potential; the difference of potential between two points of a conducting wire carrying a constant current of 1 ampere when the power dissipated between these points is equal to 1 watt; symbol V.

watt: derived SI unit; 1 joule per second or 1 volt-ampere; symbol W.

Weather protected machine: (elec) ventilated-enclosure (q.v.) machine installed in the open but protected from the weather by suitable ducting or canopies; NEMA type II is an example; anti-condensation heaters may be needed; see enclosure.

weber (Wb): unit of magnetic flux; the magnetic flux which, linking a circuit of one turn, produces in it an emf of 1 volt as it is reduced to zero at a uniform rate in 1 second.

Wheatstone bridge: (elec) network of four resistors connected in series to form a closed loop with a potential source (battery) connected across two junctions, and a galvanometer across the other two; by balancing the bridge circuit and using known or standard resistors, other resistors can be compared or calibrated; a d.c. instrument.